AF443391

Ultrathin Magnetic Structures II

B. Heinrich · J.A.C. Bland (Eds.)

Ultrathin Magnetic Structures II

Measurement Techniques and Novel Magnetic Properties

With 171 Figures

 Springer

Bretislav Heinrich
Physics Department
Simon Fraser University
Burnaby, BC, V5A 1S6
Canada
e-mail: bheinric@sfu.ca

J. Anthony C. Bland
The Cavendish Laboratory
Department of Physics
University of Cambridge
Madingley Road
CB3 0HE Cambridge
United Kingdom
e-mail: jacb1@phy.cam.ac.uk

Library of Congress Control Number: 2004104844

ISBN 3-540-21956-0 Second Printing Springer Berlin Heidelberg New York
ISBN 3-540-57687-8   First Printing Springer Berlin Heidelberg New York

This work is subject to copyright. All rights are reserved, whether the whole or part of the material is concerned, specifically the rights of translation, reprinting, reuse of illustrations, recitation, broadcasting, reproduction on microfilm or in any other way, and storage in data banks. Duplication of this publication or parts thereof is permitted only under the provisions of the German Copyright Law of September 9, 1965, in its current version, and permission for use must always be obtained from Springer. Violations are liable for prosecution under the German Copyright Law.

Springer is a part of Springer Science+Business Media

springeronline.com

© Springer-Verlag Berlin Heidelberg 1994, 2005
Printed in Germany

The use of general descriptive names, registered names, trademarks, etc. in this publication does not imply, even in the absence of a specific statement, that such names are exempt from the relevant protective laws and regulations and therefore free for general use.

Production: LE-TEX Jelonek, Schmidt & Vöckler GbR, Leipzig
Cover production: Erich Kirchner, Heidelberg

Printed on acid-free paper   57/3141/YL - 5 4 3 2 1 0

Preface

This is the second of two volumes on magnetic ultrathin metallic structures. The field is rapidly becoming one of the most active and exciting areas of current solid state research, and is relevant to longstanding problems in magnetism as well as to technologically important applications in the field of magnetic recording media, devices and sensors. The rapid growth in the field over the last decade is largely due to the coincidence of recently developed molecular beam epitaxy (MBE) techniques being applied to the growth of magnetic metal films and to the development of powerful computational methods which can be used to predict the magnetic properties of such artificial structures. An overview of the field of ultrathin magnetic structures is given in the introduction to Volume I. The reader should also refer to the introduction to Volume I for a description of some of the important concepts in ultrathin magnetic structures, of magnetic anisotropy and also of two-dimensional magnetism; an overview of film preparation and methods for investigating the magnetic properties, both theoretical and experimental, is also presented.

The book does not aim to be comprehensive but rather it is intended, as explained in the first volume, to provide an account of the underlying principles which govern the behavior of ultrathin metallic magnetic films and to describe some of the recent advances in this area, thus serving as a useful introduction to researchers entering this multidisciplinary field for the first time. The two volumes attempt to satisfy a need, perceived by the research community working in this increasingly important area, for a self-contained survey of the significant developments, key ideas and techniques and their underlying principles. In particular, the authors were encouraged to write their contributions in such a way that the tutorial material is emphasized, rather than collating the most recent research results, thus providing a treatment that contrasts with what is often found in specialized research papers and in conference proceedings. It is also hoped that the reader will benefit from the survey of techniques presented in a single treatment and so gain insight into the relative merits of the most widely used methods.

Each volume has several sections. In Volume I the basic concepts central to the field are outlined (for a brief summary, see the introduction to Volume I) and the methods for characterizing the structure of ultrathin magnetic structures are described. Volume I includes sections on. (1) the ground state of ultrathin films; (2) thermodynamic behavior of ultrathin films; (3) spin-polarized spectroscopy as a probe of ultrathin magnetic films; (4) structural studies of MBE-grown ultrathin films; and (5) magnetic studies using spin-polarized neutrons. The first volume is thus particularly concerned with general concepts and structural techniques, together with probes of magnetism using polarized particles.

Volume II includes: (1) MBE structures grown on III/V compound substrates and their magnetic properties; (2) exchange coupling and magnetoresistance; (3) RF techniques: ferromagnetic resonance (FMR), Brillouin light scattering (BLS) and nuclear magnetic resonance (NMR) applied to ultrathin structures; (4) magnetic measurements of ultrathin films using the magneto-optical Kerr effect (MOKE); and (5) Mössbauer electron conversion spectroscopic studies. The second volume is wide-ranging and aims to survey a spectrum of structures and techniques complementing those covered in Volume I.

Finally, a word about units. This is a difficulty since many magneticians tend to use Gaussian units, partly because a large body of literature now exists which is written in these units, whereas many Europeans tend to use SI units automatically (or, in some cases, because they are required to). This issue is a particular concern for those entering the field. In writing this book it was first thought that it would be best to use one system of units only. But since it is by no means clear which units to use, it was decided that it would prove more educational if the book were to make use of both units and to include a conversion table between the two systems. For this reason, some sections are written in Gaussian units and others in SI, according to the authors' preferences. The reader is therefore referred to Sect. 1.2, Vol. I by *Arrott* on units. While at first sight the reader may find it inconvenient to have to convert between units, we hope that after using this book he or she will agree that it is indeed necessary to do this and that anyone wishing to seriously read the literature in magnetism must be fully conversant with both systems.

Burnaby, Canada   B. Heinrich
Cambridge, UK   J.A.C. Bland
March 1994

List of Acronyms and Abbreviations

2D	Two-Dimensional
AED	Auger Electron Diffraction
AF	Antiferromagnetic
AMR	Anisotropic Magneto-Resistance
ASW	Augmented Spherical Wave
BLS	Brillouin Light Scattering
CEMS	Conversion Electron Mössbauer Spectroscopy
CMA	Cylindrical Mirror Analyzer
CPP	Current Perpendicular to the Planes
DOS	Density of States
DRAM	Dynamic Random Access Memory
EBS	Exchange-Biased Sandwitch
ESR	Electron Spin Resonance
FM	Ferromagnetic
FMR	Ferromagnetic Resonance
FWHM	Full Width Half Maximum
GMR	Giant Magneto-Resistance
L-L	Landau-Lifshitz
LCP	Left Circularly Polarized
LDA	Local Density Approximation
LEED	Low Energy Electron Diffraction
LON	Longitudinal
LSDA	Local-Spin-Density Approximation
LSDF	Local Spin Density Functional
MAE	Magnetic Anisotropy Energy
MBE	Molecular Beam Epitaxy
MCD	Magnetic Circular Dichroism
MFP	Mean Free Path
ML	Monolayer
MO	Magneto-Optical
MOKE	Magneto-Optical Kerr Effect
MR	Magneto-Resistance
MRAN	Magneto-Resistive Random Access Memory
NM	Noble Metal
PM	Paramagnetic

PNR	Polarized Neutron Reflection
POL	Polar
RCP	Right Circularly Polarized
RKKY	Ruderman-Kittel-Kasuya-Yosida
SAXS	Small Angle X-ray Scattering
SDW	Spin Density Wave
SEM	Scanning Electron Microscope
SEMPA	Scanning Electron Microscopy with Polarization Analysis
SL	Superlattice
SMOKE	Surface Magneto-Optic Kerr Effect
SPLEED	Spin Polarized Low Energy Electron Diffraction
SQUID	Superconducting Quantum Interference Device
STM	Scanning Tunneling Microscope
TM	Transmission Metal
UHV	Ultra High Vacuum
UMS	Ulrathin Magnetic Structures
UPS	Ultraviolet Photoemission Spectroscopy
XPS	X-ray Photoemission Spectroscopy
XTEM	Cross Section Transmission Electron Microscopy

Contents

Contributors

S.D. Bader
Argonne National Laboratory, Argonne, IL 60439, USA

P. Bruno
Institut d'Electronique Fondamentale, Bât. 220, Université Paris-Sud,
F-91405 Orsay, France

R.J. Celotta
National Institute of Standards and Technology, Gaithersburg, MD 20899,
USA

J.F. Cochran
Physics Department, Simon Fraser University, Burnaby, BC, V5A 1S6,
Canada

J.L. Erskine
Department of Physics, University of Texas at Austin, Austin,
TX 78712, USA

A. Fert
Laboratoire de Physique des Solides, Bât. 510, Université Paris-Sud,
F-91405 Orsay, France

H.A.M. de Gronckel
Department of Physics, Eindhoven University of Technology,
5600 MB Eindhoven, The Netherlands

G. Güntherodt
II Physikalisches Institut, RWTH Aachen, 52074 Aachen, Germany

K.B. Hathaway
Naval Surface Warfare Center, Silver Spring, MD 20903-5000, USA

B. Heinrich
Physics Department, Simon Fraser University, Burnaby, BC,
V5A 1S6, Canada

B. Hillebrands
Physikalisches Institut Karlsruhe, 76128 Karlsruhe, Germany

W.J.M. de Jonge
Department of Physics, Eindhoven University of Technology,
5600 MB Eindhoven, The Netherlands

K. Kopinga
Department of Physics, Eindhoven University of Technology,
5600 MB Eindhoven, The Netherlands

S.S.P. Parkin
IBM Research Division, Almaden Research Center, 650 Harry Road,
San Jose, CA 95120-6099, USA

D.T. Pierce
National Institute of Standards and Technology, Gaithersburg, MD 20899,
USA

G.A. Prinz
Naval Research Laboratory, Washington, DC 20375-5000, USA

J. Unguris
National Institute of Standards and Technology, Gaithersburg, MD 20899,
USA

J.C. Walker
Department of Physics and Astronomy, The Johns Hopkins University,
Baltimore, MD 21218, USA

1. Magnetic Metal Films on Semiconductor Substrates

G.A. PRINZ

Over the past decade, the growth of interest in magnetic metal films has been enormous, as the topics in these two volumes show. This stems from three different developments which converged during the 1980s. First, ultra high vacuum techniques were developed to carefully grow and characterize single crystal films on single crystal substrates. This field has come to be generally called "Molecular Beam Epitaxy", a term taken over from the semiconductor community. Although not strictly true for the deposition of elemental metal films, the term is nevertheless widely used and accepted in the magnetic metal film community, since the deposition sources, procedures, techniques and indeed the "MBE machines" themselves, are essentially the same. [See Appendix for a discussion of MBE techniques]. The second development was in spin-polarized electron techniques to study these new materials. These have given rise to a lexicon of "Spin-polarized" prefixed names, such as spin-polarized photo-emission, spin-polarized electron energy loss spectroscopy, spin-polarized electron microscopy, etc. These topics are discussed in detail in Volume I. Finally, the advent of the supercomputers permitted considerable progress in computational physics and specifically in the direct calculation of the electronic structure of single crystal magnetic metal slabs of finite thickness. The circle was thus closed and atomic scale structures could be grown, characterized and modeled.

Much of the experimental effort on epitaxial magnetic metal films has focused upon growth on single crystal metal substrates rather than on semiconductor substrates. The reasons for this are two-fold. First of all there is a very good lattice match between several elemental magnetic metal crystal structures and several elemental non-magnetic metals which may be readily obtained in single crystal form suitable for use as substrates. These are, respectively Fe, Co and Ni upon Cu, Ag and Au. The specific crystal structures and their lattice constants are given in Table 1.1. By far, the "workhorse" of the "industry" is Cu, but Ag is also widely used for studying bcc Fe. The second reason is both economic and cultural. Much of the work is carried out by researchers from the surface science community, often in universities, for whom the measurement techniques dominate the experiment (such as photoemission or electron scattering). In this environment, the time and effort spent on material preparation must be minimized. Furthermore, the sample generally never leaves the experimental chamber. A given substrate is used over and over, merely sputtered

B. Heinrich and J.A.C. Bland (Eds.)
Ultrathin Magnetic Structures II
© Springer-Verlag Berlin Heidelberg 1994

clean and annealed before each experiment. A given substrate thus has an infinite lifetime, often serving through several students' thesis research without ever leaving the vacuum chamber. It is important that such single crystal substrates do get such long use, since they are expensive and difficult to prepare (cutting, polishing, cleaning and annealing) in a form suitable for reliable film studies.

Unfortunately, for many important experimental magnetic characterizations (such as measurement of the magnetic moment, the magnetic anisotropy, magneto-transport, magnetic susceptibility, magnetostriction, etc.) growth on single crystal metal substrates makes the characterization difficult to impossible. Furthermore, the lack of flexibility in the lattice constants available from metal substrates has largely confined the epitaxial film studies of the 3d transition metals to the choices listed in Table 1.1. And finally, for technological applications, single crystal metal substrates are impractical even if the magnetic films have technologically useful properties.

For all of these reasons, semiconductor substrates have proven to be extremely useful. In the discussion which follows, we shall see how commonly available, inexpensive, high quality single crystal semiconductor substrates can address many of the difficulties discussed above. We shall also see that they may be used as templates upon which single crystal films of non-magnetic metals can be grown that are superior to most single crystal metal substrates used for research purposes. Finally, we shall see that the growth of ferromagnetic films on semiconductors offers the opportunity for many new technological applications.

Table 1.1. Common substrate/film combinations for some of the thermodynamically stable phases of 3d magnetic elements

Substrate (a_0 [Å])			Film (a_0 [Å])		
fcc	Cu	(3.61)	fcc	Ni	(3.52)
				β-Co	(3.55)
				γ-Fe	(3.59)
		[45° rotation on (0 0 1)]			
fcc	LiF	(4.02)	bcc	α-Fe	$\dfrac{(2.867)}{} \times \sqrt{2}$
	Al	(4.05)			
	Au	(4.07)			
	Ag	(4.09)			4.054
		[p(1/2 × 1/2) on (0 0 1)]			
fcc	NaCl	(5.64)	bcc	α-Fe	$\dfrac{(2.867)}{} \times 2$
	AlAs	(5.62)			
	GaAs	(5.65)			
	Ge	(5.66)			5.733
	ZnSe	(5.67)			

In the review which follows, we shall discuss the existing work for magnetic films (both elemental and alloys) on zincblende structures (e.g., GaAs, ZnSe, AlAs) as well as the elemental semiconductors Si, Ge and C (diamond). Potential areas of technological application will be discussed at the end of the chapter.

1.1 3d Transition Metals on Zincblende Structures

The largest body of work for magnetic 3d transition metals on semiconductors is for the zincblende compound semiconductors. This is because of the nearly factor of two relationship between the lattice constants of the metals and the semiconductors and also because of their ready availability either as bulk substrate material or as epitaxial semiconductor films upon readily available substrates. For example, as can be seen from Fig. 1.1 which displays the

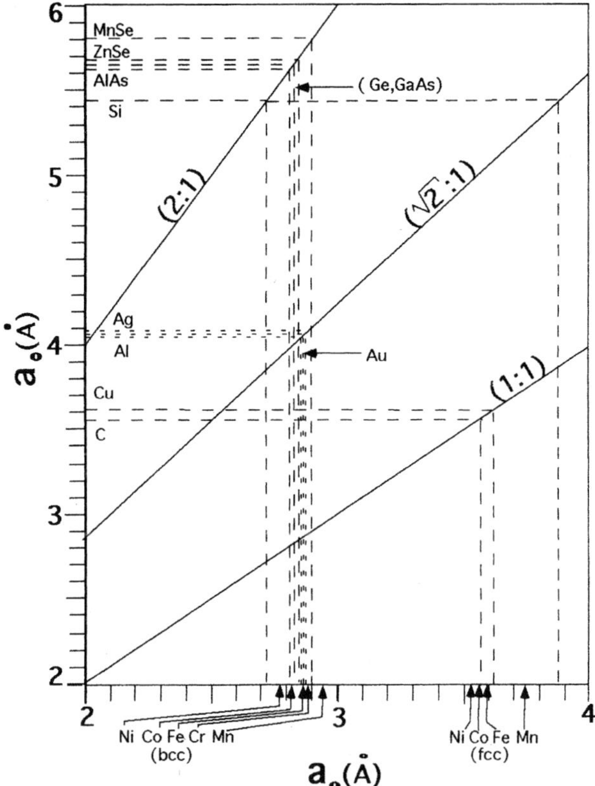

Fig. 1.1. Graphical presentation of the lattice constants of semiconductor substrates (vertical) and magnetic metal films (horizontal). The diagonal lines represent either a parallel alignment of the principal axis [2:1 and 1:1] or a rotated alignment [1:$\sqrt{2}$] of 45° on {0 0 1} or 90° on {1 1 0}

relationship between the lattice constants of substrates and metal films, Ge and GaAs are both $< 0.1\%$ mismatched to bcc Co and $\approx 1.3\%$ smaller than bcc Fe. Either of these materials can be obtained at low cost, as large polished wafers in a great variety of crystal orientations and doped to be either p-type, n-type or semi-insulating.

Using GaAs as a base, epitaxial films of AlAs, ZnSe or MnSe can be grown upon it in sufficient thickness to establish their own lattice constants. This family of compounds and their intermediate alloys thus provide substrates suitable for epitaxial growth which spans the whole range from bcc Co to bcc Mn. Indeed, the alloy system (Zn, Mn)Se itself covers the same lattice range provided by Ag, Au and Al. In fact, epitaxial films of these three metals can be obtained on these substrates and if properly prepared can serve to replace the equivalent single crystal metal substrate. Because of the utility of such films a brief digression shall be made to discuss their growth and properties.

Epitaxial growth of Al on GaAs was first reported [1.1] for GaAs(0 0 1) where it was shown to grow as a 45° rotated lattice in order to accommodate the $\sqrt{2}$ ratio in the lattice constants. Later it was shown that (1 1 0) Al would grow on (1 1 0) GaAs [1.2] where a 90° rotation was now required by the lower symmetry face. The mobility of Al on GaAs is very high and for the (1 1 0) face in particular, good quality films were only obtained for substrate temperatures below room temperature in order to avoid clustered dendritic growth. The resulting Al film can now be used as a substrate for say bcc Fe (mismatch $< 0.1\%$). However, this can only be successful in a vacuum system free of O, since Al getters O very effectively. Oxidation of Al generates a polycrystalline surface which prevents the epitaxial growth of single crystal bcc Fe. The interfacial lattice match of bcc Fe and fcc Al has been exploited to grow Fe/Al/Fe sandwiches on GaAs in order to study the coupling of the Fe moments through the Al [1.3].

Epitaxial growth of Au on GaAs(1 0 0) has also been reported [1.4], but is easily defeated by carbon contamination of the surface. Au also tends to agglomerate at etch pits on the surface and to interdiffuse and cluster when growth is carried out above room temperature. Room temperature deposition on GaAs(1 $\bar{1}$ 0) has been studied via scanning tunneling microscopy (STM) [1.5] images and here, too, clusters are formed. However, they are crystallographically oriented Au(1 1 0), eventually forming epitaxially oriented films at higher coverage.

Finally Ag growth on both GaAs(1 0 0) and GaAs(1 1 0) has been studied. Several researchers agree that room temperature growth of Ag on GaAs(1 0 0) yields Ag(1 $\bar{1}$ 0) with Ag [1 0 0] $\parallel$ GaAs [1 1 0] or GaAs [$\bar{1}$ 1 0] depending upon Ga or As termination of the substrate [1.6]. For growth above 200 °C one obtains Ag(1 0 0) [1.7]. Growth of Ag on GaAs(1 1 0) results in films which are very close to epitaxial Ag(1 1 0), however, with a small tilting of the Ag(1 1 0) plane to accommodate the lattice mismatch [1.8].

For both faces of GaAs, initial growth of Ag, like Au, proceeds via clusters rather than two-dimensional layered growth. A procedure was introduced by

Jonker et al. [1.9] by which this cluster growth of Ag could be overcome, which has since proved to have general utility in bonding weakly interacting metals to GaAs. In this procedure, 2 ML to 3 ML of bcc Fe is first deposited on the GaAs. This strongly interacting metal forms an epitaxial template of high free surface energy metal upon which a low free surface energy metal may be deposited without resulting in island formation. Because of the chemical reactivity between Fe or Co and GaAs, the seed layer is magnetically "dead" and does not therefore interfere with any subsequent magnetic measurements of a completed sample. This "seed layer" approach has also succeeded using bcc Co before growing Pd on GaAs [1.10].

1.1.1 bcc Fe

The low temperature bcc ferromagnetic phase, α-Fe, has been the most widely studied magnetic metal film grown on semiconductor substrates.

1.1.1.1 Growth and Structure

In 1979, it was found that Fe would grow epitaxially on (1 0 0) GaAs [1.11]. In 1980, it was shown that Fe would grow epitaxially on (1 1 0) GaAs and that the process could be carried out in a commercial MBE machine using BN-crucible Knudsen cells [1.12]. Following this, extensive experiments were carried out to determine the magnetic properties of these films and their dependence on thickness and growth conditions. This system provided many new insights into the properties of thin magnetic films and is now one of the best understood systems. Figure 1.2 shows the relationship between the unreconstructed surfaces of the two materials for the (1 1 0) face. Although the precise location of the Fe

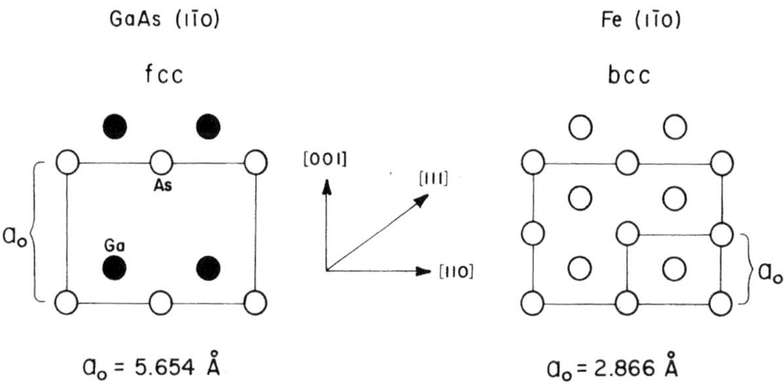

Fig. 1.2. Comparison of the (1 1̄ 0) face of GaAs and α-Fe

atoms on a GaAs(1 1 0) surface is still not certain, it is seen that to within a 1.3% mismatch, the lattice constants differ by a factor of 2, which permits the epitaxial growth.

Figure 1.2 may serve as a guide for discussing all of the zincblende substrates. Although the several different compounds listed on Fig. 1.1 differ only slightly in their lattice constants, they are chemically quite different. They range from the small band gap (1.4 eV) III–V compound GaAs which is more covalently bonded, to the large band gap (2.7 eV) II–VI compound ZnSe which is more ionically bonded. As we shall see, metallization of the surfaces of these compounds proceeds quite differently resulting in different growth modes and different magnetic properties in the resulting films.

Figure 1.3a shows the Reflection High Energy Electron Diffraction (RHEED) patterns obtained from a chemically etched and vacuum annealed surface of GaAs(1 0 0) and Fig. 1.4a represents a ZnSe(1 0 0) epitaxial film grown on GaAs. The sharp RHEED patterns from these semiconductor surfaces, indicative of smooth, terraced surfaces, contrasts dramatically with the patterns seen upon deposition of Fe. By 2 ML of Fe, the lattice spacing of bcc Fe is already established, but the broad elongated spots indicate limited long range structural order in the metal film.

It is interesting to compare these results with information obtained from a relatively new structural characterization technique, angle-resolved electron forward scattering, in the form of Auger Electron Diffraction (AED). The technique is illustrated schematically in Fig. 1.5. Electrons, escaping from the near surface region from either an Auger emission or photoemission process, are strongly scattered in the forward direction by the charge clouds surrounding atoms in their path. If these atoms are regularly arrayed as they are in a crystalline solid, this strong forward scattering acts to focus the outgoing electrons into beams along directions of high crystalline symmetry. Thus if one measures Auger emitted electrons, for example as a function of angle across a crystal surface, one observes a very non-uniform distribution, strongly peaked along lines of atoms. Furthermore, since Auger electrons are elementally specific, by collecting only those electrons emitted from the atoms being deposited upon a substrate, one can easily determine the structure those atoms are forming (fcc, bcc, etc.). As Fig. 1.5 shows, a perfect 1 ML coverage should give a featureless uniform angular distribution, a perfect 2 ML coverage will give peaks along $54°$ for a bcc (0 0 1) surface, while scanning a $\langle 1 1 0 \rangle$ azimuth. Finally, a perfect 3 ML coverage will add a central [0 0 1] peak for electrons emitted along the surface normal. This technique is especially valuable for determining the nature of the initial growth mode. For example if a [0 0 1] peak is seen for depositions which are equivalent to 2 ML or less coverage, one can be certain that the growth is not proceeding as ideal layers but rather as islands.

Examples of such data are shown in Fig. 1.6 where the AED from Fe growth on the two substrates of Fig. 1.4, GaAs(1 0 0) and ZnSe(1 0 0). For 2 ML coverage of Fe grown at 175 °C it is obvious from the 0° peak that third layer scattering is already obtained from the Fe on GaAs substrate, but almost ideal

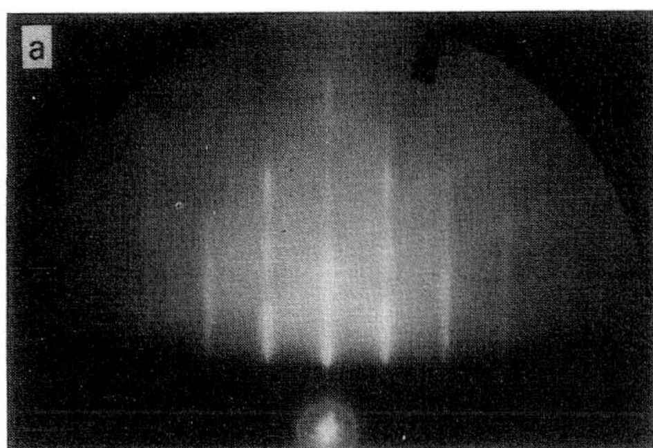

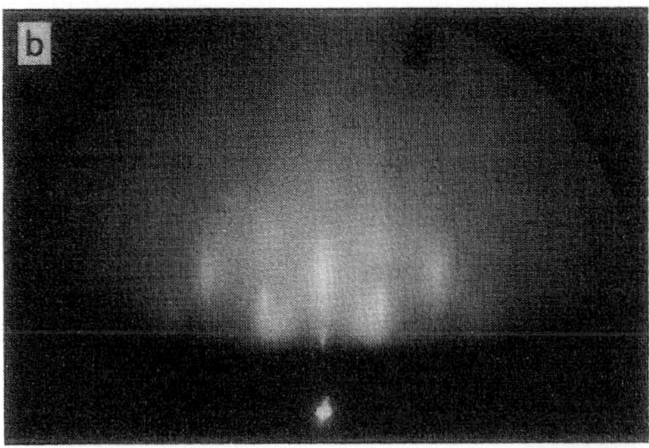

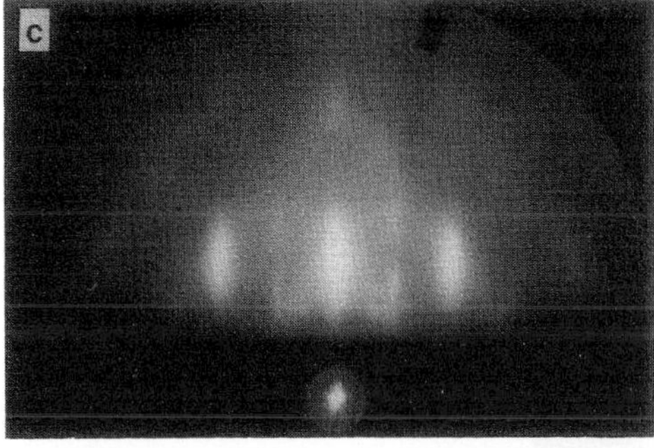

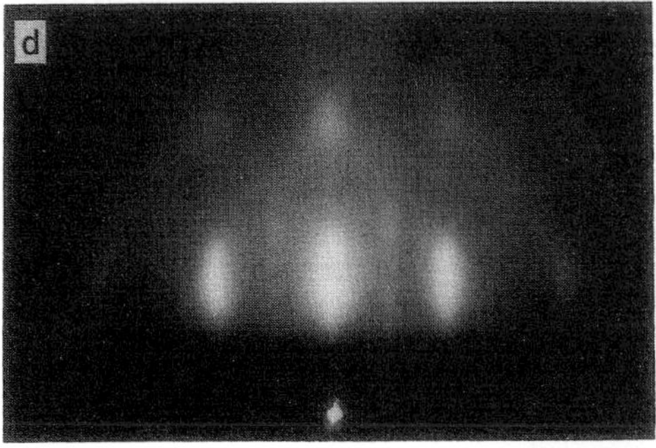

Fig. 1.3. RHEED patterns obtained along a $\langle 1\,1\,0\rangle$ azimuth at various stages of deposition at 175 °C on the GaAs (0 0 1) surface. (**a**) Clean GaAs (0 0 1) surface showing six-fold reconstruction (Ga-rich) (**b**) 1 ML Fe, (**c**) 2 ML Fe, (**d**) 3 ML Fe [1.14]

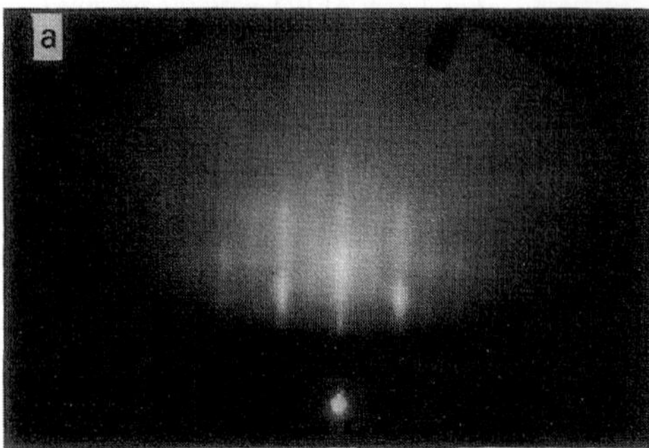
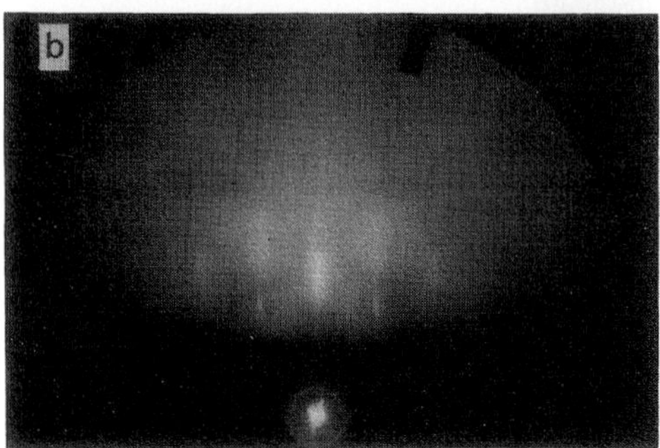
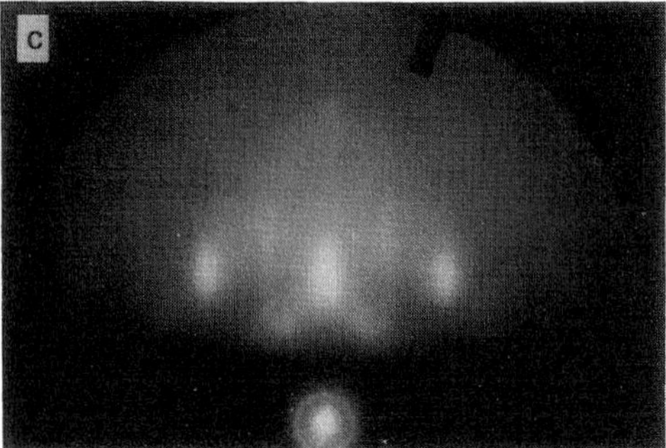
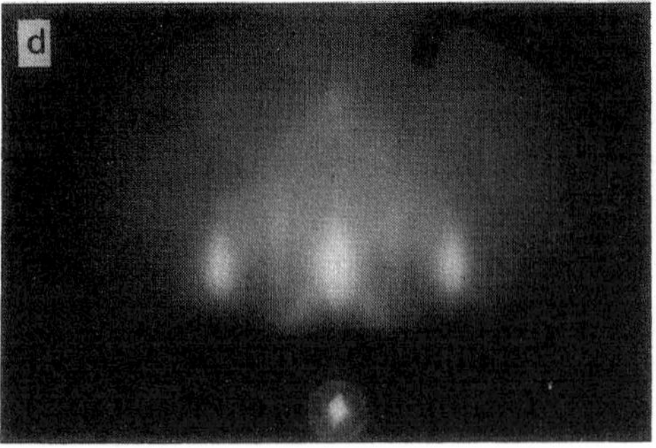

Fig. 1.4. The same deposition conditions as Fig. 1.3 on a clean ZnSe (0 0 1) surface showing two-fold reconstruction (Se-terminated) [1.14]

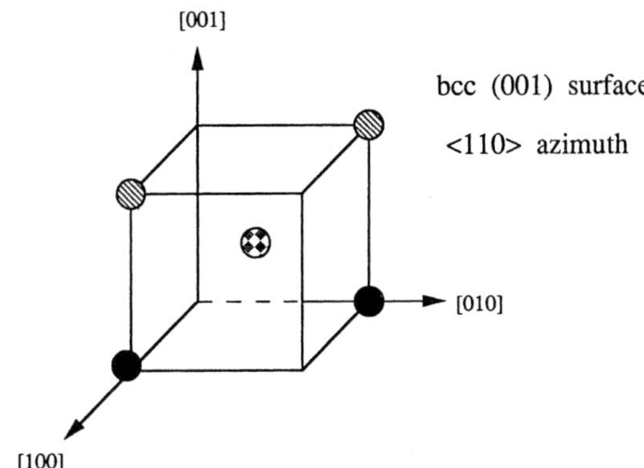

Fig. 1.5. Schematic diagram of the electron forward scattering process, illustrating a $\{1\,1\,0\}$ cross section of a bcc cell and the relative positions of first, second and third monolayer atoms for $(0\,0\,1)$-oriented growth. Auger or photoemitted electrons are focused along interatomic axes, producing characteristic peaks in the angular dependence of the emission intensity that indicate occupation of second and third monolayer sites

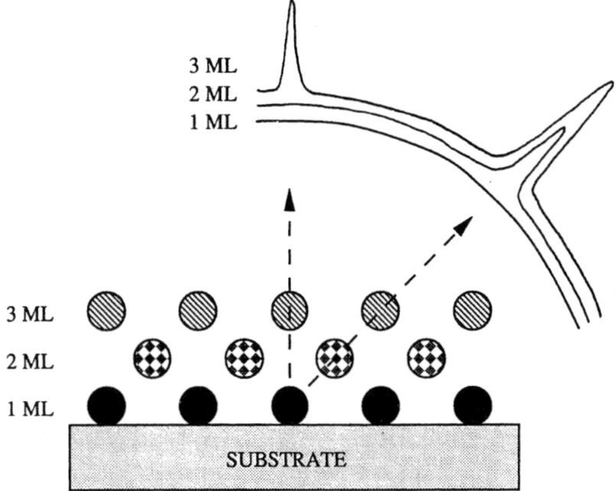

2 ML scattering from Fe on the ZnSe substrate. These definitive results are illustrative of the power of this new technique for establishing the difference between the growth modes for bcc Fe on these two zincblende surfaces: island growth on GaAs($1\,1\,0$), but layered growth on ZnSe($1\,0\,0$).

Further information on the chemistry of the interface formation can be obtained from photoemission studies. Surface sensitive X-ray photoemission spectroscopy (XPS) measurements of the Se 3d level, shown in Fig. 1.7, reveal that deposition of 1.7 ML of Fe on ZnSe results in a chemical shift of only 0.45 eV and virtually no change in line width or shape of the photoemission line. This indicates little change in the Se chemical environment. This is to be contrasted with the results seen on a GaAs surface shown in Fig. 1.8, where at similar coverages there are evident significant changes in the As 3d peak. This reactivity of Fe with the GaAs surface is also seen in the reduction of the Ga and As 3d level signals with Fe coverage, as seen in Fig. 1.9. This data shows that while there is an initial interface reaction with Ga (an interchange of Fe and Ga atoms) with the first monolayer of Fe coverage, subsequent Fe deposition covers

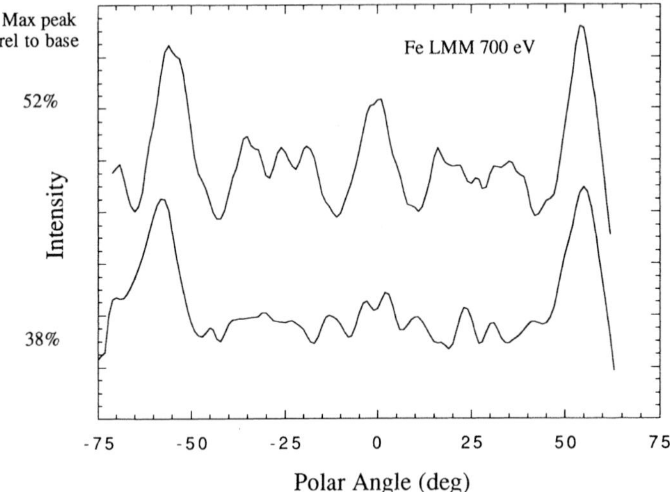

Fig. 1.6. A comparison of the Fe AED $\langle 1\,1\,0 \rangle$ azimuth polar scans for 2 ML of Fe grown on an oxide-desorbed GaAs $(0\,0\,1)$ surface (top) and the ZnSe $(0\,0\,1)$ epilayer surface (bottom) at 175 °C [1.13]

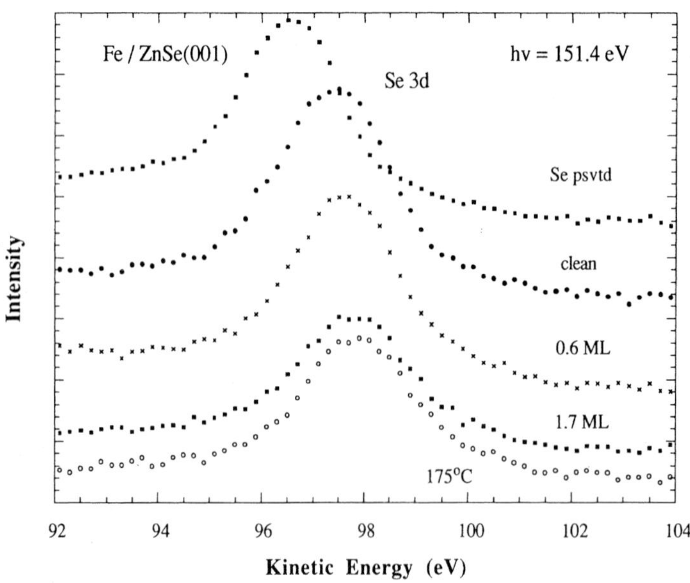

Fig. 1.7. XPS spectra ($hv = 151.4$ eV) of the Se 3d level at various stages in the Fe deposition process. The spectra have been shifted vertically for ease of comparison [1.13]

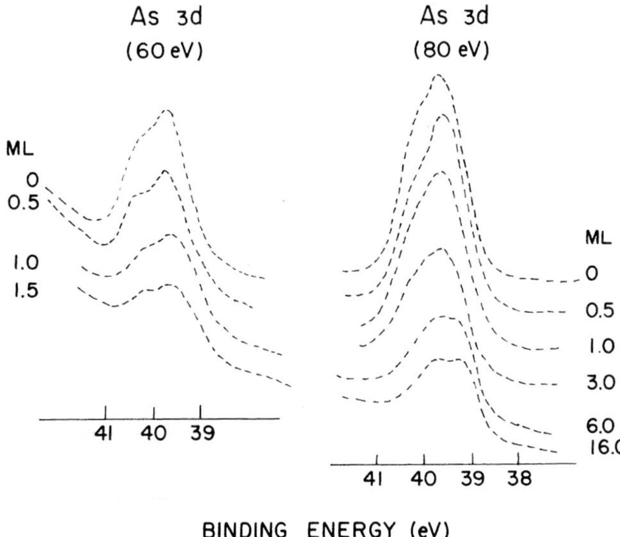

Fig. 1.8. As 3d core level spectra for $hv = 60$ and $80\,\mathrm{eV}$ at increasing stages in the Fe deposition process. Shifts induced by band bending have been subtracted out and the spectra have been normalized to comparable peak heights in order to clearly see the changes in line shape and binding energy [1.15]

the Ga in the expected exponential manner. The As signal however, persists to very high coverages, indicative of As released at the interface and persisting as a surface contaminant to very high coverages. This As surface layer acts to create the observed "extra" lines in the RHEED patterns of the Fe film surface. They may be removed by sputter cleaning the Fe film surface and annealing. The resulting RHEED pattern is then exactly as expected for bcc Fe, as illustrated in Fig. 1.10, and further Fe growth on that surface retains this pattern.

1.1.1.2 Magnetization

While the electron diffraction shows that the films being grown have the proper symmetry and spacing to correspond to a (1 1 0) face of bcc Fe, magnetic characterization showed several surprises. The first is seen in Fig. 1.11, which shows the magnetic moment/unit volume (magnetization, M) versus film thickness for a series of films grown under identical conditions. The most striking feature is the apparent decrease in M as the film thickness approaches zero. If one assumes that there is some interfacial region within which the magnetization has an exponential dependence of the form

$$M(Z) = M_0(1 - e^{-z/L_0}),$$

$$(1.1)$$

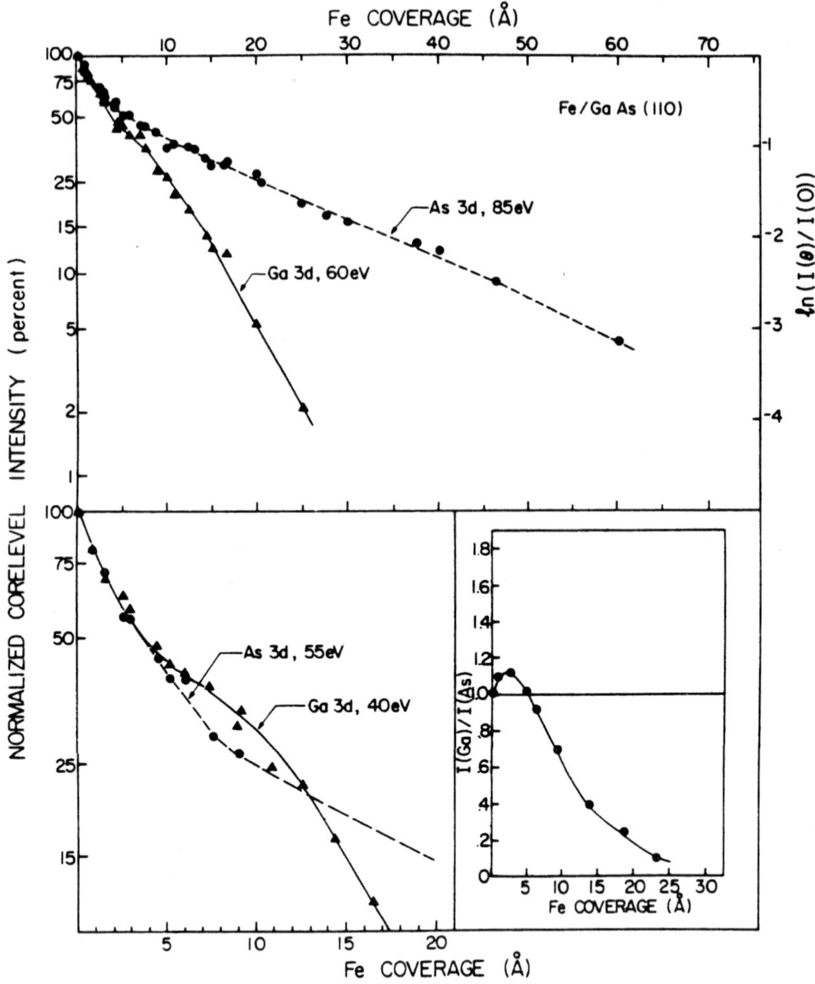

Fig. 1.9. Dependence of the integrated intensities from As 3d and Ga 3d core level transitions with increasing coverage [1.16]

where z is measured from the Fe/GaAs interface, and one integrates this over the thickness of a given film L, one obtains

$$M(L) = \frac{\int_0^L M(Z)\,\mathrm{d}Z}{\int_0^L \mathrm{d}Z} = M_0\left\{1 - \frac{L_0}{L}(1 - e^{-L/L_0})\right\}. \tag{1.2}$$

This expression has been fitted to the data for a universal value of $L_0 = 10$ Å for all of the films measured. Since similar results were obtained for films regardless of final overcoating (Al, Ge or oxide) it was concluded that the

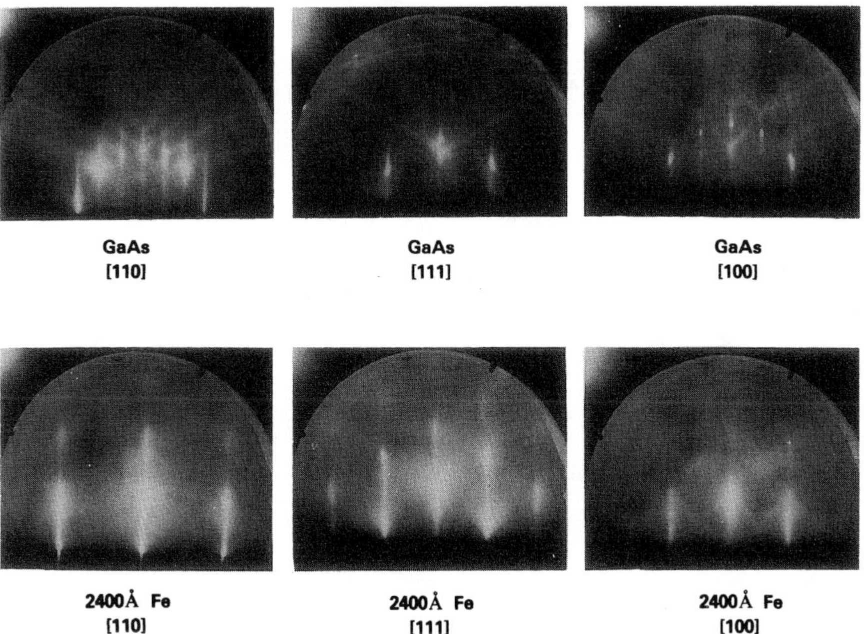

Fig. 1.10. RHEED patterns from a GaAs {1 1 0} substrate before film growth and from Fe film surface after growth for the three principal axes in the film plane [1.17]

decrease in magnetization arose from some mechanism at the Fe/GaAs interface which had an exponential decay depth of ≈ 10 Å. Although the initial Fe $\leftrightarrow$ Ga interchange could yield a magnetically dead monolayer, it is the extended presence of As in the film which is a likely source of the extended diminished magnetization. While the amount of As is too small to account for the observed magnetic effects if it merely acts as a dilutant, it has been pointed out [1.19] that an As impurity will tend to bond an Fe–Fe pair on either side of it into an antiferromagnetic alignment. In a bcc structure, therefore a single As ion could effect up to eight Fe moments. Furthermore, the Neel temperature of such compounds can be quite high, e.g., Fe_2As ($T_N = 350\,^{\circ}C$). Finally, As impurities may alter the local anisotropy near the impurity sites. This has been modeled as a random anisotropy problem [1.20]. The results indicate that the magnetic order could be strongly disrupted near the interface. Although agreement on the specific mechanism has not been reached for understanding the magnetic effects caused by As impurities in Fe films, a probable cause of the decreased magnetization appears to be present. A microscopic study of the magnetic order near the interface is required to settle the issue.

In contrast to the GaAs/Fe interface, the ZnSe/Fe interface shows much less evidence for reduced magnetization. Films of only 137 Å show the full magnetization of bulk Fe. This is consistent with the reduced chemical activity at the ZnSe/Fe interface.

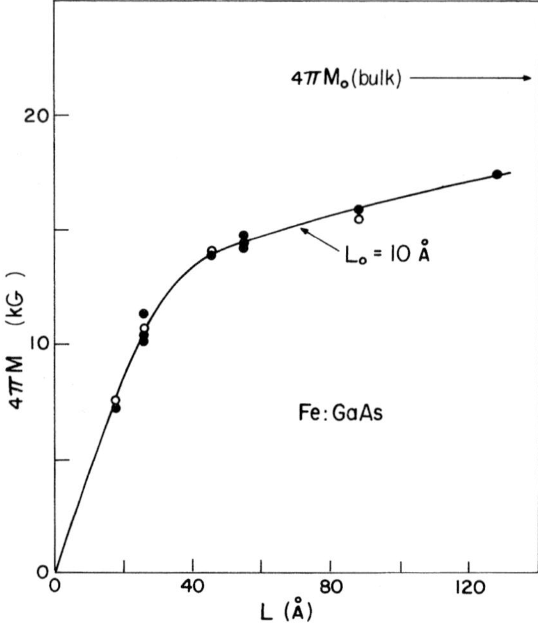

Fig. 1.11. Dependence of the magnetization upon thickness for epitaxial Fe films on (1 1 0) GaAs measured at 77 K (○) and 300 K (●) [1.18]

1.1.1.3 Ferromagnetic Resonance (FMR)

Ferromagnetic resonance (FMR) is a powerful technique for studying the magnetic properties of magnetic films. This technique is illustrated in Fig. 1.12. The magnetic moment, confined to the film plane by the demagnetizing field of surface poles, is subject to an applied field sufficient to align the moment along H. This applied field provides a restoring force such that any disturbance of M will cause it to precess about H in gyroscopic motion. The disturbance is provided by a varying microwave field of frequency ω which drives the moment into precession. At an appropriate value of field H, the restoring force will cause the natural frequency of gyroscopic motion to match the microwave frequency of the driving field and resonant absorption of energy from the radiation field will occur, which is readily observable when the sample is placed in a microwave bridge.

The role of magnetic anisotropy is to change the value of the field at which resonance occurs. For example, when M is along an easy direction, it is in a potential minimum which resists deviation of the moment. This effectively adds "stiffness" to the gyroscope and less applied field is needed to reach the resonance condition. Hence resonance occurs at a lower field value. Conversely, when the system is magnetically saturated along a hard direction, M is located

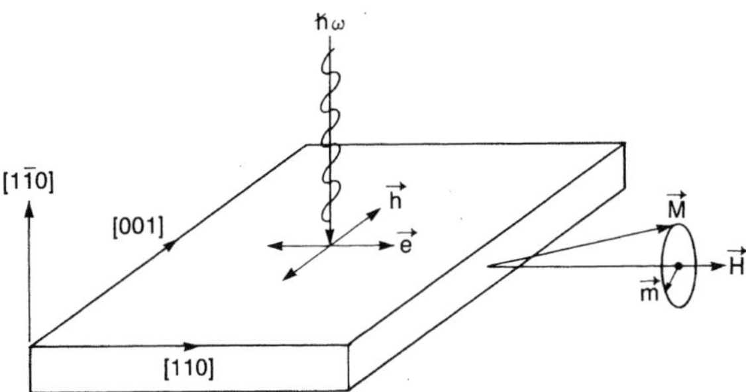

Fig. 1.12. Illustration of the experimental configuration to carry out ferromagnetic resonance for M lying in a $\{1\,1\,0\}$ plane

at an energy maximum. This contribution "softens" the restoring force, requiring a higher applied field to reach resonance. The microwave "wiggling" of M thus probes the curvature (second derivative) of the anisotropy energy surface. If the energy surface is described by powers of cosines, as shall be discussed below, the second derivative will regenerate expressions in powers of cosines. For high symmetry surfaces, which are described by simple expressions of the anisotropy, we shall see that a plot of the resonance field as a function of direction, will resemble the shape of the anisotropy energy surface itself. Quite apart from any resemblance, however, a determination of the resonance field's angular dependence readily yields an anisotropy determination.

Since the anisotropy energy is derived from effects of local environment it must have a mathematical form compatible with the symmetry of that environment. In the case of many important magnetic metals (e.g., bcc Fe, fcc Ni, fcc Co) this means that the form must be invariant to operations of the cubic group. Thus, if we express the energy contributed by the magnetic moment as a function of its direction as a general expansion of the form

$$E = \kappa_2(\alpha_1^2 + \alpha_2^2 + \alpha_3^2) + \kappa_4(\alpha_1^4 + \alpha_2^4 + \alpha_3^4) + \kappa_6(\alpha_1^6 + \alpha_2^6 + \alpha_3^6) + \cdots, \quad (1.3)$$

where $\alpha_1, \alpha_2, \alpha_3$ are the direction cosines of M with respect to the cubic axes x, y, z, only even powers are permitted by symmetry. This may be simplified by dropping terms which are merely additive constants (e.g., $\alpha_1^2 + \alpha_2^2 + \alpha_3^2 = 1$) and by expressing the terms in second powers only, one obtains the form most conventional in the literature

$$E_{\text{an}} = K_1(\alpha_1^2\alpha_2^2 + \alpha_2^2\alpha_3^2 + \alpha_3^2\alpha_1^2) + K_2(\alpha_1^2\alpha_2^2\alpha_3^2) + \cdots. \quad (1.4)$$

The terms get progressively smaller both because of changes in K_i as well as the decline in the values of the cosine products.

For the remainder of this discussion we shall assume $K_i = 0$ for $i > 1$ and just deal with the generally dominant cubic term. Generalization to lower symmetry systems (e.g., hcp Co) is straightforward. Figure 1.13 illustrates the effect of the K_1 term upon the formerly spherical energy surface. Figure 1.13a is the surface

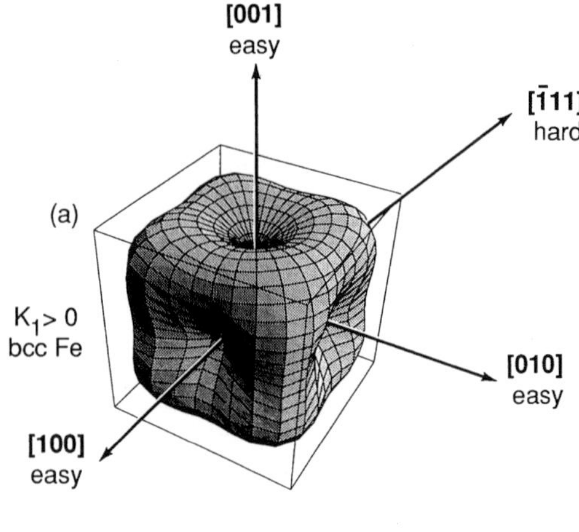

Fig. 1.13. (a) The anisotropy energy surface introduced by $K_1 > 0$, typical for a cubic system like bcc Fe; (b) the anisotropy energy surface introduced by $K_1 < 0$, typical for a cubic system like fcc Ni

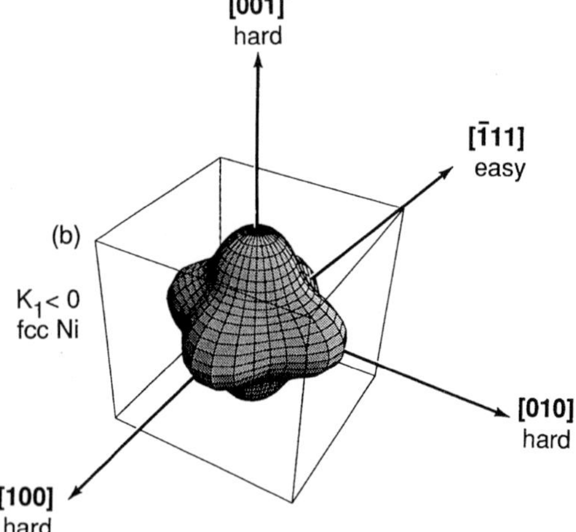

obtained by the addition of a cubic anisotropy with $K_1 > 0$. This characterizes bcc Fe and we see that the energy minima created along $\langle 0\,0\,1 \rangle$, $\langle 0\,1\,0 \rangle$ and $\langle 1\,0\,0 \rangle$ make them the magnetically "easy" axes. That is, in the absence of an applied field the moment will lie along one of these directions. In fact, in a bulk single crystal sample, different sections of the sample will in general be magnetized along each of these directions (i.e. magnetic domains) due to dipoles created on the sample surface, and the sample may possess a zero net macroscopic moment. The case for $K_1 < 0$ is illustrated in Fig. 1.13b, and here we see that the $\langle 1\,1\,1 \rangle$ directions now locate the minima on the energy surface. This is the case for fcc Ni where $\langle 1\,1\,1 \rangle$ are the magnetic "easy" directions. In contrast to these energy minima, the energy maxima in both cases are called the magnetically "hard" axes, since it demands the application of an external magnetic field to pull M into those directions.

In order to see how this discussion applies to epitaxial films one must first recognize that the geometric shape of a film introduces a profound anisotropy. This arises from the same mechanism that creates domains in bulk crystals. Any region of the sample surface which is perpendicular to M will have a net magnetic pole density which serves as a source of magnetic field B. This field B, passes back into the sample generating a $+M\cdot B$ contribution, which raises the energy of the system. The magnetization M will always orient itself to minimize this energy. In a film this results in the moment lying in the plane of the film, minimizing the dipole filled area and the resulting B field. We ignore in this discussion any anisotropy arising from termination at the surface, commonly called "surface anisotropy". These effects, generally seen in very thin films (a few atomic layers), are discussed in Chap. 2 of Volume I.

When discussing films, therefore, one need only look at the anisotropy energy contribution for M lying in different directions in the plane of the film. This is easily done from the three-dimensional surfaces of Fig. 1.13 by "cutting" the figure with a plane parallel to the surface of the crystalline film. This is done in Fig. 1.14 for $K_1 > 0$ and a $(0\,0\,1)$ film. The intersection of these two surfaces

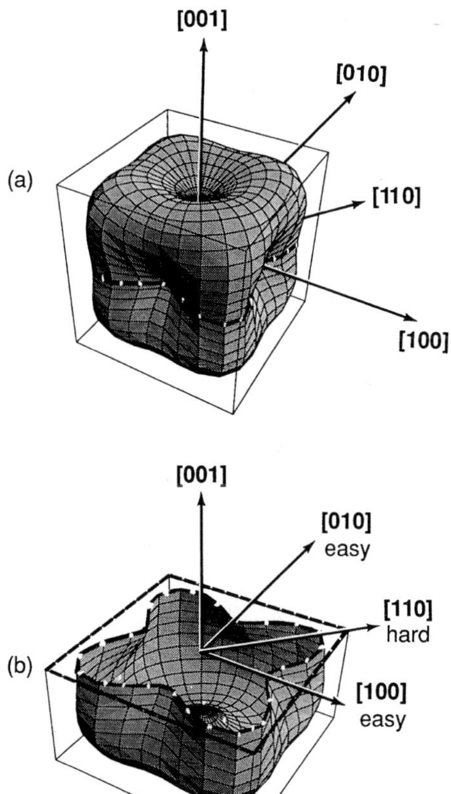

Fig. 1.14. (a) Dashed line indicates the intersection of $(0\,0\,1)$ plane with the $K_1 > 0$ cubic anisotropy energy surface; (b) exposed edge of energy surface reveals the angular dependence of the anisotropy energy in the $(0\,0\,1)$ plane

reveals a curve which possesses four-fold symmetry with the explicit form

$$K_1^{(0\,0\,1)} = \frac{K_1}{8}(1 - \cos 4\phi). \tag{1.5}$$

To see how this relates to FMR measurements, we give the ferromagnetic resonance condition

$$\left(\frac{\omega}{\gamma}\right)^2 = (H_0 + \alpha_H)(H_0 + \beta_H), \tag{1.6}$$

where α_H and β_H are the anisotropy contributions to the resonance field H_0 (see Chapter 3, Sect. 3.1). For a $(0\,0\,1)$ surface, they are [1.19]

$$\alpha_H^{(0\,0\,1)} = 4\pi M + \frac{K_1}{2M}(3 + \cos 4\phi) \quad \text{and} \quad \beta_H^{(0\,0\,1)} = \frac{2K_1}{M}\cos 4\phi. \tag{1.7}$$

In the absence of anisotropy, $\omega = \gamma H_0$, independent of direction. Here $\gamma = g(e/2mc)$ is the gyromagnetic ratio.

The terms containing K_1 derive from the cubic crystalline anisotropy. The "shape anisotropy" is $4\pi M$ and derives from the fact that we are dealing with a film geometry. When the gyroscopic motion carries the magnetic moment out of the plane of the film, there is a restoring force proportional to the magnetization created by the B field of the surface poles. If the sample were spherical this term would be zero. However, in a film it dominates the anisotropy, since (expressed in common units) $4\pi M = 2.2 \times 10^4$ Oe while $K_1/M = 2.5 \times 10^2$ Oe, for Fe. Because of this, we can approximate $\alpha \approx 4\pi M$, yielding

$$\left(\frac{\omega}{\gamma}\right)^2 = (H_0 + 4\pi M)(H_0 + \beta_H). \tag{1.8}$$

Finally, at the resonance fields we shall be discussing ($\approx 6 \times 10^3$ Oe at 35 GHz) we shall make the further simplifying approximation to let

$$H_0 + 4\pi M = 4\pi M\left(\frac{H_0}{4\pi M} + 1\right) \doteq 4\pi M. \tag{1.9}$$

At resonance, we then approximately have the simplified expression

$$H_0 = \left(\frac{1}{4\pi M}\right)\left(\frac{\omega}{\gamma}\right)^2 - \beta_H. \tag{1.10}$$

We see that all of the angular dependence now lies in β and at a fixed frequency ω, for a measured value of M, the angular dependence of the resonance field H_0 will yield K_1. Specifically

$$H_0^{(0\,0\,1)} = \left(\frac{1}{4\pi M}\right)\left(\frac{\omega}{\gamma}\right)^2 - \frac{2K_1}{M}\cos 4\phi. \tag{1.11}$$

This mimics the angular dependence of the anisotropy in (1.5).

An example of such data is shown in Fig. 1.15, for a 200 Å film of Fe grown epitaxially on (0 0 1) GaAs. Inspection of the figure shows that, while there is a dominant four-fold symmetry, the $\langle 1\,1\,0 \rangle$ directions are not equivalent. This is made dramatically evident in Fig. 1.15b in which the difference between minimum and maximum values in Fig. 1.15a are plotted. This result can easily be represented mathematically with the introduction of a lower order symmetry term, an in-plane uniaxial anisotropy given by

$$E_{\mathrm{u}} = K_{\mathrm{u}} \cos^2(\phi - \phi_{\mathrm{u}}), \tag{1.12}$$

where ϕ_{u} represents the direction in the plane for which this energy is maximized. This representation is presented pictorially, for an appropriate choice of K_{u}, in Fig. 1.16, where it clearly illustrates the observed angular dependence in the resonance field in Fig. 1.15b. This result raises the question of why a cubic material (bcc Fe) grown upon a closely matched ($\approx 1.3\%$) cubic substrate (GaAs) should exhibit evidence of a uniaxial distortion in its magnetic anisotropy. Although the answer is not yet known, it undoubtedly arises from the fact that the (0 0 1) surface of the zincblende structure of GaAs is not four-fold symmetric.

The (1 0 0) surface of GaAs is illustrated in Fig. 1.17a for a Ga-terminated surface and Fig. 1.17b for an As-terminated surface. Although there is four-fold symmetry for the atomic locations, the dangling bonds have only two-fold symmetry. Note that they are oriented along [1 1 0] in Fig. 1.17a but along [1 $\bar{1}$ 0] in Fig. 1.17b. When Fe atoms are first deposited on (1 0 0) it is likely that they preferentially satisfy these bonds, and depending upon the nature of the (1 0 0) surface only one type may be available. For example, GaAs(1 0 0) surfaces that are prepared by chemical etching and subsequent vacuum annealing (a standard practice) result in a Ga-terminated surface as in Fig. 1.17a. Hence, the initial growth of Fe on this surface may yield a grain structure oriented along [1 1 0]. Subsequent growth may trap this grain structure as oriented defects which could serve to relieve or trap an oriented strain in the final film. Through

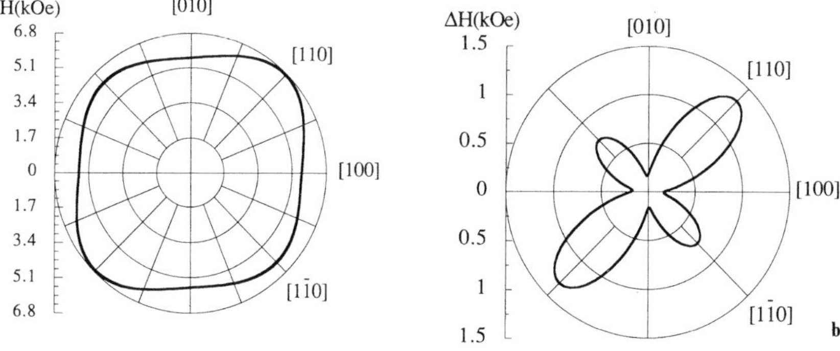

Fig. 1.15. (a) Resonance field value in the (0 0 1) plane of a 200 Å Fe film grown on GaAs obtained at 35 GHz [1.19]; (b) variation of the resonance field between maximum and minimum values

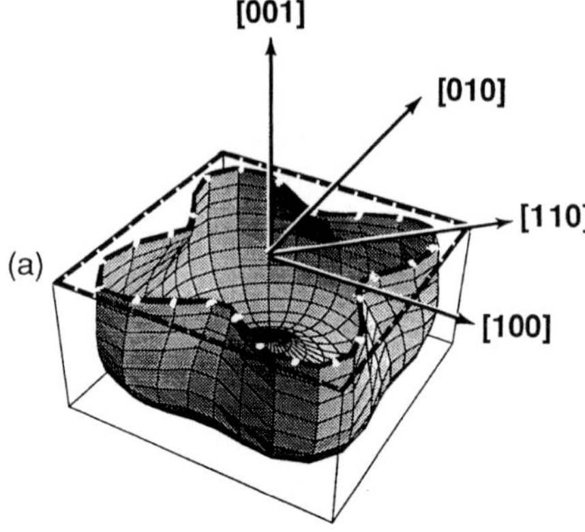

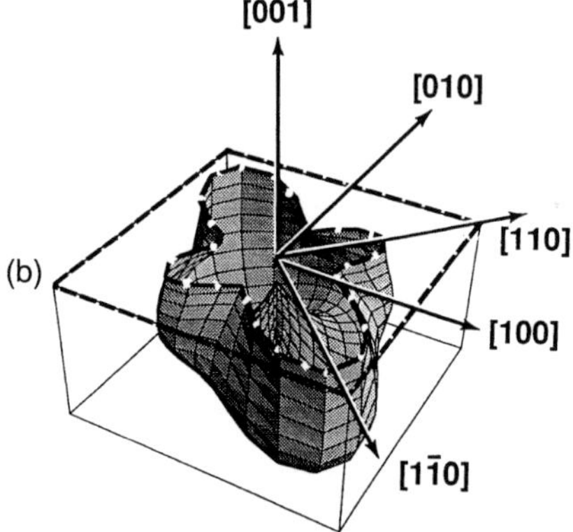

Fig. 1.16. (a) Angular variation of the $K_1 > 0$ anisotropy in $(0\,0\,1)$ plane; (b) effect of adding a uniaxial anisotropy energy term of the form $K_u \cos^2(\phi - \phi_u)$ to the cubic anisotropy

the mechanism of magnetostriction, strains can reveal themselves in the magnetic anisotropy.

Although an explanation based upon this mechanism is plausible, the nucleation, growth and subsequent defect structure of Fe films on $(1\,0\,0)$ GaAs has not been directly observed microscopically. This would be an ideal topic to be addressed in an ultra-high vacuum (UHV) high resolution electron microscope fitted to execute carefully controlled growth studies.

GaAs provides one additional example of this interplay between structure and anisotropy: the growth of Fe on $(1\,1\,0)$ GaAs. The $(1\,1\,0)$ surface of Fe is readily illustrated by "cutting" the three-dimensional anisotropy surface as described earlier, but this time the plane is vertical and contains three important axes: $\langle 0\,0\,1 \rangle$; $\langle 1\,1\,1 \rangle$; and $\langle 1\,1\,0 \rangle$, as illustrated in Fig. 1.18. Figure 1.18b shows the intersection of the surface with the plane, and reveals it to have the outline of butterfly wings. Therefore one would expect if a single crystal film of $(1\,1\,0)$ oriented Fe were grown, it would exhibit a variation in magneto-crystalline energy given by this curve. The minimum in energy is again along the $[0\,0\,1]$ "easy" axis, $[\bar{1}\,1\,1]$ is the maximum energy ("hard" axis), and $[\bar{1}\,1\,0]$ is a local

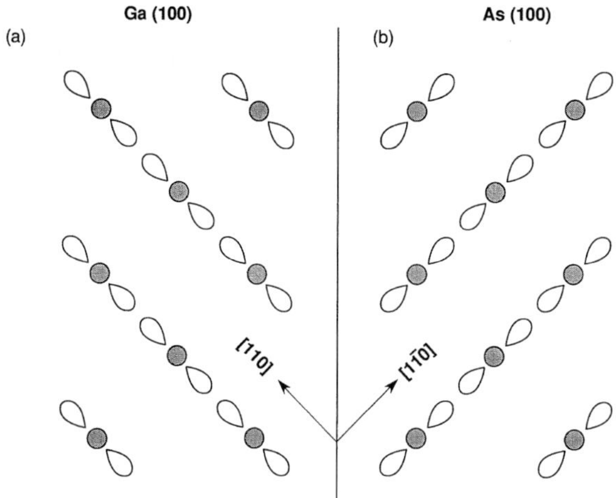

Fig. 1.17. (**a**) Ga-terminated GaAs (0 0 1) surface with dangling bonds indicated as lobe-shaped elements extending out of plane; (**b**) As-terminated GaAs (0 0 1) surface with dangling bonds indicated

minimum called the "intermediate" axis. Mathematically the curve is described by

$$K_1^{(1\,1\,0)} = \frac{K_1}{32}[7 - 4\cos 2\theta - 3\cos 4\theta], \tag{1.13}$$

where θ is measured from the [0 0 1] axis. It is interesting to note that the angular dependence in this lower symmetry plane exhibits both a four-fold and a two-fold rotational symmetry term.

For this surface, the ferromagnetic resonance solution again takes the form

$$\left(\frac{\omega}{\gamma}\right)^2 = (H + \alpha)(H + \beta),$$

where now [1.22]

$$\alpha^{(1\,1\,0)} = 4\pi M + \frac{K_1}{M}[2 - 7\sin^2\theta + 3\sin^4\theta], \tag{1.14}$$

and

$$\beta^{(1\,1\,0)} = \frac{K_1}{M}[2 - 7\sin^2\theta + 3\sin^4\theta]. \tag{1.15}$$

We again approximate $\alpha = 4\pi M$, as before and take $H_0 + 4\pi M \doteq 4\pi M$, to obtain

$$H_0^{(1\,1\,0)} = \left(\frac{1}{4\pi M}\right)\left(\frac{\omega}{\gamma}\right)^2 - \frac{K_1}{M}[2 - 13\sin^2\theta + 12\sin^4\theta]. \tag{1.16}$$

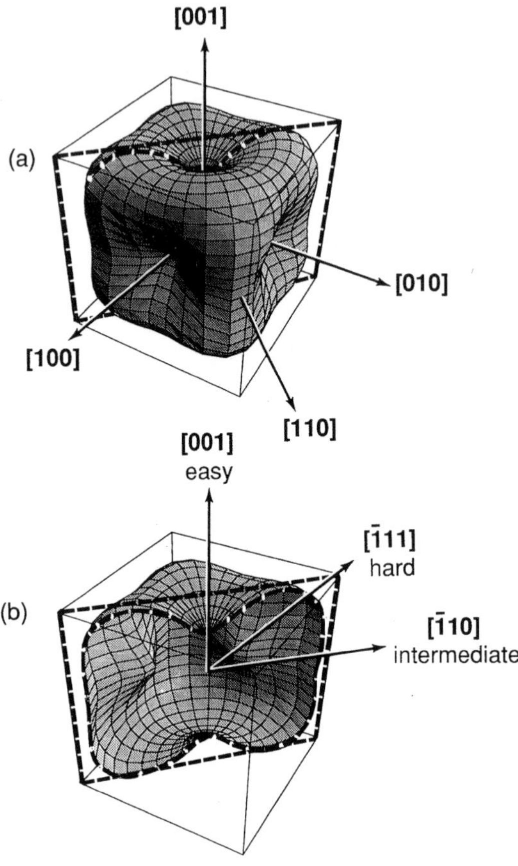

Fig. 1.18. (a) Dashed line indicates intersection of (1 1 0) plane with $K_1 > 0$ cubic anisotropy energy surface; (b) exposed edge of energy surface reveals angular dependence of anisotropy energy in (1 1 0) plane

This can be recast to the form

$$H_0^{(1\,1\,0)} = \left(\frac{1}{4\pi M}\right)\left(\frac{\omega}{\gamma}\right)^2 - \frac{K_1}{M}[\cos 2\theta + 3\cos 4\theta]. \tag{1.17}$$

We see that this contains the same angularly dependent terms as the anisotropy energy surface given by (1.13) except that the resonance experiment, through the second derivative process, emphasizes the higher order terms by generating larger relative prefactors.

As discussed earlier, this angular dependence can be readily measured using ferromagnetic resonance and already has been for Fe films grown epitaxially upon GaAs [1.21]. The results of that study are illustrated in Fig. 1.19, which shows the angular dependence of the ferromagnetic resonance field in the (1 1 0) plane for Fe films grown to three different thicknesses on GaAs(1 1 0). The curve labeled "200 Å" is quite similar to what one would expect for (1 1 0) Fe described by (1.17), except that the energy minimum in the [0 0 1] direction is deeper. The

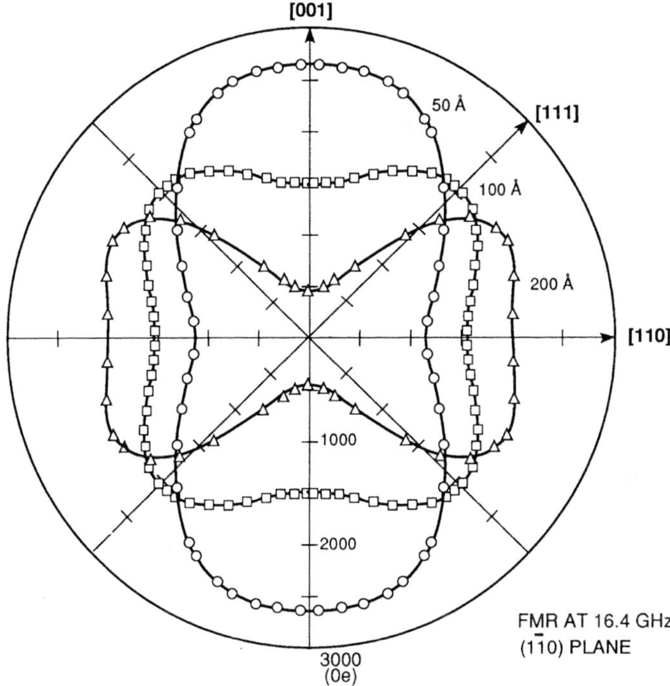

Fig. 1.19. Angular dependence of the ferromagnetic resonance field at 16 GHz in the (1 1 0) plane for Fe films grown on GaAs for different thicknesses [1.21]

curve for the 100 Å film shows almost perfect four-fold symmetry, again not the dependence given in (1.17). Finally, the curve for the 50 Å is dramatically altered to the extent that the [0 0 1] axis is now an energy maximum. Thus we see the evolution from 200 Å to 50 Å thickness reverses the role of easy axis to hard axis.

This behavior is easily described by again introducing an in-plane uniaxial anisotropy of the form

$$E_{\mathrm{u}} = K_{\mathrm{u}}\cos^2\theta. \tag{1.18}$$

It is necessary, however to allow both K_1 and K_{u} to vary with the film thickness in order to describe the observed behavior. This is illustrated in Fig. 1.20 for a series of samples ranging from 20 Å to 120 Å in thickness. The data indicates that "bulk" behavior is approached for K_1 once films exceed a thickness of ≈ 100 Å. This is in agreement with the common experience that a 1% strain in an epitaxial film is typically relieved at these thicknesses, and also that strain is a common mechanism for altering the anisotropy through magnetostriction. However the presence and variation of the uniaxial anisotropy, represented by K_{u}, is completely unexpected. Its effect on the three-dimensional anisotropy energy surface is illustrated in Fig. 1.21, where values were chosen for K_1 and K_{u} to reproduce the data of Fig. 1.19.

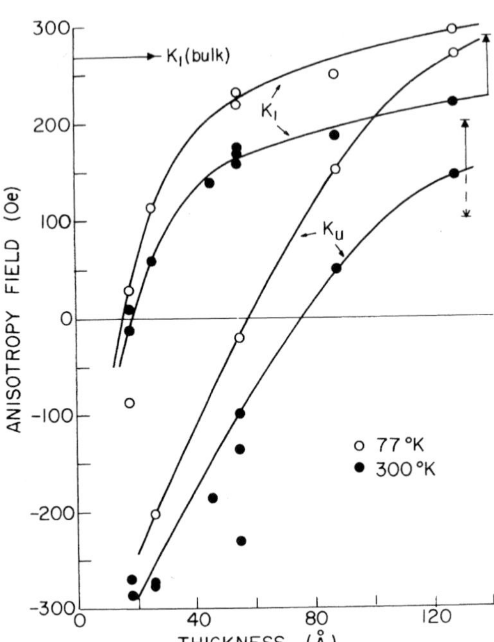

Fig. 1.20. Dependence of the magnetic anisotropy constants upon film thickness. Arrows indicate the calculated temperature shifts expected [1.23]

In order to understand the origin of the uniaxial anisotropy, we again return to the GaAs/Fe interface to search for a clue. Figure 1.22 shows the atomic locations for Ga and As in the (unreconstructed) (1 $\bar{1}$ 0) face. Notice that they form zig-zag ridges, with troughs separating them, parallel to the [1 1 0] direction. Each of these atoms has a dangling bond projecting up out of the surface perpendicular to these ridges. In a recent study of the growth of Fe on (1 1 0) GaAs by STM [1.24], images of this surface with 0.1 Å coverage were obtained which indicated that the initial growth was in the form of small clusters elongated along the [1 1 0] direction. Although this first work may not be definitive, it could indicate some anisotropy in the initial nucleation and growth.

1.1.2 bcc Co

Body-centered cubic cobalt is not a known thermodynamic phase occurring naturally. The first clue that such a phase might exist was found in the Fe–Co alloy phase diagram [1.25] which supports the phase up to 75% Co. An estimate of the lattice constant for a possible bcc phase of Co was obtained by extrapolation, as shown in Fig. 1.23. This lattice constant of 2.819 Å is within 0.4% of that of GaAs divided by two (i.e. 2.825 Å), which had already been shown to support the growth of bcc Fe ($a_0 = 2.867$). The growth of bcc Co on GaAs has now been studied in a number of laboratories using a variety of

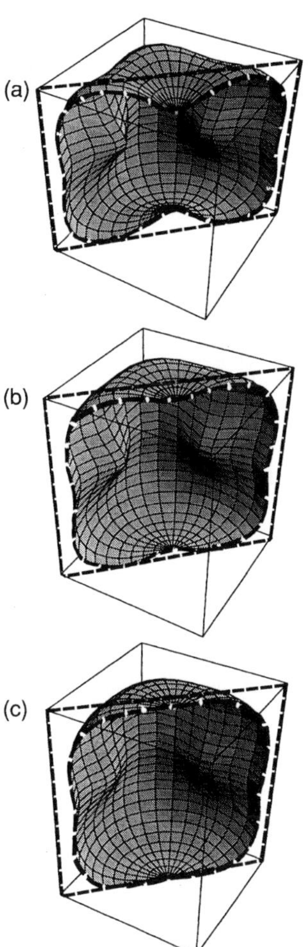

Fig. 1.21. Variation in the anisotropy energy surface as a function of K_1 and K_u with the $(1\,1\,0)$ edge exposed: (**a**) $K_1 = 1.0$, $K_u = 0$; (**b**) $K_1 = 0.5$, $K_u = -0.1$; (**c**) $K_1 = 0.3$, $K_u = -0.1$

techniques, establishing it as one of the most studied of the 3d magnetic metastable phases, and has stimulated much of the theoretical search for new phases.

RHEED data taken during growth are shown in Fig. 1.24 along with the analogous RHEED patterns for the GaAs$(1\,1\,0)$ substrate and a Fe$(1\,1\,0)$ film. One immediately sees that the film is a single crystal, of approximately the same lattice spacing as bcc α-Fe and possessing the same in-plane symmetry and orientation. Also shown is the diffraction pattern obtained when a critical thickness is exceeded and the film transforms to the low energy stable phase α-Co(hcp).

X-ray diffraction measurements, made to obtain the interplanar atomic spacing perpendicular to the film plane are shown in Fig. 1.25. One sees the $(1\,1\,0)$ Co diffraction peak lying on the shoulder of the GaAs$(2\,2\,0)$ exhibiting

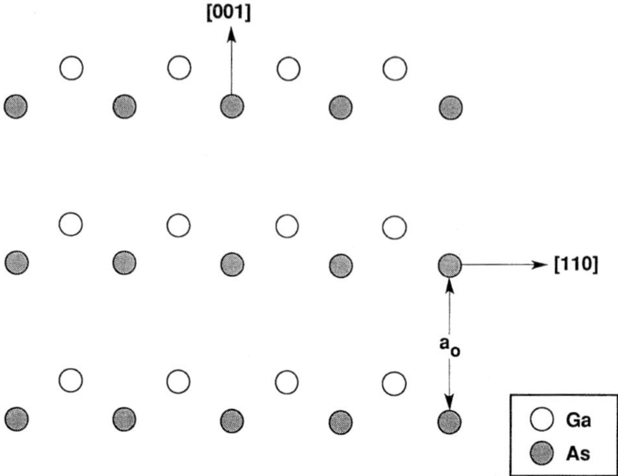

Fig. 1.22. Unreconstructed (1 $\bar{1}$ 0) face of GaAs

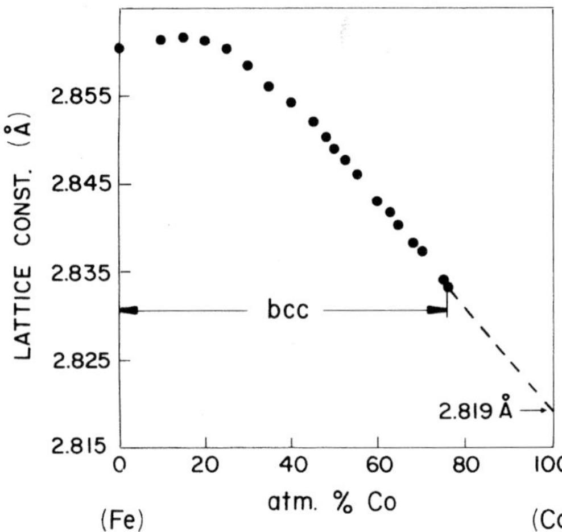

Fig. 1.23. Lattice constant versus composition for the Fe–Co alloy system [1.26]

only a 0.1% contraction of the lattice planes compared to GaAs. Extended X-ray Absorption Fine Structure (EXAFS) exploiting an electron conversion technique especially suitable for thin films, yielded real space atomic distributions shown in Fig. 1.26, where bcc Fe, hcp Co and fcc Cu data are provided for comparison. It is clear that the Co/GaAs structure is bcc and the determined lattice constant (2.82 $\pm$ 0.001 Å) compares well with both the X-ray and extrapolated alloy values. Finally, AED data, shown in Fig. 1.27, were obtained for Co deposition on GaAs(0 0 1) and ZnSe(0 0 1). The pattern on GaAs exhibits a

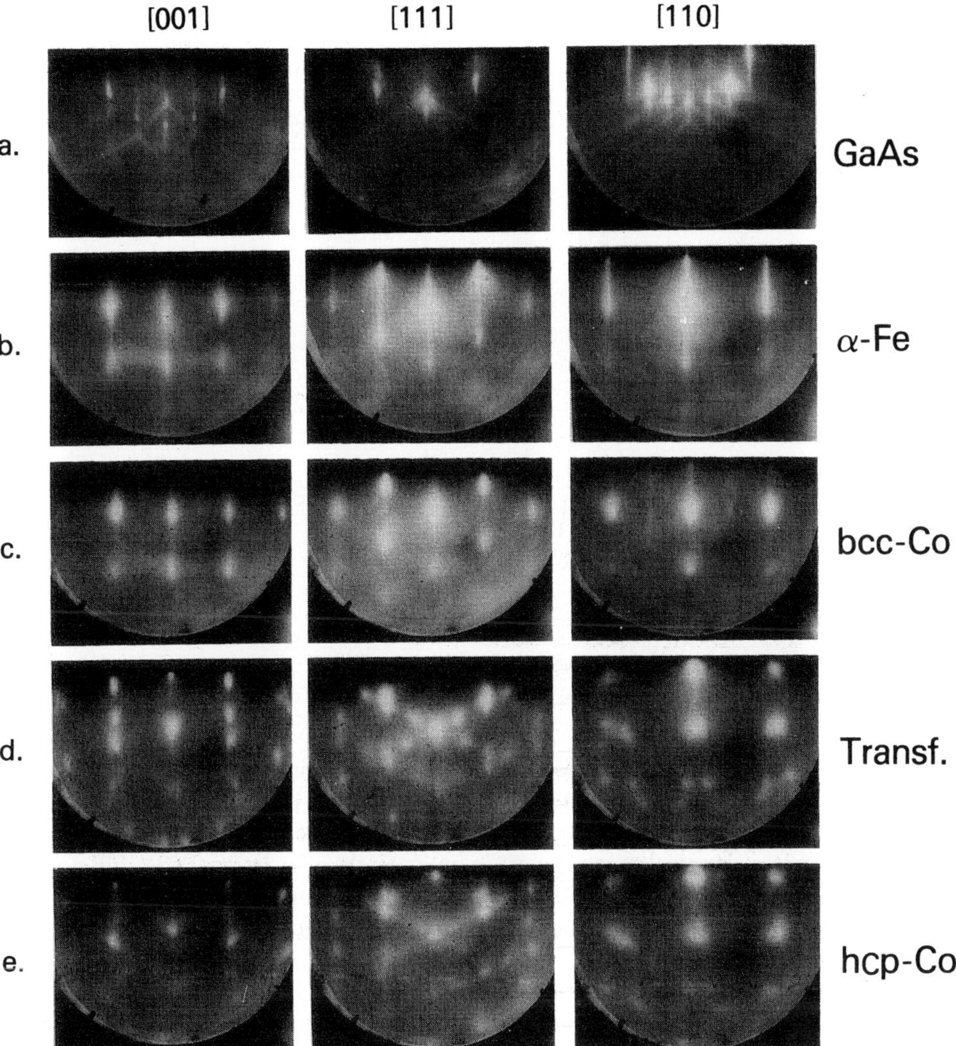

Fig. 1.24. RHEED patterns along the principal axes in the (1 $\bar{1}$ 0) face of: (a) GaAs; (b) α-Fe; (c) bcc-Co; (d) twinning transformation; (e) hcp-Co with (11·0) [00·1] Co ∥ (1 1 0) [1 1 1] GaAs [1.27]

structure characteristic of a bcc (0 0 1) surface, with strong forward scattering peaks at normal emission (0°) and $\pm$ 45° at only 4 ML coverage. These scans reveal a predominantly three-dimensional mode of growth, similar to that observed for Fe on GaAs discussed previously. In contrast to this bcc single-crystal growth, deposition of Co on ZnSe(0 0 1) epilayers, pseudomorphic with a GaAs substrate, results in a poorly ordered multicrystalline-phase. While it exhibits some features suggestive of cubic ordering, its peaks are poorly defined,

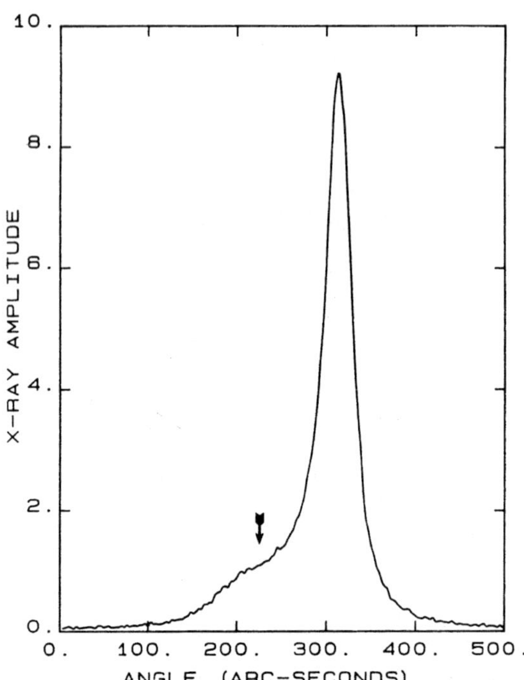

Fig. 1.25. Double-crystal X-ray rocking curve for a 357 Å Co film on a GaAs substrate. The arrow indicates the position of the Co (1 1 0) peak on the shoulder of the GaAs (2 2 0) peak [1.28]

with a high level of high-angle emitted electrons. Similar indications are obtained from the RHEED pattern of Co on ZnSe(0 0 1), shown in Fig. 1.28. XPS data from the Se 3d level for Co deposition on ZnSe, depicted in Fig. 1.29, reveals even less chemical reactivity of Co on ZnSe than for Fe on ZnSe, and no evidence for compound formation even after 250 °C annealing. These results suggest that chemical bonding is important at the interface in order to establish bcc Co at a 175 °C growth temperature.

Theoretical treatment of bcc Co was first reported [1.30] in 1983, with the band structure calculated for $a_0 = 2.77$ Å. After experimental synthesis of this phase was demonstrated [1.31] in 1985, total energy calculations illustrated in Fig. 1.30 yielded the bcc phase with a minimum at $a_0 = 2.82$ Å, some 70 meV/atom above the minimum for the fcc phase. More recent work has shown that this bcc minimum is stable only to uniform cubic strain. It is unstable in the presence of a volume conserving tetragonal strain, which carries the system to the fcc minimum [1.33]. Epitaxial growth, by pinning the structure at the interface, stabilizes the bcc phase until it reaches a critical thickness at which point the accumulated energy difference between the bcc and fcc phases exceeds the energy provided by the interface. The system then structurally relaxes. This thickness depends upon growth conditions and interface preparation, but is typically 30 ML to 40 ML, which corresponds to an accumulated energy of 2–3 eV. This is a typical binding energy/atom for a metal on GaAs.

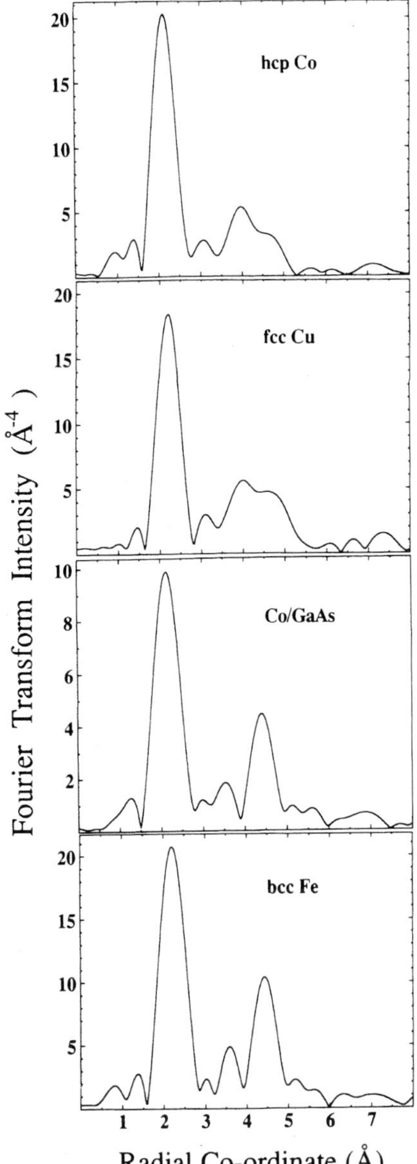

Fig. 1.26. Fourier transform of the isolated fine structure for an hcp Co rod; a fcc Cu foil; a Co film deposited on GaAs (1 1 0); and a thick film of bcc Fe [1.29]

Magnetic characterization of bcc Co has yielded a magnetic moment of $1.53\mu_B$ per atom [1.31], lower than the theoretical value of $1.7\mu_B$ predicted from first-principles total energy calculations [1.32]. Subsequent nuclear magnetic resonance measurements obtained a broad resonance line width (75 MHz) centered at a frequency corresponding to the observed magnetic moment [1.28].

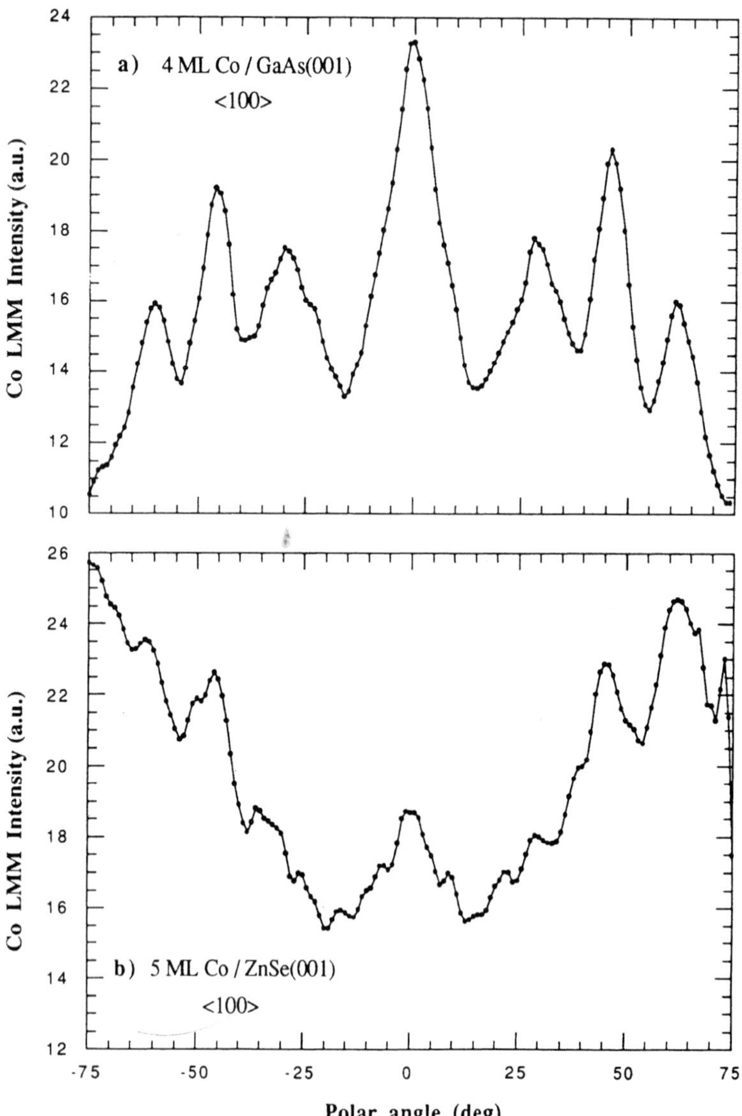

Fig. 1.27. (a) AED polar scan along a $\langle 001 \rangle$ azimuth for 4 ML of Co grown on GaAs (0 0 1) at 175 °C. The structure is that expected from a bcc overlayer. (b) The same scan obtained from 5 ML Co coverage on ZnSe (0 0 1) [1.14]

More recent NMR studies on (0 0 1) films confirm this low average value for the moment, but Polarized Neutron Reflection (PNR) studies of those films show a gradient in the moment [1.34]. At the GaAs interface the moment falls to $1.0\mu_B$ but rises to $1.7\mu_B$ at the center of a 100 Å thick film. This center value matches the theoretical prediction and supports the speculation that chemical reactivity

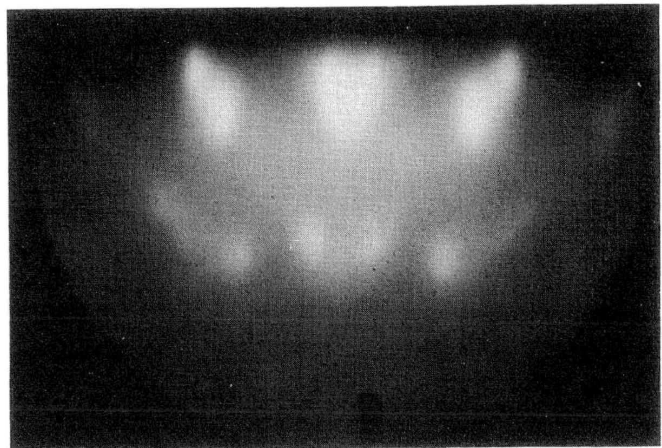

Fig. 1.28. RHEED pattern obtained along the ⟨1 1 0⟩ azimuth from 5 ML of Co deposited at 175 °C on the ZnSe (0 0 1) surface, corresponding to AED data in Fig. 1.27b [1.14]

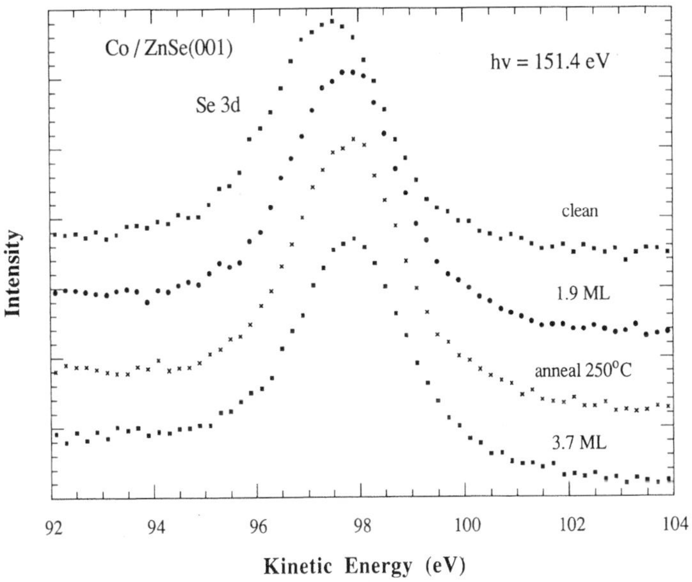

Fig. 1.29. XPS spectra (hv = 151.4 eV) of the Se 3d level at various stages in the deposition of Co on ZnSe (0 0 1). The spectra have been shifted vertically for clarity

at the GaAs interface decreases the magnetic moment in Co films as it does in Fe films.

Detailed ferromagnetic resonance studies confirmed the prediction obtained by extrapolation of Fe/Co alloy properties that the cubic anisotropy constant

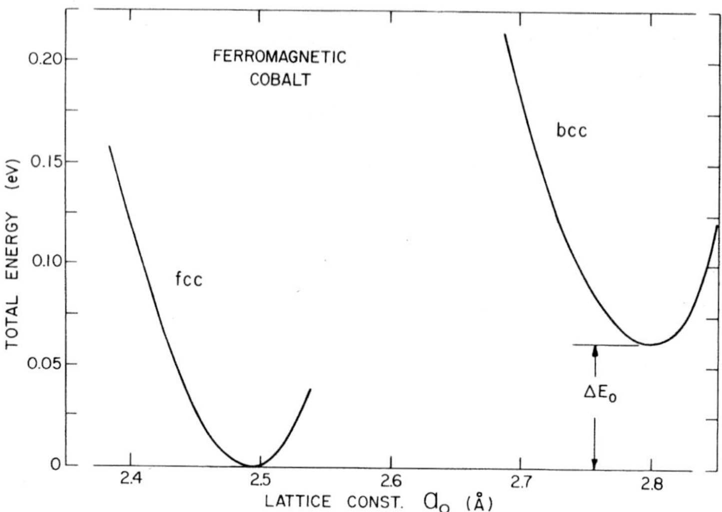

Fig. 1.30. Relative energies of the two ferromagnetic cubic phases of Co as a function of lattice constant, adapted from [1.32]. The scale represents the true lattice constant of a bcc structure, but $(\sqrt{2})^{-1}$ times the true lattice structure of the fcc structure

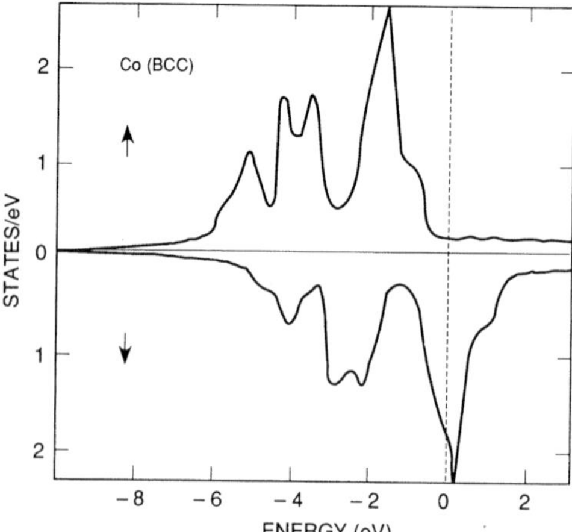

Fig. 1.31. Spin resolved density of states for bcc Co [1.37]

K_1 in bcc Co should be opposite in sign and twice the magnitude of that found in bcc Fe [1.35]. Finally, recent Brillouin light scattering studies have yielded a value for the exchange stiffness constant in bcc Co of $D = 2.6$ (± 0.3) $\times 10^{-9}$ Oe cm^2 [1.36]. This value can be compared to that obtained from hcp Co of $D = 4.04 \times 10^{-9}$ Oe cm^2. Thus D scales from one structure to another for spin wave excitations near $k = 0$, exactly as the number of near neighbors.

One of the most interesting properties of bcc Co is its electronic structure. The density of states (DOS) obtained from total energy calculations is shown in Fig. 1.31. The majority states are completely filled and there exists a high DOS for minority spins at the Fermi level. This description has been confirmed by spin-polarized photoemission [1.38] and spin-polarized electron energy loss spectroscopy [1.39]. One should expect, from this electronic state distribution, that bcc Co is energetically unstable. However, captured as an epitaxial structure on GaAs, it is available for basic research as well as technological exploitation.

1.1.3 τ-MnAl

In addition to the elemental ferromagnetic metals, bcc Fe and bcc Co, a compound ferromagnetic, τ-MnAl, has been successfully grown on a zinc-blende substrate [1.40]. τ-MnAl is a metastable phase obtained in the bulk by quenching $M_{55}Al_{45}$ from the high temperature hexagonal ε-phase through the eutectoid point at 870 °C. The epitaxial growth of these films was possible because the unit cell of τ-MnAl ($a_0 = 2.77$ Å) has nearly a factor of two relationship to that of AlAs ($a_0 = 5.66$ Å). Its crystal structure is similar to the ordered cubic CsCl structure exhibited by the transition metal aluminides (FeAl, CoAl, NiAl), which have all been successfully grown on III–V semiconductors. However, τ-MnAl is tetragonally distorted ($c/a = 1.28$).

The epitaxial growth of this compound is more sophisticated than a simple elemental metal film. First, an AlAs(0 0 1) epitaxial film is grown on GaAs(0 0 1) to establish an Al-compound lattice matched surface. Then a thin (< 50 Å) amorphous Mn–Al film is deposited at room temperature. This amorphous film is then annealed at 100–300 °C in order to crystallize it into the τ-MnAl phase. This surface now serves as a template to carry out further growth of τ-MnAl by co-deposition of Mn and Al at 200 °C. Following growth a post-deposition annealing at 400 °C improves the crystal structure. Magnetic characterization of these ferromagnetic films indicates a strong uniaxial anisotropy along the tetragonal axis perpendicular to the film plane. A coercive field of 5 kOe was observed at room temperature and a very square hysteresis loop with high remanence is obtained in optimally grown films.

1.2 3d Transition Metals on the Diamond Structure

The group IV elements, C, Si and Ge are all found naturally in the diamond structure, although for diamond itself this is a metastable phase. The diamond structure is obtained from the zincblende structure by merely allowing all of the atomic sites to contain the same species. Thus diamond, like zincblende, is formed by nesting together two fcc lattices, which are displaced from each other along a body diagonal by 1/4, the unit cell containing eight atoms. The unit cell face can thus be pictured as a face-centered square of length 3.567 Å(C), 5.43 Å(Si) or 5.657 Å(Ge). The first of these in Fig. 1.1 is seen to be well-matched to the fcc transition metals Fe, Co and Ni while the last is a good two-fold match to the bcc transition metals Fe, Co, Cr. Si falls between these two and is not well-matched to either group.

1.2.1 fcc Ni on C

The only published work of a magnetic transition metal on diamond [1.41] is for fcc Ni ($a_0 = 3.5238$ Å), yielding a lattice mismatch of 1.2% (Ni under tensile stress). Deposition at a substrate temperature of 500 °C in a vacuum of 10^{-8} Torr yielded a highly textured Ni (0 0 1) film with faceted morphology on C (0 0 1), indicative of three dimensional growth without long-range order. In contrast, growth in a commercial MBE system, utilizing electron beam (e-beam) sources, in a vacuum $< 10^{-9}$ Torr at 500 °C yielded two-dimensional growth. No magnetic characterization of the films was provided.

1.2.2 Cu on Si

Si does not provide a close lattice match to any of the elemental magnetic metals. Considered as a two-fold lattice match to the bcc structures, Fig. 1.1 shows it to be 4% smaller than bcc Co and 5% smaller than bcc Fe. If one treats it as a 45° rotated lattice, then it is 6% too large to match fcc Cu and 8% too large for the worst case, fcc Ni. Nevertheless, there are reports describing the growth of fcc Cu on Si in 10^{-7} Torr yielding oriented, aligned films of Cu(0 0 1) on Si(0 0 1) with Cu [0 1 0] ∥ Si [0 0 1], or Cu(1 1 1) on Si(1 1 1) [1.42]. These films, effectively a replacement for single crystal Cu substrates, have then been used to grow fcc metals Ni, Co, Rh, Ir, Pd, Au, Ag, Pt and Al as well as bcc metals Fe, Cr, V, Mo and W [1.43]. Magnetic characterization of films of fcc Co and NiFe alloys grown this way indicate magnetic anisotropies characteristic of oriented crystalline films [1.44]. Although much of this work still remains unconfirmed in other laboratories, this use of Si as a substrate material could prove important both scientifically and technologically. An important issue in the metallization of Si is the reaction of transition metals with the clean Si surface to form silicides. The most thoroughly studied are $NiSi_2$ and $CoSi_2$ which form single crystal epitaxial films on Si in the CaF_2 cubic crystal structure. The lattice mismatch of these two silicides with Si are 0.4% and 1.2% respectively. Although these silicides do not appear to be useful magnetically (they are both paramagnetic) they are of interest to the semiconductor community since their high conductivity provides epitaxial conductors on Si.

1.2.3 Fe on Ge

From Fig. 1.1 one sees that Ge has essentially the same lattice constant as GaAs and therefore is an attractive candidate for use as a substrate for epitaxial growth of the bcc 3d transition metals. Furthermore, it is a material readily available commercially in extremely high purity, as single crystals with very low defect density and in large format sizes as polished wafers. Unfortunately, Ge forms a large number of alloys with Fe [1.45]. Studies of epitaxial growth carried out at $\approx 150\,^{\circ}$C showed excellent RHEED patterns, however films up to 100 Å exhibited no magnetic moment [1.46]. Alloys in which the Ge concentration equals or exceeds the Fe concentration are all antiferromagnetic, and the high rate of interdiffusion at the interface creates a magnetically "dead" region. This is particularly unfortunate since Ge offers an elemental material which is closely lattice-matched to bcc Fe and bcc Co and could thus serve to grow superlattice structures of alternating magnetic metal/semiconductor layers. GaAs and the other zincblende compounds are not suitable for this because of the anti-phase problem (i.e. the inability of an elemental surface to define the order of constituents in a compound overlayer, which results in nucleated islands that are equally likely to be in-phase or out-of-phase when their growth fronts come into contact). Lower temperature growth of Fe on Ge decreases the interface interdiffusion but results in highly defected films with rough surfaces.

1.3 Rare Earths

After the 3d metals, the other family of magnetic metals are the 4f rare earths. These metals, generally occurring in the hexagonal close packed (hcp) structure, exhibit a rich variety of magnetic ordering. These include both ferromagnetic and antiferromagnetic order, with the latter manifesting themselves in both collinear as well as various helical arrays. The earliest work centered on rare earth metal films grown on Al_2O_3 substrates [1.47]. In order to prevent an oxidizing reaction between the rare earth metals and the Al_2O_3, a buffer layer of Nb was first deposited and then a layer of Y metal to provide a template for the subsequent deposition of the rare earth metal. A typical example would be the deposition of Dy on Y. Quite good epitaxial single crystal films were obtained this way, however the misfit of 1.6% does introduce strains. The magnetic properties of the rare earths with very high anisotropy are particularly sensitive to strain. This may explain the shifts in the ferromagnetic ordering temperature with decreasing film thickness seen for the $Dy/Y/Nb/Al_2O_3$ film system.

In 1988 *Farrow* et al. [148] reported a new approach to growing rare earth metal films employing GaAs $(\bar{1}\,\bar{1}\,\bar{1})$ substrates. In this case a buffer layer of LaF_3 was deposited first to act both as a chemical buffer as well as crystalline template. LaF_3 is hexagonal with a unit cell base dimension $a_0 = 4.148$ Å. This is approximately 1/2 of the face diagonal of the GaAs unit cell $\sqrt{2}a_0 = 7.995$ Å.

Excellent single crystal growth of LaF_3 was obtained in a commercial MBE machine by sublimation from the compound source material. Subsequent growth of Dy on LaF_3 results in films whose magnetic transitions at 178 and 115 K occur at the same temperature as bulk crystals over a broad range of film thickness (3000–75 Å). This choice of the rare earth fluorides for the buffer layers seems especially promising since solid solutions of LaF_3 with other rare earth fluorides allows one to "tune" the buffer layer lattice constant to match the various rare earth metals. While there is very little evidence of chemical reaction between the buffer and the underlying GaAs substrate, there is unfortunately considerable interaction between the rare earths themselves. At the interface there is a Dy $\leftrightarrow$ La interchange, resulting in a diminished magnetic moment in finished Dy films [1.49].

1.4 Applications

The growth of magnetic films on semiconductors has not only made important contributions to fundamental physics studies, but has also opened up opportunities for technological applications. This of course stems from the fact that the prodigious developments of planar electronics now make semiconductor substrates the starting point for virtually all electronic devices. The incorporation of magnetic elements into those devices remains a largely unexplored field of research. However, some recent developments are underway, and a few of these new directions shall now be discussed.

1.4.1 Non-Volatile Magnetic Memory

The first example is the demonstration of a commercial non-volatile ferromagnetic memory concept [1.50]. This memory, called MRAM (Magneto-resistive Random Access Memory) is based upon arrays of ferromagnetic sandwich structures (Fig. 1.32). Each sandwich consists of two layers of ferromagnetic permalloy separated by an exchange-breaking layer between them. The two ferromagnetic layers are magnetized in the film plane, but antiparallel to each other. The fringing fields from each layer are captured by the other, thus effectively eliminating coupling between neighboring memory elements. This represents an important advance over earlier thin-film ferromagnetic memory concepts, which were plagued by cross-coupling between elements, preventing high packing densities. A unique easy axis of magnetization is formed in the magnetic layers by depositing them under the influence of a biasing magnetic field. The two anti-aligned layers define a "0" bit when oriented in one sense along this axis and a "1" bit when they are both reversed. This reversal is accomplished by a current pulse through an overlaid conducting line. The sense of which bit state is stored is accomplished by measuring the change of

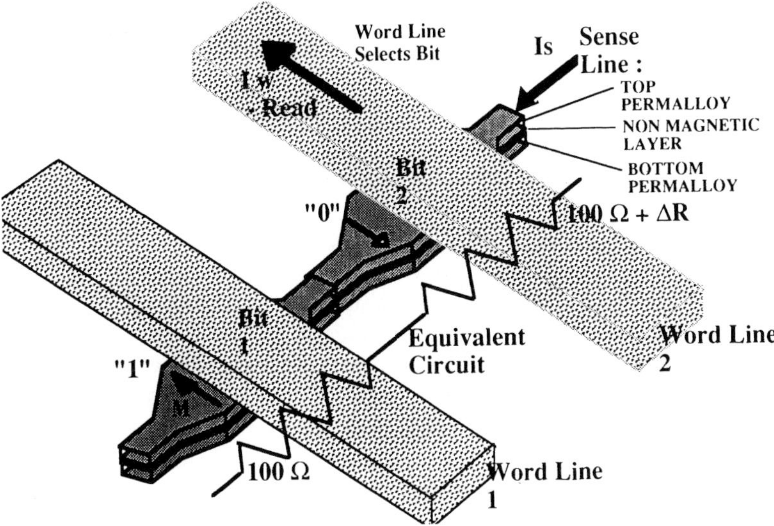

Fig. 1.32. Magneto-resistive Random Access Memory (MRAM) elements. Two elements are shown, one set as "0", the other as "1", with the overlaid conducting lines used for reading and writing [1.50]

resistance in the sandwich structure under the influence of a low current in the same overlay line. The change is caused by altering the relative orientation of the magnetization and the sense current direction through the sandwich. These magneto-resistance changes are of the order of $\approx 2\%$. In order to improve the performance of this device, it is important to increase the magnitude of this change. An increase in the magneto-resistance shows up proportionally as an increase in the measured signal. One may use this increase to lower the device power or increase its speed. Since the frequency bandwidth is proportional to the square of the signal, a factor of 10 increase in signal leads to a factor of 100 increase in speed of operation. MRAMs would then be faster than current Dynamic Random Access Memory (DRAM) with the added advantages of being cheaper to fabricate (fewer processing steps) and being intrinsically non-volatile as well as radiation-hard.

1.4.2 Microwave Devices

One of the most useful properties of magnetic materials is the fact that they can couple to a radiation field. In the microwave region of the spectrum this coupling occurs when the magnetization vector M_0 is driven by the h component of the radiation. In the most common case with a thin film, the magnetization M_0 lies in the plane of the film. Plane-polarized radiation propagating normal to the plane of the film, as illustrated in Fig. 1.12, exerts a torque on the magnetization vector when $h \perp M_0$ which causes it to exhibit gyroscopic rotation. As discussed previously, this system is driven into resonance when

$$2\pi v_0 = \gamma \sqrt{H_{\text{eff}}(H_{\text{eff}} + 4\pi M_0)}, \qquad (1.19)$$

where v_0 is the frequency of the radiation, γ is the gyromagnetic ratio, and H_{eff} is the externally applied magnetic field plus any internal anisotropy field.

For Fe films, where $4\pi M_0 = 21\,500$ Oe, this resonance occurs near 10 GHz when $H_{\text{app}} = 0$. Here $H_{\text{eff}} = H_{\text{an}} + H_{\text{app}}$ and $H_{\text{an}} \doteq 500$ Oe. This is a very useful frequency regime for many microwave devices. An example is shown in Fig. 1.33, a schematic drawing illustrating a microwave stripline device concept. Striplines are an effective way to carry microwave signals in planar circuits. They are guided wave devices into which, in this illustration, a microwave signal is injected on the left from a coaxial cable. The n$^+$-GaAs layer acts as a ground plane and the structure sustains an electric field between the ground plane and the conducting strip located above the insulating layer of GaAs. The radiation field is thus plane-polarized with its electric vector e vertical and corresponding h vector horizontal and it propagates down the device structure from left to right, emerging on the other end where it is again picked up by another coaxial cable. If a ferromagnetic metal section replaces the conducting strip, as shown, then in that region the radiation field can transfer energy into the ferromagnetic metal if the magnetization vector M is parallel to the direction of propagation. In this orientation, the ferromagnetic resonance condition can be satisfied that is ($M \perp h$), and the radiation field can couple to the magnetization. It will, however, couple effectively only at the ferromagnetic resonance frequency $\omega_0 = 2\pi v_0$ given above, so that a display of power transmitted versus frequency will show a sharp decrease in transmission at v_0. This device is called a "notch"

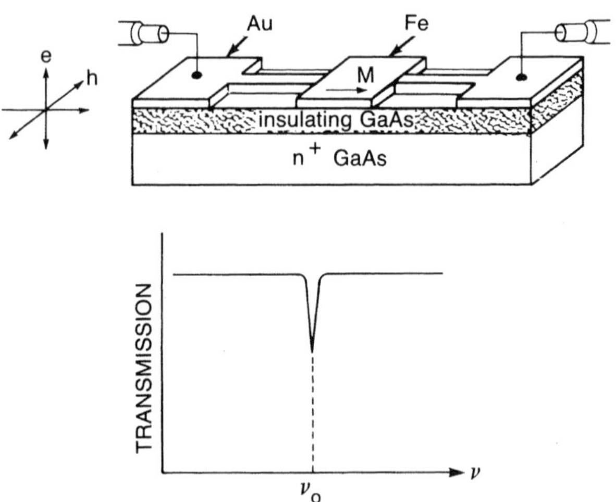

Fig. 1.33. Stripline microwave "notch" filter, where v_0 is defined by (1.17)

filter. It should be noted that the location of the "notch" can be moved by applying an external magnetic field, as indicated by 1.19, parallel to the magnetization direction M. Furthermore, if an external magnetic field is applied perpendicular to the direction M, as shown in Fig. 1.33, and has sufficient size to reorient the direction of M perpendicular to the direction of signal propagation, the radiation field can then no longer couple to M, since now $h \parallel M$. This will effectively turn the filter off. As discussed earlier, this requires an applied field of < 10 Oe, easily achieved in a device configuration. The device described here is not intended to be a prototype design, but rather a pedagogical example to illustrate the concepts involved. Real devices are currently under study and have been reported [1.51].

The width of the "notch" in the filter device described is just the line width of the ferromagnetic resonance observed in the film. This line width is a measure of the quality of the films, insofar as there are variations in M or in the internal anisotropy field caused by variations in strain, thickness or composition. In this regard, the epitaxial Fe films grown on ZnSe have shown the narrowest linewidths ever observed for a ferromagnetic metal (45 Oe at 35 GHz) [1.52] which become comparable to those observed in ferrimagnetic insulators (ferrites). Ferromagnetic metals were abandoned years ago for high frequency device applications because their observed linewidths were too large, compared to ferrites. These recent results for epitaxial metal films suggest that those conclusions should be reconsidered. This is especially true for higher frequency applications where the low $4\pi M$ of ferrites (≈ 2 kOe) keeps their zero field resonance frequency low, well below the range of interest for most high frequency planar device applications.

1.4.3 Spin Injection Devices

One of the most interesting uses of ferromagnetic metal films is as a source of spin-polarized carriers. If one considers the simple Stoner picture of an ideal rigid-band saturated ferromagnetic metal, the majority spin d states lie below the Fermi level and are all filled, but the minority states are filled only up to the Fermi level. Assuming the s- and p-state electrons to be unpolarized, one might expect that some fraction of the carriers will be polarized to the extent that the minority d states contribute to the conductivity. One could exploit this condition to obtain polarized carriers by using a ferromagnetic metal film as an electrical contact on a non-magnetic metal or semiconductor.

The pioneering work on this concept was reported in 1970 by *Meservey* et al. [1.53] in which they carried out tunneling experiments from a ferromagnetic metal film through an Al_2O_3 barrier into a superconducting metal film (Al). The superconducting film, with its sharply peaked density of states at the superconducting gap, acted as a spin-polarized electron detector in the presence of an applied magnetic biasing field. The tunneling current, originating from the ferromagnetic film, proved to be highly polarized ($+44\%$ for Fe, $+34\%$ for Co and $+11\%$ for Ni) [1.54] where the polarization is defined as $P = (n_+ - n_-)/(n_+ + n_-)$, $n_{\pm}$ being the number of carriers of each spin character.

This was a surprisingly high degree of polarization, but even more surprising was that the polarization direction was that for the majority spins. For the simple band structure assumed, carriers would only be available from the minority spin-states near the Fermi level. After several years, an explanation was finally provided by *Stearns* [1.55], who pointed out that while most of the d electrons are tightly bound, a small number of d electrons possess a more itinerant character. These itinerant d states are the main component of the tunneling current. The relative number of majority and minority spins in these itinerant d-bands is proportional to the magnetic moment of the material and thus accounts for the observed sign and magnitude of the polarization of the tunneling current.

A more recent achievement in this field was the injection of spin-polarized carriers into a paramagnetic metal by *Johnson* and *Silsbee* in 1985 [1.56]. This is illustrated in Fig. 1.34 where the paramagnetic metal (Al) is shown as a long bar with two pads of ferromagnetic metal deposited upon it to act as a source and drain of carriers. These carriers will be polarized, since they originate in a magnetized material as shown in the figure. The injected carriers travel through the bar and are picked up at the second ferromagnetic pad. The physics is illustrated, again in a simple Stoner picture, as minority electrons entering the paramagnet, going into unoccupied minority states at the Fermi level; traveling through the metal, under the influence of the applied electric field; finally, leaving the metal by entering unoccupied minority states in the second ferromagnetic pad. The pads may thus be thought of as polarizing filters acting on the current passing through the circuit. If this picture is correct, the polarized current in the paramagnet should act to unbalance the equally populated up- and down-spin bands, inducing a net magnetization. Because there will be spin-flip scattering events during the transit through the paramagnet, the magnitude

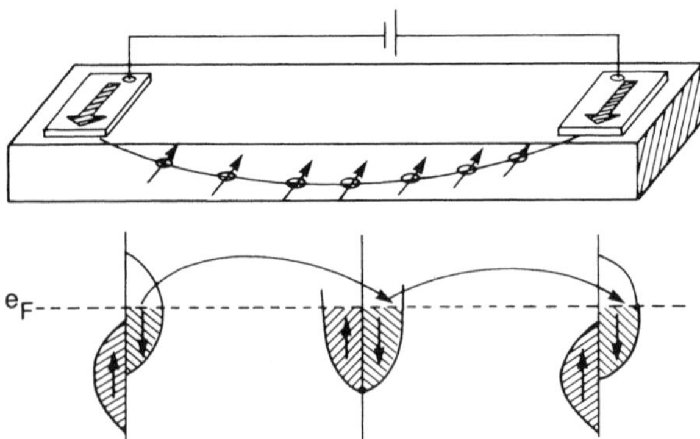

Fig. 1.34. Injection of spin-polarized current into a paramagnetic bar from ferromagnetic pads [1.57]

of the induced magnetization should decrease along the path. Its measurement would thus yield a determination of the spin-flip scattering length. This measurement was carried out by putting thin-film SQUID pick-up loops down on the surface of the bar, in a regular array between the two pads, and measuring the induced magnetic moment. It was found that the polarization relaxed with a decay length of 100 μm at 40 K [1.57]. It should be noted that, since the two pads act as polarizing filters for the current, by reversing one of them the polarized current can be blocked, much in the same way that crossed optical polarizers can block light passing through them.

This analogy was recently invoked for a proposed device which would apply the spin injection concept to ferromagnetic metal films on semiconductors. *Datta* and *Das* [1.58] have suggested the construction of a spin-polarized field effect transistor (Spin-FET), as depicted in Fig. 1.35. The current carrying medium would be an inversion layer formed at the heterojunction between InAlAs and InGaAs. The two-dimensional electron gas in that layer would provide a very high mobility, free of spin-flip scattering events. The spin-polarized carriers are injected and collected by ferromagnetic metal pads as discussed above. However, one can expect that the strong internal electric field present in the inversion layer heterostructure interface region, oriented perpendicular to the layer, will cause the spins of the carriers to precess, due to spin–orbit coupling. This precession will rotate them out of alignment with the magnetization of the second ferromagnetic pad, decreasing the transmitted current of the device. If a gate electrode is deposited on top of the device, one can apply a gate voltage V_g to increase or decrease the effective electric field causing the spin precession. This will serve to control the alignment of the carriers' spin with respect to the magnetization vector in the second pad, thus permitting modulation of the current passing through the device. Although this proposed device demands carefully controlled material growth and lithography, its fabrication is well within the reach of current technology. Devices such as these, which

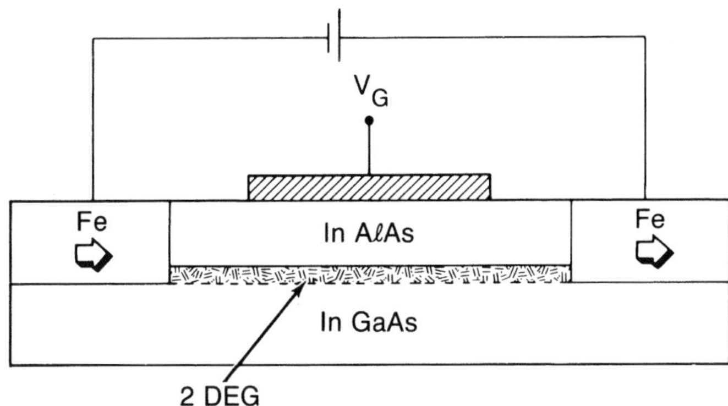

Fig. 1.35. Proposed spin-polarized field effect transistor [1.58]

distinguish between spin-up and spin-down carriers, essentially possess an added dimension of device parameter space which, up until now, has not been exploited.

Appendix

When using the term "MBE", we generally mean vacuum systems with a base pressure of $\approx 10^{-11}$ Torr and during deposition the pressure remaining $\approx 10^{-10}$ Torr. We also imply sources of reasonable capacity (1 cm^3 or greater) capable of sustained monitored deposition at very well-controlled rates. The sources are generally Knudsen cells, which are crucibles surrounded by resistively-heated windings, enclosed in radiation shields. The immediate surrounds of the sources are cooled by either liquid N_2 or chilled H_2O, as are also the surfaces facing the source openings. Crucibles commonly used are pyrolytic BN, Al_2O_3 and BeO. Although these sources have served for temperatures up to 1300 °C when properly constructed, various forms of e-beam heated sources have also been incorporated for evaporating high temperature materials. These include open hearths, e-beam heated metal crucibles and e-beam heated center wire electrodes. In all cases, the challenge is to carry off the heat generated in the evaporation process in order to maintain a low background pressure. The final essential element of an "MBE machine" is in situ monitoring of crystal growth via RHEED. This technique, with its glancing-angle trajectory for incident and diffracted electrons, does not geometrically interfere with the deposition process and therefore allows one to monitor the crystalline structure of the film during deposition.

References

1.1   R. Ludeke, G. Landgren: J. Vac. Sci. Technol. **19**, 667 (1981)

1.2   G.A. Prinz, J.M. Ferrari, M. Goldenberg: Appl. Phys. Lett. **40**, 155 (1982)

1.3   C.J. Gutierrez, J.J. Krebs, M.E. Filipkowski, G.A. Prinz: J. Magn. Magn. Mat. **116**, L305–L310 (1992)

1.4   S. Chang, I.M. Vitomirov, L.J. Brillson, C. Mailhiot, D.F. Rioux, Y.J. Kime, P.D. Kirchner, G.D. Pettit, J.M. Woodall: Phys. Rev. B **45**, 13438 (1992)

1.5   R.M. Feenstra: Phys. Rev. Lett. **63**, 1412 (1989)

1.6   T.C. Nason, L. You, T.-M. Lu: Appl. Phys. Lett. **60**, 174 (1992)

1.7   J.J. Massies, P. Delescluse, P. Etienne, N.T. Linh: Thin Solid Films **90**, 113 (1982)

1.8   B.M. Trafas, Y.-N. Yang, R.L. Siefert, J.H. Weaver: Phys. Rev. B **43**, 14107 (1991)

1.9   B.T. Jonker, J.J. Krebs, G.A. Prinz: Phys. Rev. B **39**, 1399 (1989)

1.10 J.M. Slaughter, Brad N. Engel, M.H. Wiedmann, Patrick A. Kearney, Charles M. Falco: "Thin Films, Surfaces and Interfaces" in *Science and Technology of Nanostructured Magnetic Materials*, ed. by G.C Hadjipanayis, G.A. Prinz (Plenum Press, New York, 1991) pp. 67–70

1.11 J.R. Waldrop, R.W. Grant: Appl. Phys. Lett. **34**, 630 (1979)

1.12 G.A. Prinz, J.J. Krebs: Appl. Phys. Lett. **39**, (1981)
1.13 B.T. Jonker, G.A. Prinz: J. Appl. Phys. **69**, 2938 (1991)
1.14 B.T. Jonker, G.A. Prinz, Y.U. Idzerda: J. Vac. Sci. Technol. B **9** 2437 (1991)
1.15 C. Carbone, B.T. Jonker, K.-H. Walker, G.A. Prinz, E. Kisker: Solid State Comm. **61**, 297 (1987)
1.16 M.W. Ruckman, J.J. Joyce, J.J. Weaver: Phys. Rev. B **33**, 7029 (1986)
1.17 S.B. Qadri, M. Goldenberg, G.A. Prinz, J.M. Ferrari: J. Vac. Sci. Technol. B **3**, 718 (1985)
1.18 T.J. Mc Guire, J.J. Krebs, G.A. Prinz: J. Appl. Phys. **55**, 2505 (1984)
1.19 J.J. Krebs, B.T. Jonker, G.A. Prinz: J. Appl. Phys. **61**, 2596 (1987)
1.20 J.R. Cullen, K.B. Hathaway, J.M.D. Coey: J. Appl. Phys. **63**, 3649 (1988)
1.21 G.A. Prinz, G.T. Rado, J.J. Krebs: J. Appl. Phys. **53**, 2087 (1982)
1.22 J.J. Krebs, B.T. Jonker, G.A. Prinz: J. Appl. Phys. **61**, 3744 (1987)
1.23 J.J. Krebs, F.J. Rachford, P. Lubitz, G.A. Prinz: J. Appl. Phys. **53**, 8058 (1982)
1.24 R.A. Dragonset, P.N. First, J.A. Stroscio, D.T. Pierce, R.J. Celotta: "Characterization of Epitaxial Fe on GaAs (1 1 0) by Scanning Tunneling Microscopy" in *Growth, Characterization and Properties of Ultrathin Magnetic Films and Multilayers*, ed. by B.T. Jonker, J.P. Heremans, E.E. Marinero (Materials Research Society, Vol. 151, 1989) p. 193
1.25 M. Hansen: *Constitution of Binary Alloys* (McGraw-Hill, New York, 1958)
1.26 W.C. Ellis, E.S. Greiner: Trans. Am. Soc. Met. **29**, 415 (1941)
1.27 G.A. Prinz: "Metallic Epitaxy of Transition Metals on Semiconductors", Mat. Res. Soc. Sym. Proc. Vol. 56, 139 (1986)
1.28 P.C. Reidi, T. Dumelow, M. Rubinstein, G.A. Prinz, S.B. Qadri: Phys. Rev. B **36**, 4595 (1987)
1.29 Y.U. Idzerda, B.T. Jonker, W.T. Elam, G.A. Prinz: J. Vac. Sci. Technol. A **8**, 1572 (1990)
1.30 D. Bagayoko, A. Ziegler, J. Callaway: Phys. Rev. B **27**, 7046 (1983)
1.31 G.A. Prinz: Phys. Rev. Lett. **54**, 1051 (1985)
1.32 V.L. Moruzzi, P.M. Marcus, H. Schwarz, P. Mohn: J. Magn. Magn. Mat. **54–57**, 955 (1986)
1.33 A.Y. Liu, D.J. Singh: Phys. Rev. B **47**, 8515 (1993)
1.34 J.A.C. Bland, R.D. Bateson, P.C. Reidi, R.G. Graham, H.J. Lauter, J. Penfold, C. Shackleton: J. Appl. Phys. **69**, 4989 (1991)
1.35 G.A. Prinz, C. Vittoria, J.J. Krebs, K.B. Hathaway: J. Appl. Phys. **57**, 3672 (1985)
1.36 J.M. Karanikas, R. Sooryakumar, G.A. Prinz, B.T. Jonker: J. Appl. Phys. **69**, 6120 (1991)
1.37 K. Schwartz, P. Mohn, P. Blaha, J. Kübler: J. Phys. F. **14**, 2659 (1984)
1.38 G.A. Prinz, E. Kisker, K.B. Hathaway, K. Schröder, K.-H. Walker: J. Appl. Phys. **57**, 3024 (1985)
1.39 Y.U. Idzerda, D.M. Lind, D.A. Papaconstantopoulos, G.A. Prinz, B.T. Jonker, J.J. Krebs: Phys. Rev. Lett. **61**, 1222 (1988)
1.40 T. Sands, J.P. Horbison, M.L. Leadbeater, S.J. Allen, Jr., G.W. Hull, R. Ramesh, V.G. Keramidas: Appl. Phys. Lett. **57**, 2609 (1990)
1.41 T.P. Humphreys, Hyengtag Jeon, R.J. Nemanich, J.B. Posthill, R.A. Rudder, D.P. Malta, G.C. Hudson, R.J. Markunas, J.D. Hunn, N.R. Parikh: Mat. Res. Soc. Symp. Proc. Vol. 202, 463 (1991)
1.42 Chin-Au Chang: J. Appl. Phys. **67**, 566 (1990)
1.43 Chin-Au Chang: Surf. Sci. Lett. **237**, L421 (1990)
1.44 Chin-Au Chang: J. Magn. Magn. Mat. **109**, 243 (1992)
1.45 M. Richardson: Acta Chem. Scand. **21**, 2305 (1967)
1.46 G.A. Prinz (unpublished)
1.47 M.B. Salamon, Shantanu Sinha, J.J. Rhyne, J.E. Cunningham, R.E. Erwin, J. Borchers, C.P. Flynn: Phys. Rev. Lett. **56**, 259 (1986)
1.48 R.F.C. Farrow, S.S.P. Parkin, V.S. Speriosu: J. Appl. Phys. **64**, 5315 (1988)
1.49 R.F.C. Farrow, M.F. Toney, B.D. Hermsmeier, S.S.P. Parkin, D.G. Wiesler: J. Appl. Phys. **70**, 4465 (1991)
1.50 A.V. Pohm, J.S.T. Huang, J.M. Daughton, D.R. Krahn, V. Mehra: IEEE Trans. Mag. **24**, 3117 (1988)
1.51 V.S. Liau, T.Wong, W. Stacey, S. Ali, E. Schloemann: IEEE MTT-S Digest **DD-3**, 957 (1991)

1.52 G.A. Prinz, B.T. Jonker, J.J. Krebs, J.M. Ferrari, F. Kovanic: Appl. Phys. Lett. **48**, 1756 (1986)
1.53 R. Meservey, P.M. Tedrow, P. Fulde: Phys. Rev. Lett. **25**, 1270 (1970)
1.54 P.M. Tedrow, R. Meservey: Phys. Rev. B **7**, 318 (1973)
1.55 M.B. Stearns: J. Mang. Magn. Mat. **5**, 167 (1977)
1.56 M. Johnson, R.H. Silsbee: Phys. Rev. Lett. **55**, 1790 (1985)
1.57 M. Johnson, R.H. Silsbee: Phys. Rev. B **37**, 5326 (1988)
1.58 S. Datta, B. Das: Appl. Phys. Lett. **56**, 665 (1990)

2. Magnetic Coupling and Magnetoresistance

In this chapter, the related subjects of magnetic coupling and magnetoresistance in ultrathin film structures are discussed in detail. Antiferromagnetic exchange coupling can occur between two ultrathin ferromagnetic films (e.g., Fe) separated by a non-magnetic spacer layer (e.g., Cr) of the correct thickness. Antiparallel alignment of the adjacent ferromagnetic layer magnetizations in Fe/Cr multilayers gives rise to the phenomenon of giant magnetoresistance via the so-called spin valve effect, although indirect exchange coupling is only one of several ways in which such an antiparallel alignment, and hence giant magnetoresistance, can be obtained. Oscillatory coupling has been found to occur in which the coupling strength oscillates as a function of thickness of the spacer layer, and in appropriate ferromagnetic/non-magnetic multilayer systems, this is accompanied by an oscillatory magnetoresistivity. In this chapter we survey both theoretical and experimental aspects of coupling and magnetoresistivity in magnetic multilayers. In the first section, a range of theoretical models proposed to explain exchange coupling are discussed by *Hathaway*. This is followed by a review by *Fert* and *Bruno* of the experimental results and theoretical models for interlayer coupling and magnetoresistance. *Pierce, Unguris* and *Celotta* discuss studies of exchange coupling using scanning electron microscopy with polarization analysis. This study focuses chiefly on epitaxial films. The reader is referred to Volume 1, Chap. 4 for a discussion of spin-polarized electron spectroscopy techniques. Finally, *Parkin* concludes the chapter with a discussion of giant magnetoresistance and coupling in polycrystalline transition metal multilayers. The reader is referred to this final section for a comparison of MBE-grown and sputtered films.

2.1 Theory of Exchange Coupling in Magnetic Multilayers

K.B. Hathaway

Magnetic transition metals separated by thin layers of non-magnetic metals exhibit an exchange coupling which oscillates with a period of approximately 10 Å and which decays with increasing thickness of the spacer layer. For some systems, notably Fe/Cr, a short-period oscillation (~ 2 Å), like that which would be expected from the response function of an electron gas, is observed superim-

B. Heinrich and J.A.C. Bland (Eds.)
Ultrathin Magnetic Structures II
© Springer-Verlag Berlin Heidelberg 1994

posed on the long-period oscillation. Most recently, biquadratic coupling, which leads to 90° relative orientations of the magnetizations in adjacent magnetic layers, has been observed in some systems for certain ranges of interlayer thickness. In attempting to explain this exchange coupling, theorists have proposed a wide variety of models, in many cases drawing on earlier work. The antecedents of the proposed models range from theories successfully used to explain coupling between dilute magnetic impurities in metals in the 1950's and 1960's, to the continuously evolving theories of itinerant magnetism.

What follows is an attempt to classify the various proposed theories of exchange coupling according to those aspects of the experimental behavior that each attempts to explain. The theoretical picture at the time this is written is still evolving as new experimental results are reported. This article is an attempt to provide a conceptual framework within which to understand the strengths and weaknesses of various theoretical approaches. Detailed discussions of the experimental observations are available elsewhere in this volume and references to experiment will only be included where results are not well known. However the theoretical references are intended to be comprehensive.

2.1.1 RKKY-Like Models

Many of the models proposed for exchange coupling between two ferromagnet (FM) layers through a paramagnet (PM) layer follow from earlier work on the coupling between magnetic impurities in a host metal. The most widely applied of these early models is the Ruderman–Kittel–Kasuya–Yosida (RKKY) coupling, first proposed for nuclear spins but later applied to both transition metal magnetic impurity systems and exchange in rare earth metals.

The RKKY interaction was calculated by *Ruderman* and *Kittel* [2.1] for the indirect exchange coupling of two nuclear spins via their hyperfine contact interaction with the conduction electrons. *Kasuya* [2.2] and *Yosida* [2.3] proposed a similar coupling between two localized d (or f) electrons via their mutual coupling to the conduction (s) electrons. But they could not simply assume a simple s–d contact interaction, as the localized states have finite spatial extent. Rather they were forced to derive the s–d interaction from the general many-body Hamiltonian. We give here a simple discussion of the s–d Hamiltonian and the resulting coupling between d spins which glosses over many of the mathematical subtleties. More complete discussions can be found in several review articles [2.4–6]. We follow the general derivations of *Kondo* [2.4] and *Freeman* [2.5].

The s–d exchange interaction results when the electrostatic electron–electron interaction term in the Hamiltonian

$$V_{ij} = \sum_{i>j} \frac{1}{r_{ij}} \tag{2.1}$$

is treated explicitly rather than in an average or mean field approximation.

V_{ij} (which can be generalized to include screening) is a two body interaction, so its matrix elements involve four single-electron states. We want to extract a magnetic interaction, so we consider only transitions between states very close in energy, i.e., those processes in which a conduction electron is scattered and a localized state is occupied both before and after the scattering. The relevant matrix element thus involves different initial and final states of the conduction electron $|k\rangle$ and $\langle k'|$ and the same initial and final localized state $|n\rangle$ and $\langle n|$. Writing the required terms using creation and annihilation operators (a^*, a) for the localized electrons and creation and annihilation operators (c^*, c) for the conduction electrons with spin indices s, s' we have:

$$\sum_{kk'nss'} [a_{ns'}^* a_{ns'} c_{ks}^* c_{k's} \langle kn|V_{ij}|k'n\rangle - a_{ns'}^* a_{ns} c_{ks}^* c_{k's'} \langle kn|V_{ij}|nk'\rangle]. \tag{2.2}$$

The first term is a Coulomb scattering term because electron 1 remains in a conduction state and electron 2 remains in a localized state. The second term contains the desired exchange interaction because electrons 1 and 2 change places. The matrix element V is given in terms of the localized electron state (ϕ) and conduction electron states (ψ) by

$$\langle kn|V_{ij}|nk'\rangle = \iint d^3r_1\, d^3r_2\, \psi_k(r_1)\phi_n(r_2) V_{12}\phi_n(r_1)\psi_{k'}(r_2). \tag{2.3}$$

Note that since V is independent of spin the initial and final total spin is constant. Writing out the exchange terms with the spins shown explicit for spin $\frac{1}{2}$ gives

$$-\sum_{kk'n} [\tfrac{1}{2}(a_{n\uparrow}^* a_{n\uparrow} - a_{n\downarrow}^* a_{n\downarrow})(c_{k\uparrow}^* c_{k'\uparrow} - c_{k\downarrow}^* c_{k'\downarrow}) + a_{n\downarrow}^* a_{n\uparrow} c_{k\uparrow}^* c_{k'\downarrow}$$

$$+ a_{n\uparrow}^* a_{n\downarrow} c_{k\downarrow}^* c_{k'\uparrow}]\langle kn|V_{ij}|nk'\rangle. \tag{2.4}$$

If we assume that (1) the localized states, ϕ, are non-overlapping atomic-like states centered at sites R_n, and (2) the conduction electron states, ψ, are Bloch functions, the matrix element of V can be written to define an s–d exchange coupling J_{sd}

$$\langle kn|V_{ij}|nk'\rangle = \frac{1}{N} e^{i(k-k')\cdot R_n} J_{sd}(k, k'). \tag{2.5}$$

If we also assume that the localized state is always occupied by a single electron, then

$$\tfrac{1}{2}(a_{n\uparrow}^* a_{n\uparrow} - a_{n\downarrow}^* a_{n\downarrow}) = S_n^z, \qquad a_{n\uparrow}^* a_{n\downarrow} = S_n^+ \quad \text{and} \quad a_{n\downarrow}^* a_{n\uparrow} = S_n^-. \tag{2.6}$$

(If the localized orbital is degenerate, within certain restrictions S_n may still be used to denote the total spin of the localized electrons.) Inserting these relations into expression (2.4) we finally obtain the s–d *Hamiltonian*:

$$H_{sd} = -\frac{1}{N}\sum_{kk'n} J_{sd}(k, k') e^{i(k-k')\cdot R_n}[S_n^z(c_{k\uparrow}^* c_{k'\uparrow} - c_{k\downarrow}^* c_{k'\downarrow})$$

$$+ S_n^+ c_{k\downarrow}^* c_{k'\uparrow} + S_n^- c_{k\uparrow}^* c_{k'\downarrow}]. \tag{2.7}$$

H_{sd} can be rewritten in the form of a contact interaction between the localized spin S_n and the itinerant spin s_i only if J depends on the scattered wave vector in electron scattering $q = k - k'$. In that case we can use the relations $c_{ks} = (1/\sqrt{N}) \sum_i e^{i k \cdot r_i} c_{is}$ and the definitions analogous to (2.6) for the c_{is} operators, then perform the sums over k and q to get

$$H_{sd} = -\sum J_{sd}(|r_i - R_n|) S_n \cdot s_i. \tag{2.8}$$

Now we can use the s–d interaction embodied in H_{sd} to obtain the RKKY interaction between two localized spins at n and n'. We treat H_{sd} up to second order in perturbation theory, considering the virtual excitation of an electron in state ks into an empty state $k's$ or $k'-s$. (We ignore some subtleties concerning the treatment of diagonal terms in second rather than first order, [2.7].)

$$H_{dd} = -\frac{1}{N^2} \sum_{kk'} |j_{sd}(k, k')|^2 \frac{f_k(1 - f_{k'})}{\varepsilon_{k'} - \varepsilon_k} e^{i(k' - k) \cdot (R_n - R_{n'})}$$

$$\times [2S_n^z S_{n'}^z + S_n^- S_{n'}^+ + S_n^+ S_{n'}^-], \tag{2.9}$$

where the f's are the Fermi functions. Rearranging the sum and noting that the term $f_k f_{k'} = 0$, we get

$$H_{dd} = -\frac{1}{N^2} \sum_{kk'} |j_{sd}(k, k')|^2 \frac{f_k - f_{k'}}{\varepsilon_{k'} - \varepsilon_k} e^{i(k' - k) \cdot (R_n - R_{n'})} S_n \cdot S_{n'}. \tag{2.10}$$

If j_{sd} depends only on $q = k - k'$ and the susceptibility of the non-interacting electron gas is

$$\chi(q) = \frac{1}{N} \sum_k \frac{f_k - f_{k+q}}{\varepsilon_{k+q} - \varepsilon_k}, \tag{2.11}$$

we can write

$$H_{dd} = -\frac{2}{N} \sum_q |j_{sd}(q)|^2 \chi(q) e^{iq \cdot (R_n - R_{n'})} S_n \cdot S_{n'} \equiv J(R_n - R_{n'}) S_n \cdot S_{n'}. \tag{2.12}$$

Thus the interaction $J(R_n - R_{n'})$ between two localized spins at l and l' is given by the Fourier transform of the q-dependent s–d interaction (squared) times the q-dependent susceptibility of the electron gas. Calculating $J(R_n - R_{n'})$ for real systems requires either assumptions or information regarding $\chi(q)$. If $j_{sd}(q)$ equals a constant and $\chi(q)$ is given by the susceptibility of the three-dimensional electron gas

$$J(R) = 9\pi \frac{j}{\varepsilon_F} \left[\frac{\cos 2k_F R}{(2k_F R)^3} - \frac{\sin 2k_F R}{(2k_F R)^4} \right]. \tag{2.13}$$

The term in square brackets is referred to as the RKKY range function.

The first extension of the RKKY model to the case of two interacting magnetic layers was done by *Yafet* [2.8], to explain the coherence of magnetization of Gd layers separated by non-magnetic Y layers in Gd/Y superlattices. He treated the Gd layers as localized spins interacting via the measured bulk exchange interaction, and used the computed wave vector dependent susceptibilities of Gd and Y to obtain an effective Gd–Y q-dependent exchange interaction. The resulting exchange coupling, $J(d)$, is obtained by integrating over components parallel to the layer planes. $J(d)$ shows a well defined oscillation for d between 4 and 11 monolayers of Y. For large d irregular oscillations reflect the structure in $\chi(q)$ with interference occuring between many wave vectors with roughly equal contributions to the susceptibility. Magnetic ordering in rare earth superlattices was also considered by *Fairbairn* and *Yip* [2.9] who calculated the RKKY coupling using wave functions computed for a superlattice of square well potentials.

Oscillations in the exchange coupling observed for transition metal magnetic/nonmagnetic multilayers suggested that the RKKY model might be applied to those materials as well. If we take the RKKY range function to describe the interaction between every pair of spins in a layered structure, we can calculate an exchange energy by simply summing these interactions. As an illustration we consider a FM/PM/FM layered structure in which all three layers have the same lattice structure and spacing. The spins within each FM layer are rigidly coupled by direct exchange and we assume the indirect exchange coupling between spins in different FM layers to be mediated by an

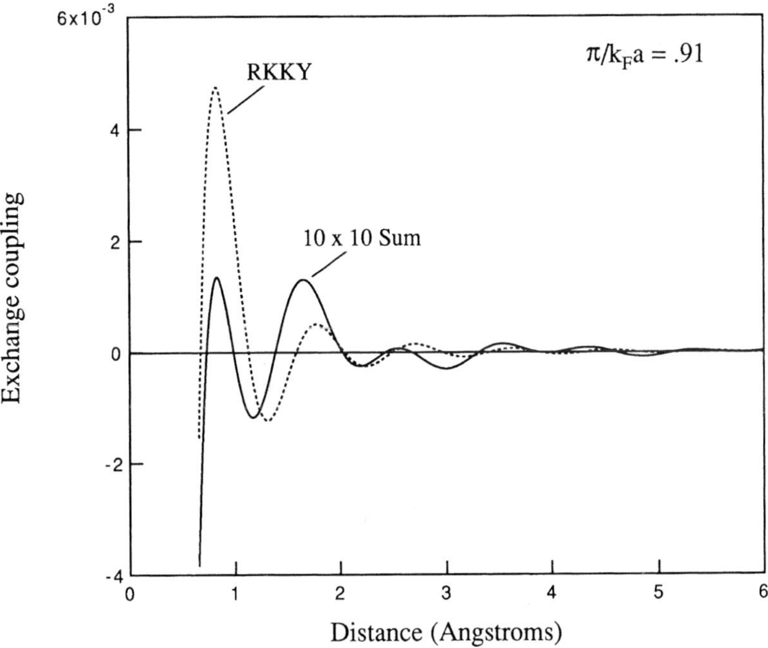

Fig. 2.1. Exchange coupling between two ferromagnetic layers of discrete spins obtained by summing the three-dimensional RKKY range function for every pair of spins in different layers as a function of separation in units of lattice constant, a. Calculated for $\pi/k_F = 0.91a$, (solid line) summed out to 10th neighbors in the plane. Dashed line is RKKY range function for single spin pair

electron gas that is continuous across the interfaces. Since the RKKY range function oscillates with period π/k_F, if the lattice spacings are not significantly smaller than π/k_F the discrete sum will produce irregular oscillations in the net exchange coupling. The net exchange coupling for two single atomic layer FM layers, calculated by summing out to 10th neighbors in the plane for $(\pi/k_F) = 0.91a$ is shown in Fig. 2.1. In the asymptotic limit of the layer thickness, $d > a$ the oscillations in J become smooth.

An alternative way to obtain smooth oscillations in J is to eliminate the discreteness of the spins in the FM layers by replacing them with a continuous sheet of constant spin density. This model, which is more appropriate for itinerant magnets, we dub the continuous planar RKKY model. It has been investigated by *Baltensperger* and *Helman* [2.10], among others. In their calculation the (semi-infinite) ferromagnetic (FM) layers are treated as continuous spin sheets and the non-magnetic or paramagnetic (PM) metal layer is representated as a free electron gas, with results shown in Fig. 2.2. The asymptotic large d limit of the range function for this model is

$$\left[\frac{\sin 2k_F R}{(2k_F R)^2} - 5\frac{\cos 2k_F R}{(2k_F R)^3}\right], \tag{2.14}$$

which is similar to the form of the three-dimensional RKKY range function but reflects the two-dimensional nature of the multilayer structure by decaying as d^{-2} at large distances rather than d^{-3}. This type of RKKY model can be improved for transition metal systems, as was done by *Yafet* [2.8] for rare earths, by using susceptibilities measured or calculated for the specific material instead of that for the uniform electron gas.

While it is clear that the two-dimensional or layer RKKY model does indeed give an oscillating exchange coupling, it has some obvious shortcomings. The first is that it is derived from perturbation theory which assumes that the FM layers have only a weak effect on the electrons in the PM. The interaction is either modelled directly as a contact interaction between localized spins and conduction electrons or comes from a perturbation treatment of the full Hamiltonian which takes the form of a contact interaction (i.e. the s–d Hamiltonian), whereas for transition metals strong hybridization is anticipated, with no clear

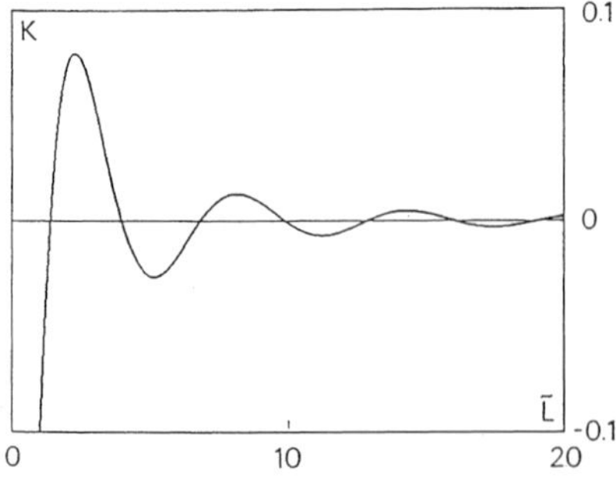

Fig. 2.2. Separation dependence of the exchange coupling between two ferromagnetic layers treated as sheets of constant spin density. From [2.10]

separation between the ferromagnetic moments and the itinerant electrons comprising the Fermi sea. These approximations (perturbation approach, contact interaction) yield an oscillating range function whose form is independent of the properties of the FM, i.e. the FM properties show up only in the coefficient of the coupling. Second, the predicted oscillation periods for free electrons do not reproduce the long period oscillations observed for many materials. Third, there is no possibility of an antiferromagnetic or ferromagnetic non-oscillatory background, which seems to exist in some systems. Fourth, the form assumed for a contact interaction leads directly to a Heisenberg form for the exchange, which cannot explain the recently observed biquadratic coupling. In the sections which follow, modifications or alternatives to the RKKY picture which address each of these shortcomings will be discussed.

2.1.2 Non-Perturbation Calculations for Strongly Hybridized Systems

The most straightforward approach to go beyond perturbation theory to treat the effects of strong coupling (due to large exchange splitting, or strong ferromagnetism in the FM layers) is to determine the wavefunctions for the interacting FM/PM system exactly. Approaches using first-principles electronic structure techniques involve the fewest approximations but are limited to relatively thin layers – results of such calculations will be discussed in Sect. 2.16. The simplest approximation is to assume free-electron wave functions in both the FM and PM layers with perfectly flat interfaces between them. Our calculation based on this approximation [2.11] will be discussed in some detail to provide a framework for evaluating other approaches. A "matched" exchange-split FM/PM free-electron band structure is shown in Fig. 2.3. This figure introduces the two important parameters of the model, the Fermi wave vector k_F (identical for the PM and the FM minority spin band) and the FM exchange splitting, $2h_0$. This matching of the FM minority spin band and the PM band is appropriate for FM/PM materials in which the PM has fewer electrons than the FM (for elemental layers, the PM element lies to the left of the FM element in the periodic table). It is a reasonable choice for a free-electron representation of Fe/Cr and Co/Ru as can be seen for Fe/Cr from the densities of states shown in Fig. 2.4. Note that when the FM and PM have the same crystal structure the rigid band picture is not too bad, and the bands match not only their densities of states, but also the individual symmetries of energy states as well. Thus this approximation is applicable when the electrons in the PM and FM have the same symmetry, both s-like or both d-like, for example. This picture is consistent with that of *Stearns* [2.12] for transition metal magnets, that there exist highly polarized, but free-electron-like d states with effective masses only a few times unity. This choice of band matching (in addition to simplifying the calculation) provides the largest difference between the discontinuities in the potential encountered by the minority and majority spins in traversing the FM/PM interface. Note that the potential is not self-consistent, i.e. the exchange splitting

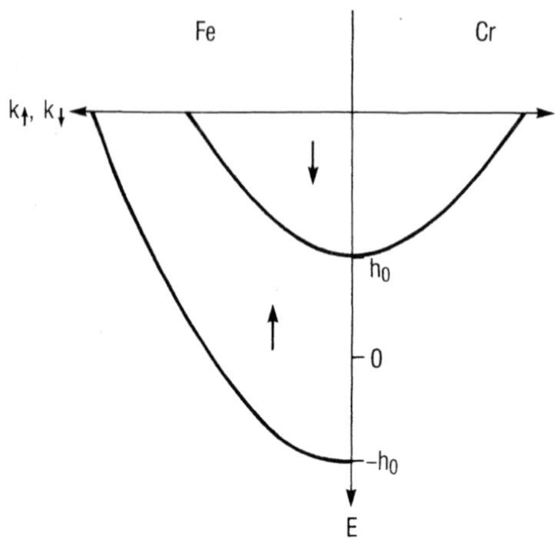

Fig. 2.3. Free-electron band structures for a ferromagnet (FM)/paramagnet (PM) interface in which the PM band is assumed to exactly match the minority-spin FM band. $k_\uparrow$ and $k_\downarrow$ are majority- and minority-spin wave vectors in the FM, and k is the wave vector in the PM. The FM exchange splitting is $2h_0$

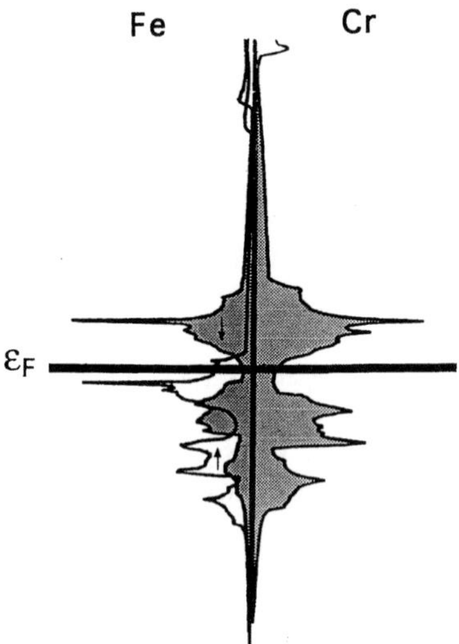

Fig. 2.4. Calculated densities of states for Fe and Cr, from *D.A. Papaconstantopoulos, Handbook of the Band Structure of Elemental Solids.* (Plenum, NY, 1986), with their respective Fermi energies, ε_F, aligned. The shaded portions (Fe minority-spin density of states on the left and Cr total electron density of states on the right) show the matching leading to the free-electron band structure of Fig. 2.3

is assumed to maintain the bulk value over the entire FM layer, even though the spin polarization varies near the interfaces.

Figure 2.5 shows the one-dimensional potentials for majority and minority spins in a FM/PM/FM sandwich structure with this band structure when the magnetizations in the FM layers are parallel or antiparallel. Note that the parallel magnetization case corresponds to scattering of the majority spin electrons by a barrier potential. Wave functions may be determined for these two cases by matching amplitudes and derivatives at the interfaces. The wave functions can then be used to compute the total energies for the two cases, with the difference yielding the Heisenberg exchange coefficient. We adopt a slightly

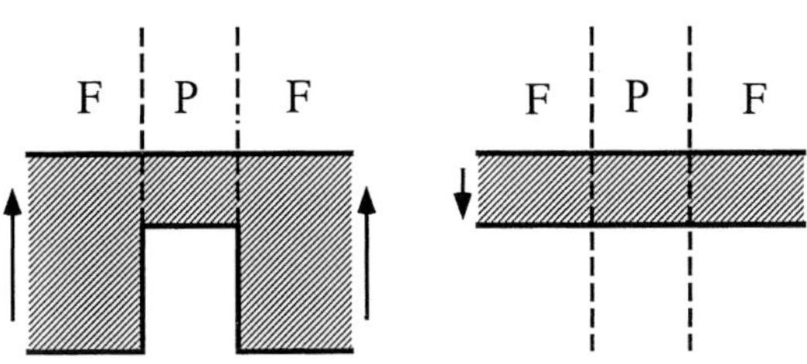

Fig. 2.5. One-dimensional potentials (normal to layer plane) for majority- and minority-spin electrons for different alignments of the ferromagnet layer magnetizations: (**a**) parallel alignment, producing a barrier potential for majority-spin electrons, and (**b**) antiparallel alignment producing step potentials for both majority- and minority-spin electrons

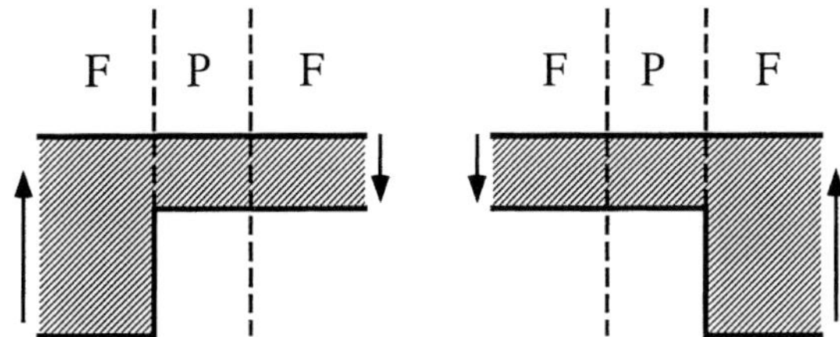

different approach due to *Slonczewski* [2.13] which allows for the possibility of non-Heisenberg exchange, and which is also intuitively related to the physical picture of electrons scattered by potential discontinuities. We assume two semi-infinite FM layers, with unit magnetizations m_1 and m_2 in the plane of the layers related by the angle ϕ_{12} such that $m_1 \cdot m_2 = \cos \phi_{12}$. The time rate of change of m_1 is given by the torque exerted on m_1 by m_2 which is proportional to $m_1 x m_2$. We define J such that the classical torque on m_1, for both m_1 and m_2 in the $(x\text{–}z)$ plane of the layers, is

$$J(\cos \phi_{12})m_1 x m_2 = J(\cos \phi_{12})\sin \phi_{12} y. \tag{2.15}$$

J is only identical to the classical exchange coupling defined by $E = Jm_1 \cdot m_2$ in the Heisenberg limit; in general, the energy is obtained from the torque by integration over ϕ_{12}. We would like to be able to calculate the torque from the quantum mechanical wave functions which diagonalize the Hamiltonian, H, of the system for a specified ϕ_{12}. We can calculate

$$\frac{\partial \langle S(t) \rangle}{\partial t} = \langle [H, S] \rangle \equiv \langle \dot{S} \rangle, \tag{2.16}$$

where S is the electron spin. We find that $\langle S(t) \rangle$ is related to a tensor spin current, j, through the conservation equation:

$$\frac{\partial \langle S(t) \rangle}{\partial t} = -\frac{\hbar}{2} \int_V dV \, \nabla \cdot j(x, t), \tag{2.17}$$

where j is defined by

$$j_{\alpha\beta} \equiv \frac{\hbar}{2mi}\left\{\psi^*(r,t)\sigma_\beta\frac{\partial\psi(r,t)}{\partial r_\alpha} - \frac{\partial\psi^*(r,t)}{\partial r_\alpha}\sigma_\beta\psi(r,t)\right\}, \tag{2.18}$$

where σ is the vector of Pauli spin matrices. In order to compare with the classical torque, we need to calculate the time derivative of the y-component of the spin S:

$$\begin{aligned}
\frac{\partial\langle S_y(t)\rangle}{\partial t} &= -\frac{\hbar}{2}\int_V dV\, \nabla\cdot j(r,t)\cdot y = -\frac{\hbar}{2}\int_V dA\, j_{yy}(r,t) \\
&= -\frac{\hbar}{2}Aj_{yy}(y)\Big|_{y=0}, \tag{2.19}
\end{aligned}$$

where we have used the fact that the current is uniform over the surface defined by the FM/PM interface (with area A) which is normal to the y-direction. The total spin, S_1, in FM(1) can be obtained by summing over occupied states and its time derivative is related to the total spin current, j_{yy}^{T}, obtained by summing j_{yy} in the same way. We can finally equate the quantum and classical torques to obtain

$$J(\cos\phi_{12})\sin\phi_{12} = -\tfrac{1}{2}A\hbar j_{yy}^{\mathrm{T}}. \tag{2.20}$$

Thus the spin current provides an alternative to computing total energies as a way of calculating exchange once the electron wave functions are known. A nice feature of this approach is that any ϕ_{12} dependence of J derives naturally from the ϕ_{12} dependence of the wave functions.

We now turn to a calculation of the spin current for our trilayer system. The wave functions for the trilayer are obtained by matching free-electron plane waves (with appropriate real and imaginary wavenumbers) for each spin polarization at both interfaces in the usual way. The different polarization axes in the two FM's are accounted for by applying a spinor transformation to the two-component spinor wave functions at one of the interfaces. Thus, for example, an electron which is represented by a minority-spin plane wave in one FM will, if the magnetizations of the FM's are not colinear, be resolved into a two-component spinor in the other FM with its components having different wave numbers. The spin current, which is constant only in the PM, constitutes a flow of off-diagonal spin which corresponds to the classical precession expected for non-colinear alignments of the FM magnetizations. Calculation of j_{yy} in the PM yields the following forms for the contributions from wave functions for majority spins and minority spins incident at the infinite boundaries:

For $E < h_0$

$$j_{yy}^{\uparrow} = -\frac{\hbar}{\pi^2 m}\int_0^{k_0} dk_\uparrow \int_0^{\sqrt{k_{\uparrow F}^2 - k_\uparrow^2}} k_\parallel\, dk_\parallel$$

$$\times \frac{4\kappa k_\uparrow^2(\kappa^2 - k_\uparrow^2)e^{-2\kappa d}\sin\phi_{12}}{\left|(k_\uparrow + i\kappa)^2 - (k_\uparrow - i\kappa)^2 e^{-2\kappa d}\cos^2\dfrac{\phi_{12}}{2}\right|^2} \tag{2.21a}$$

and for $E > h_0$

$$j_{yy}^{\downarrow} = -\frac{\hbar}{\pi^2 m} \int_0^{k_{\downarrow F}} dk_{\downarrow} \int_0^{\sqrt{k_{\downarrow F}^2 - k_{\downarrow}^2}} k_{\parallel} \, dk_{\parallel}$$

$$\times \frac{k_{\downarrow} k_0^4 \sin(2k_{\downarrow}d)\sin\phi_{12}}{\left| (k_{\uparrow} + k_{\downarrow})^2 - (k_{\uparrow} - k_{\downarrow})^2 e^{-2ik_{\downarrow}d}\cos^2\dfrac{\phi_{12}}{2} \right|^2}. \tag{2.21b}$$

$k_{\uparrow}$ and $k_{\downarrow}$ are the majority and minority spin wave numbers which are related by $k_{\uparrow}^2 - k_{\downarrow}^2 = k_0^2 = 4mh_0/\hbar^2$. For $E < h_0$ we define an imaginary wave vector in the PM barrier $\kappa = ik_{\downarrow} = \sqrt{k_0^2 - k_{\uparrow}^2}$.

Note that the integrand in (2.21a) has an exponential decay with spacer layer thickness, d, while that in (2.21b) has an oscillatory behavior. Performing the energy integrations of (2.21) gives a cancellation of the exponentially decaying contribution, analogous to that which occurs for the spin disturbance in a PM caused by a single FM interface, shown years ago by *Bardasis* et al. [2.14] using this same matched band structure. This cancellation can be shown quite generally for all values of ϕ_{12} [2.15]. For purposes of illustration here we show how the cancellation occurs in the asymptotic (large d) limits of the two contributions to the spin current evaluated at $\phi_{12} = \pi$.

$$j_{yy}^{\uparrow} = \frac{\hbar^2 k_0^2 k_{F\downarrow}^2}{8\pi^2 m}\left[\frac{1}{(k_0 d)^3} + \cdots \right],$$

$$j_{yy}^{\downarrow} = \frac{\hbar^2 k_0^2 k_{F\downarrow}^2}{8\pi^2 m}\left[\frac{\sin 2k_{F\downarrow}d}{(k_{F\downarrow}d)^2} - \frac{1}{(k_0 d)^3} + \cdots \right]. \tag{2.22}$$

Clearly the lowest order non-oscillating terms cancel. Details of the cancellation for all d and ϕ_{12} are given in [2.15], in which it is shown that (2.21a) can be rewritten in the same form as (2.21b) via extension by contour integration in the complex plane.

Note that if the exact cancellation between the slowly decaying evanescent states with energies just below h_0 and the long wavelength travelling wave states with energies just above h_0 were somehow disrupted, there would in fact be an exponentially decaying term in the spin current and thus in the exchange coupling. Any discreteness in the density of states which prevents the energy integration from proceeding smoothly through the top of the barrier can be expected to yield such a term. *Slonczewski* [2.13] has shown this to be the case for an insulating barrier where the energy integration stops at e_F, below the top of the barrier. A similar "superexchange" coupling arises when discrete states are included along with continuum states in a perturbation calculation. Such models have been investigated by several authors and will be discussed in Sect. 2.1.5.

The cancellation of the non-oscillatory contribution in our free-electron model and integration of the remaining oscillatory terms over momentum states parallel to the interface, leaves the following expression for the exchange coupling:

$$J(\phi_{12}) = \frac{A\hbar^2 k_0^4}{2\pi^2 m} \int_{z_F}^{\infty} dz$$

$$\times \frac{z(z - \sqrt{1 + z^2})^4 (z^2 - z_F^2)\sin(2k_0 dz)}{1 - 2(z - \sqrt{1 + z^2})^4 \cos(2k_0 dz)\cos^2\frac{\phi_{12}}{2} + (z - \sqrt{1 + z^2})^8 \cos^4\frac{\phi_{12}}{2}},$$

$$(2.23)$$

where $z_F = k_F/k_0$. The terms in ϕ_{12} in the denominator give the non-Heisenberg contributions of the exchange coupling, and $J(\phi_{12})$ can be written quite generally as a power series in $\cos\phi_{12}$, as will be shown in Sect. 2.1.7. For purposes of illustration we first consider the behavior of the Heisenberg-like part of the exchange. If we restrict consideration to the region $\phi_{12} \sim \pi$, i.e. we calculate the torque for small deviations from antiparallel alignment of the FM moments, the terms in ϕ_{12} in the denominator disappear, and the exchange constant is independent of ϕ_{12}. J can then be evaluated numerically for particular ratios of k_0/k_F with the results shown in Fig. 2.6. Note that in the limit $k_0/k_F \to 0$ (weak coupling) J reduces exactly to the planar RKKY range function shown in Fig. 2.2 for all d. The asymptotic (large d) behavior of this function in the limits $k_0/k_F \to 0$ (weak coupling) and $k_0/k_F \to \infty$ (strong coupling) are given by

$$J \cong -k_0^4\left[\frac{\sin 2k_F d}{4(2k_F d)^2} - \frac{5\cos 2k_F d}{4(2k_F d)^3}\right] \quad \text{"weak"} \qquad (2.24a)$$

$$J \cong -2k_F^4\left[\frac{\sin 2k_F d}{(2k_F d)^2} + \frac{6\cos 2k_F d}{(2k_F d)^3}\right] \quad \text{"strong"} \qquad (2.24b)$$

As the exchange splitting ($\sim k_0^2$) increases, the small-d behavior of the range function changes, reflecting the interplay between the two length scales, k_F^{-1} and k_0^{-1}. The magnitude of the exchange (which is always ferromagnetic at $d = 0$) is proportional to k_0^4. Each exchange coupling function, $J(d)$, displayed in Fig. 2.6 is normalized to its value, J_0, at $d = 0$. The value for J_0 is given by $(\hbar^2 k_0^4/8\pi^2 m)(\pi/32) = (21)(2h_0)^2$ erg/cm^2, where $2h_0$ is the exchange splitting in the FM in eV. For an exchange splitting of 1 eV and for moderate coupling ($k_0/k_F \sim 1$) this gives a magnitude of the first antiferromagnetic peak of ~ 2 erg/cm^2.

For large k_0 the envelope of the oscillations decays much more rapidly than for RKKY, and the positions of the initial peaks in units of k_F^{-1} are moved toward the origin. Quantitative comparisons between these results and data for Co/Ru and Fe/Cr [2.11] confirm these trends in the shape of the range function: Fe/Cr behaves like $k_0/k_F = 0.6$ (moderate coupling) and Co/Ru behaves like $k_0/k_F = 3.0$ (strong coupling).

This model does not fit the behavior of Co/Cu. This is not surprising since the assumed band structure of Fig. 2.3 is inappropriate as a free-electron approximation for Co/Cu and other systems for which the PM has more nearly

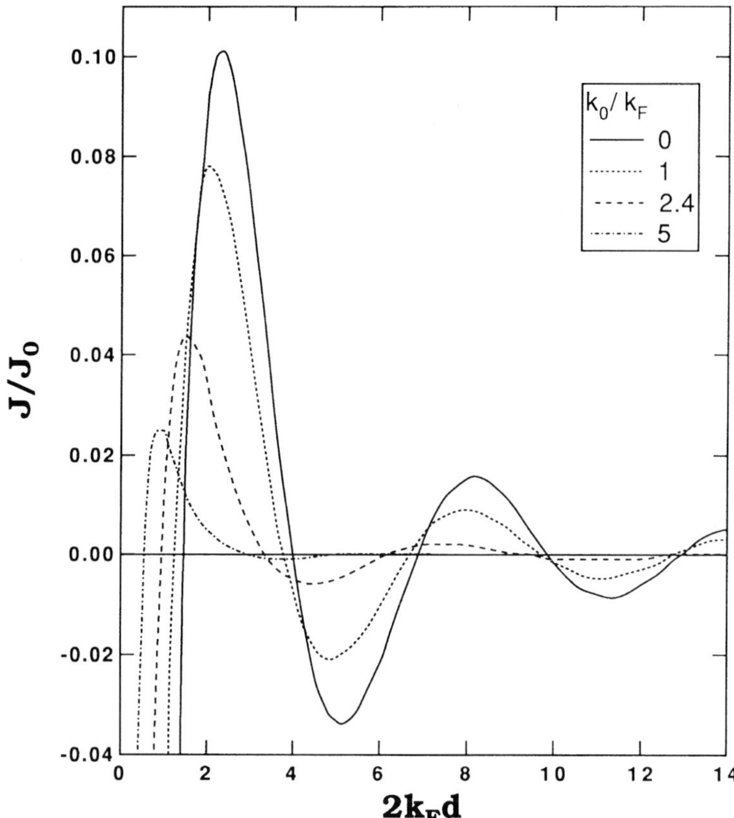

Fig. 2.6. The exchange coupling, J, as a function of the thickness of the paramagnetic layer, d, times twice its Fermi wave vector, k_F for various values of k_0/k_F where k_0 is the wave vector corresponding to the ferromagnetic exchange energy, $k_0 = (4h_0 m/\hbar^2)^{1/2}$. The J's have been normalized to 1 at $d = 0$ by division J_0 (see text) thus inverting their relative magnitudes

filled bands than the FM. *Barnas* [2.16] has treated the alternate "matched" band structure in which the PM band matches the FM *majority*-spin band, (more appropriate for Co/Cu) shown in Fig. 2.7. He performed numerical calculations of wavefunctions and total energies for an infinite superlattice structure, with the results shown in Fig. 2.8. Note that in contrast to our previous results for minority-spin matched bands where the $d = 0$ coupling was always ferromagnetic, the initial phase (sign) of the exchange coupling for this case may be either ferromagnetic or antiferromagnetic. For this band structure, parallel magnetization alignment corresponds to scattering of minority-spin electrons by a potential well, allowing the possibility of bound states for certain ranges of the parameters. Although one cannot reach general conclusions from the selected numerical results presented in *Barnas*' paper, it is likely that the bound states account for the variability in the initial phase.

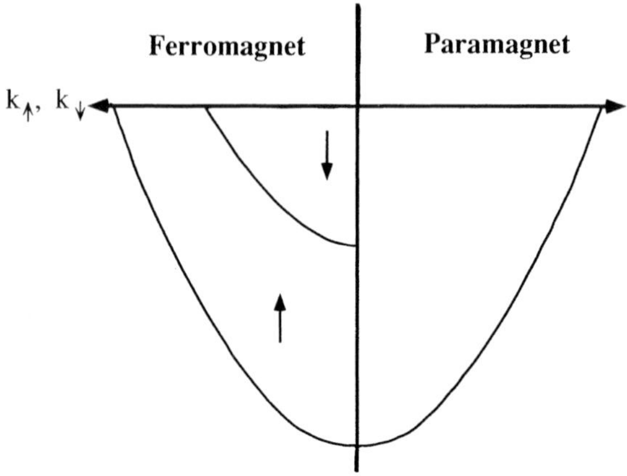

Fig. 2.7. Free-electron band structure with band matching between ferromagnet majority-spin and paramagnet bands, approximating transition metal/noble metal systems. Investigated in [2.16]. Notation as in Fig. 2.3

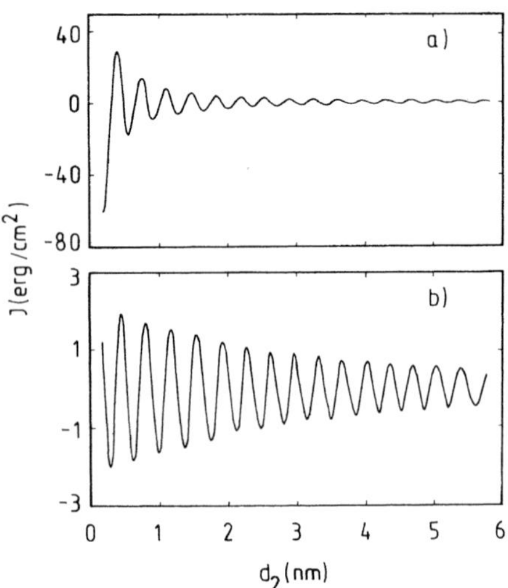

Fig. 2.8. Numerical calculations of exchange coupling for infinite ferromagnet/paramagnet superlattice with the band structure of Fig. 2.7 as reported in [2.16]. Note the variation in magnitude, the decay rate of the envelope function, and the initial phase for two different values of k_0/k_F, (**a**) 1.3, and (**b**) 0.4

Barnas has also considered explicitly the effects of varying the thicknesses of the FM layers. His calculations show oscillations in the exchange coupling for fixed d, but increasing FM thickness. Since most experimental studies have maintained constant FM thicknesses, until a careful study investigating the effect of this parameter is reported we can only speculate that these effects may be showing up as variability in the magnitudes of J between different studies.

Huberman [2.17] has calculated the energy of a free-electron model for a FM/PM/FM trilayer (for finite FM thicknesses) where the PM band lies between the minority-spin and majority-spin bands in the FM. For this unmatched band structure both barrier scattering and well scattering contribute to the energy. He has only considered the case of ferromagnetic alignment of the

FM magnetizations, for which he finds that both the total energy and total number of electron states differ from the corresponding bulk energies and number of states. These differences, which arise from both the tunneling and quantized well states, oscillate with d and approach finite constants when d becomes very large. This result indicates the sensitivity of this class of calculations to the existence of discrete well states.

We have shown that the free-electron approximation allows calculation of the exchange coupling without resorting to perturbation theory, and reduces exactly to the simple free-electron RKKY result in the weak coupling limit. However, at least some of the d states of a transition metal ferromagnet are probably better represented by tight-binding bands. Tight-binding model Hamiltonians have been investigated by *Edwards* et al. [2.18] and by *Deavon* et al. [2.19]. Tight-binding wave functions also have an advantage of including the lattice structure in the calculation from the beginning (since the calculations are performed for atoms at lattice sites). The inclusion of lattice periodicity in the calculation, which is absent from the free-electron model discussed earlier is important in obtaining long-period oscillations, as discussed in the following section. The band matching assumed by *Edwards* et al. is the same as that assumed in our free-electron calculation discussed above, with the FM minority band exactly matching the PM band. They treat a simple cubic single-band model with an infinite U parameter (repulsion in the Hubbard Hamiltonian) which corresponds approximately to the strong coupling limit of the free-electron model. In fact, the exchange coupling they obtain in the asymptotic (large d) limit is identical to that obtained for $k_0/k_F \to \infty$ (strong coupling, (2.24b)) in the free-electron calculation. The oscillating exchange functions obtained for choices of ε_F near the middle of the band (Fig. 2.9) are similar to those shown for free electrons in Fig. 2.6. However, when ε_F is near the band edges (i.e. near the Brillouin zone boundary) the range functions exhibit long-period oscillations due to the "beating" between k_F and various reciprocal lattice vectors, as discussed in the following section. Another difference between the free-electron and tight-binding calculations is that tight-binding bands give a description for the hole states as well as for the electrons. A tight-binding model requires that when the electrons of one spin confront a barrier in crossing the paramagnet, the corresponding picture for holes is that of states trapped in a one-dimensional quantum well. In the hole picture the model of *Edwards* et al. is more like the band structure treated by *Barnas* where the PM band matches the majority spin band in the FM. Thus the tight-binding calculation makes a connection between the two matched band structure cases treated for free-electrons.

Deavon et al. [2.19] also performed a non-perturbation calculation using a simple tight-binding model with a tight-binding spectrum in the direction perpendicular to the layer planes and a free-electron density of states in the layer planes. The magnetization in the FM was set by a local exchange parameter independent of the hybridization and thus did not have the strong effect on the exchange coupling observed in our free-electron calculation. Long-period oscil-

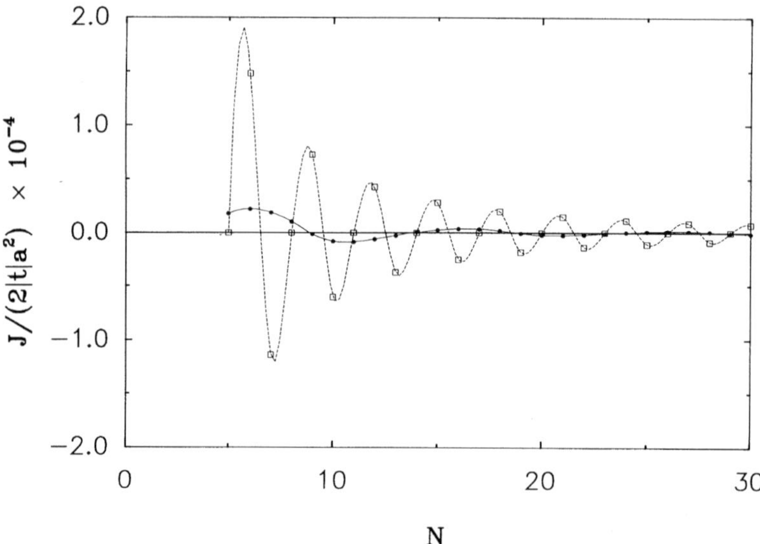

Fig. 2.9. The exchange coupling J as a function of the number of atomic planes N in the paramagnet layer for the model of [2.18], for two different choices of ε_F relative to the band edges $(-3.0,\ -1.0)$: $\varepsilon_F = -2.5$ (squares) and $\varepsilon_F = -1.05$ (circles). Note the larger oscillation period for ε_F near the band edge (circles)

lations were observed here as well for particular values of the tight-binding parameters, and were interpreted physically as the result of "aliasing", as discussed in the following section.

2.1.3 Oscillation of the Exchange Coupling with Interlayer Thickness, d

Since the discovery of oscillations in the exchange coupling, the long period of these oscillations, which appears inconsistent with a free-electron RKKY description, has occupied most of the attention of theorists. The free-electron wave functions implicit in RKKY calculations contain no information about the lattice structure. Introducing the lattice periodicity into the calculation provides an additional length scale, which can lead to long-period oscillations. This can be accomplished in two ways: using wave functions that meet the Bloch conditions of lattice periodicity (or their derived susceptibilities), or by imposing conditions of lattice periodicity on the locations of the magnetic layers, which we discuss first. Introducing lattice periodicity into free-electron calculations in this way, often called "aliasing", has been proposed by several authors [2.19–21]. Aliasing can be understood by imagining a short period oscillating exchange coupling which is only sampled by FM layers located at discrete lattice plane positions. When the PM Fermi wave vector, which determines the short oscillation periods, is close to the wave vector corresponding to the spacing, a, of the lattice planes (i.e. a reciprocal lattice vector) "beating" occurs between the two corresponding frequencies, giving an apparent long-period oscillation, as shown in Fig. 2.10. Aliasing can be applied in a *post facto* manner to free-electron calculations and occurs naturally for calculations where the FM planes are separated by an integral number of lattice spacings. However, as oscillatory exchange coupling has now been observed for many systems with oscillation periods very close to 10 Å, it appears unlikely that k_F and π/a could have nearly the same relationship for so many materials, and it is likely that this picture is too simple. Real metals indeed do not have a single value of k_F (corresponding to a spherical Fermi surface) but rather complex Fermi surfaces, as well as various sets of lattice planes. Recent calculations by *Bruno* and *Chappert* [2.22] and by *Stiles* [2.23] have examined the bulk Fermi surface in relationship to the reciprocal lattice for several elemental PMs and find the potential for several oscillation periods as shown for Cu in Fig. 2.11. Since the relevant Fermi wave vectors are those perpendicular to the growth plane, and it is in this direction that the bulk Fermi surface is most disrupted by the finite thickness of the FM, a better understanding of the Fermi surfaces of finite systems may be of use here.

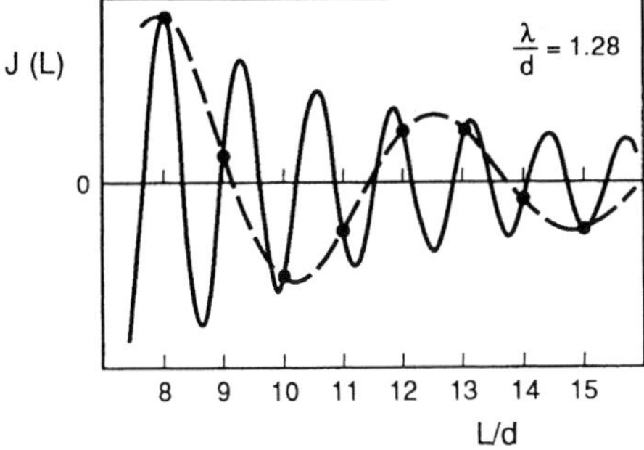

Fig. 2.10. RKKY-like oscillating exchange coupling with period λ (solid line) showing the longer period oscillation (dashed line) obtained by sampling the function only at integral values of the spacing, a, between atomic planes, i.e. "aliasing". From [2.21]

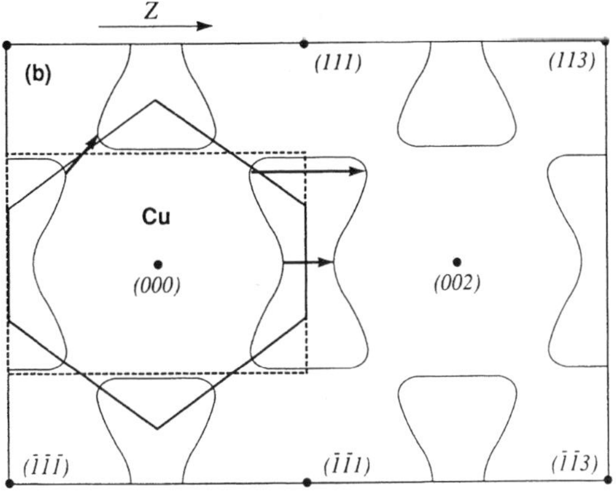

Fig. 2.11. Fermi surface of Cu in the (1 0 0) plane in the extended zone scheme. Arrows indicate values of $2(k_F - G)$ for reciprocal lattice vectors G which can give rise to oscillations with periods greater than π/k_F

Calculations of regions of large densities of states in k-space and "nesting vectors" can indicate what periods might be observed: a complete calculation of exchange coupling requires matrix elements for the exchange interaction as a function of k, as well. This sort of undertaking is of the same order of difficulty as calculating the exchange coupling directly from first-principles band structures, to be discussed in Sect. 2.18. However, a step in this direction is the incorporation of realistic (theoretical or empirical) susceptibility functions for the PM into the perturbation formalism. This was done by *Yafet*, as mentioned earlier, for Gd/Y superlattices by using calculated $\chi(q)$ functions for Gd and Y. *Wang* et al. [2.24] have incorporated the calculated $\chi(q)$ for Cr into their perturbation calculation for Fe/Cr, as discussed in the next section.

Herman and *Schreiffer* [2.25] have presented a study which illuminates the difference between imposing the lattice periodicity on the location of the magnetic planes (aliasing) vs. using wave functions for the PM which incorporate the lattice periodicity via the Bloch periodicity condition. They compare the effects of different types of interface roughness in the two cases. The long-period oscillations produced by aliasing are washed out by roughness in which the interface magnetic atoms are slightly displaced from lattice sites. (These atoms "sample" the exchange at non-integral numbers of atomic plane spacings.) But when Bloch wave functions are used for the PM the *same* interference condition occurs between the oscillations induced by the lattice potential and the Fermi-surface induced oscillations, also producing long periods. This interference mechanism is robust with respect to off-site disorder confined to a few planes near the interface and appears to require only that there be enough well-ordered layers in the PM to establish a periodic potential leading to a Bloch periodicity condition for the wave functions.

2.1.4 Non-Oscillatory Exchange Terms and Anderson-Like Models

We turn now to the possibility that the exchange coupling contains not only an oscillatory term but also a slowly decaying, non-oscillatory contribution. As mentioned earlier *Slonczewski* [2.13] predicted such an exchange coupling (with no oscillatory term) for two ferromagnets separated by an insulator (Fig. 2.12) with the form

$$J = \frac{(U_0 - \varepsilon_F)}{8\pi^2 d^2} \frac{8\kappa^2 (\kappa^3 - k_\uparrow k_\downarrow)(k_\uparrow - k_\downarrow)^2 (k_\uparrow + k_\downarrow) e^{-2\kappa d}}{(\kappa^2 + k_\uparrow^2)^2 (\kappa^2 + k_\downarrow^2)^2}, \tag{2.25}$$

where $k_\uparrow$, and $k_\downarrow$ and d are defined as in Sect. 2.1.2, U_0 is the height of the barrier, and $\kappa = [2m(U_0 - \varepsilon_F)/\hbar^2]^{1/2}$. For large U_0, (2.25) requires no integration over energy, as all of the wave vectors may be approximated by their values at the Fermi energy. The exponential decay of this function arises from the exponentially decaying overlap of the FM wave functions extending into the insulating barrier. The important thing to note here is that ε_F lies in this exponentially decaying region and that the maximum contribution comes from

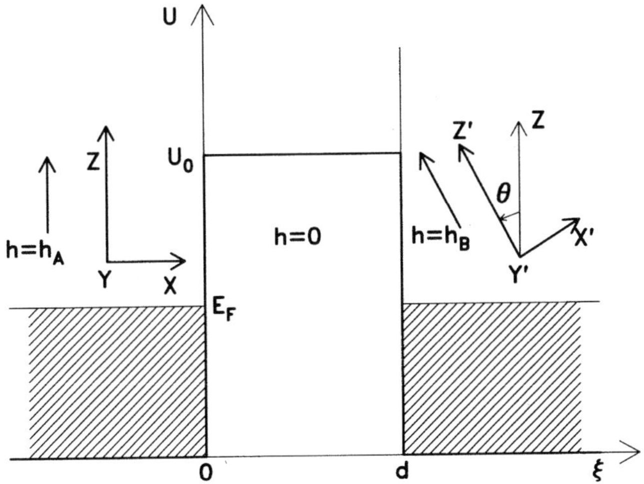

Fig. 2.12. Potential diagram for a ferromagnet/insulator/ferromagnetic structure (cf. Fig. 2.5), used for calculation of [2.13]

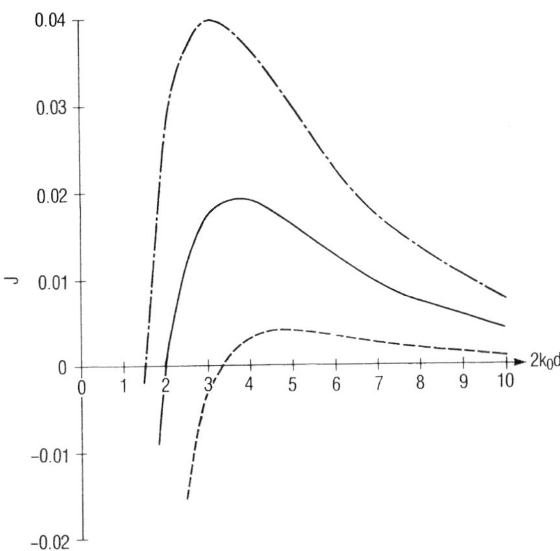

Fig. 2.13. Non-oscillatory contribution to the exchange coupling in the free-electron model, arising from evanescent (decaying) states in the paramagnet with energies $< h_0$. This contribution is cancelled in the full calculation but is representative of superexchange terms

energies near ε_F. In our non-perturbation free-electron model of Sect. 2.1.2, the non-oscillatory term in the coupling arising from exponentially decaying states with energies below the barrier height was exactly cancelled by an equivalent term from states lying just above the top of the barrier. The crucial element producing this cancellation is continuous integration through the energy of the top of the barrier. Discreteness in the density of states of either the FM or PM would disrupt this cancellation and could yield a "superexchange" term similar to that obtained for an insulating barrier. The form of this term can be assumed to be similar to the non-oscillatory contribution to the coupling from states with $E < h_0$ in our free-electron model, as shown in Fig. 2.13 for various ratios of k_0/k_{F}.

One would like to have a model for exchange coupling which includes both itinerant and discrete states in an integrated treatment. This is also the problem considered by *Anderson* [2.26, 27] in evaluating the stability/formation of magnetic moments of transition metal impurities in host metals. Because the "*Anderson* model" underlies many of the treatments of exchange coupling discussed below, we will outline it briefly. The major utility of the *Anderson* model is that it treats a Hamiltonian that is simple enough that the effect of s–d mixing on the configuration and energy levels of the localized states can be evaluated. The *Anderson* model, like the real many-body Hamiltonian, can be treated in the mean field approximation when electron correlations are not too important. The *Anderson* Hamiltonian has the form

$$H = H_{\mathrm{s}} + H_{\mathrm{d}} + H_{\mathrm{mix}}, \tag{2.26}$$

where

$$H_{\mathrm{s}} = \sum_{k,s} \varepsilon_k c_{ks}^* c_{ks},$$

$$H_{\mathrm{d}} = \sum_{n,s} E_n a_{ns}^* a_{ns} + \frac{1}{2}(U - J) \sum_{m \neq n,s} a_{ms}^* a_{ms} a_{ns}^* a_{ns} + U \sum_{m,n} a_{ms}^* a_{ms} a_{n-s}^* a_{n-s}, \tag{2.27}$$

$$H_{\mathrm{mix}} = \sum_{k,m,s} V_{km} c_{ks}^* a_{ms} + V_{mk} a_{ms}^* c_{ks}.$$

U is a Coulomb interaction between the localized (d) states and J is the exchange integral. H_{mix} is reminiscent of the Coulomb electron–electron terms which couple s and d states in the derivation of the s–d Hamiltonian. The broadening of the d electron levels is determined by the interplay between H_{d} and H_{mix}. When H_{mix} is small the local levels will be narrow. We have seen that the degree to which the FM levels are localized can dramatically affect the contribution of a non-oscillatory term to the coupling.

Discussions of the *Anderson* model [2.28] provide some important physical insights into how and when localized states should be included in theories of exchange coupling. We first consider when a mean field or band approach is valid versus when correlations must be treated explicitly. (Recall that the s–d interaction came only from an explicit treatment of correlations in the electron–electron Hamiltonian.) The equation for H_{mix} contains terms which allow a band electron to hop on and off a (partially occupied) local ion. The partially screened Coulomb interaction energy U of two electrons on the local ion tends to keep the electrons apart. The lifetime of an extra electron on the ion is given by the inverse of the width of the virtual level, $\hbar/\Delta$, which is proportional to V^2.

If the level is broad $\Delta > U$, and $\hbar/\Delta < \hbar/U$ (the time for the electrons to interact via U), then the electrons hop on and off so fast that they do not have time to experience the repulsion due to U which would otherwise tend to correlate their motion. In this case a mean field or Hartree Fock approach is valid and there is a Stoner-like criterion for the formation of a stable moment on the local ion $(U + 4J)/\Delta > 1$. It is however in the other limit $(U > \Delta)$ where correlations are large and the local level is narrow that local moments are most likely to exist and the *Anderson* Hamiltonian is most useful. In this limit H_{mix} can be treated by perturbation theory. Thus we conclude that narrow localized levels may be treated by perturbation theory including the effects of correlation directly, and very broad levels may be treated as bands using one-electron approaches.

Schreiffer [2.29] has shown that by applying a canonical transformation to the *Anderson* Hamiltonian he can restructure it as a power series in H_{mix} with no linear term. Explicit evaluation of the term in H_{mix}^2 for an s state ion shows that it consists of two contributions, a direct one-body potential which scatters a conduction electron without spin-flip, and a spin–spin exchange interaction. The exchange interaction has exactly the form of the s–d Hamiltonian derived earlier. This process of transforming the *Anderson* Hamiltonian into the s–d Hamiltonian leads to another important physical insight: in the derivation it is apparent that for ionic transitions within a given multiplet only conduction electrons of the same symmetry are exchange scattered. For example a transition from the ionic state $M = 5/2$ and conduction electron state $klm\text{-}s$ to the ionic state $3/2$ and electron state $k'lms$ proceeds via a conduction electron hopping onto the ion and one of the ion's electrons hopping off. Since the transition is to the same atomic multiplet, $S = 5/2$, only $l = 2$ conduction electrons give non-zero contributions. Thus free-electron-like states with d symmetry should make the largest contributions to the coupling between d localized states.

Goncalves da Silva and *Falicov* [2.30] several years ago extended the *Anderson* s–d Hamiltonian to calculate exchange in rare earth metals (with localized f-electrons) via their itinerant sp electrons. In fourth order perturbation theory their model gave both an *Anderson*-type superexchange and, for metallic itinerant bands, an oscillating RKKY exchange. A similar model, but for localized d- rather than f-electrons, has recently been proposed to apply to exchange coupled FM/PM mulilayers. *Lacroix* and *Gavigan* [2.31] discuss an illustrative example in which an *Anderson*-like Hamiltonian is used to calculate the exchange coupling between Co layers separated by Cu. They model Cu as a single free-electron band and Co as an exchange-split narrow d-band, with a mixing interaction between them, V. In order to make the simplest possible calculation they narrow the d-bandwidth to a single exchange-split level with energies $\varepsilon_{\mathrm{F}} - \Delta$ and $\varepsilon_{\mathrm{F}} + \Delta'$, and they do not treat the two-dimensional multilayer geometry. This yields a model identical to that of *Goncalves da Silva* and *Falicov*. Applications of fourth order perturbation theory yields essentially identical results for both calculations for the oscillatory part of the exchange coupling.

From *Lacroix* and *Gavigan* (which we specialize to $\Delta = \Delta'$) we obtain the asymptotic perturbation energy:

$$\Delta E = -\frac{4V^4}{(2\pi)^3 \Delta^2}(2mk_F)\frac{\cos 2k_F d}{d^3}.$$
(2.28)

This is the usual form for a three-dimensional RKKY interaction with a R^{-3} decay. The two references evaluate the superexchange terms in different limits, with *Goncalves da Silva* and *Falicov* obtaining an exponentially decaying term in the limit of an empty PM conduction band:

$$J(d) \propto V^4 \frac{\exp(-2k_0 d)}{(k_0 d)^2},$$

$$k_0 = \left[\frac{2m\Delta}{\hbar^2}\right]^{1/2},$$
(2.29)

which is similar to *Slonczewski*'s result for an insulating barrier and our result shown in Fig. 2.12. *Lacroix* and *Gavigan* obtain a large superexchange contribution when the product $\alpha = dk_F(\Delta/\varepsilon_F)$ is small, i.e. when d is small or the d-electron levels (Δ, Δ') are close to the Fermi level, with the form

$$\Delta E_2 \propto \frac{V^4}{\Delta^3}\frac{\sin^2 k_F d}{d^2} Ln^2 \frac{\alpha}{2}.$$
(2.30)

If Δ is large, i.e. if the discrete energy levels are far from the Fermi energy the superexchange contribution is small. In both calculations these non-oscillatory contributions to the exchange coupling are antiferromagnetic. Thus we see that the introduction of discrete energy levels can in principle explain the existence of an antiferromagnetic exchange coupling background superimposed on the oscillating RKKY coupling.

Wang et al. [2.24] have proposed that the discrete states which lead to non-oscillatory coupling are interface states associated with Fe atoms in Cr at the (interdiffused) interfaces of Fe/Cr multilayers, with energies lying below e_F. Their results, obtained by approximating these interface states as discrete levels at $\varepsilon_F - \Delta = 0.04$ Ry, are shown by the dotted line in Fig. 2.14. In this calculation the PM (Cr) has been modelled by a susceptibility $\chi(r)$ calculated from bulk *ab initio* band structures, and the two-dimensional nature of the structure is accounted for in the calculation.

The next level of complexity in treating coupling between discrete states in the FM layers is to include the effects of broadening of these states by hybridization with the itinerant electron bands. *Bruno* [2.32] has considered coupling in a system described by the *Anderson* Hamiltonian following the method used by *Caroli* [2.33] to treat coupling between magnetic impurities in a non-magnetic host metal, but including the two-dimensional nature of the multilayer geometry. The energy levels of the magnetic layers are only localized with respect to the direction in k-space corresponding to momenta perpendicular to the interfaces, and are continuous bands in the layer planes. The mixing

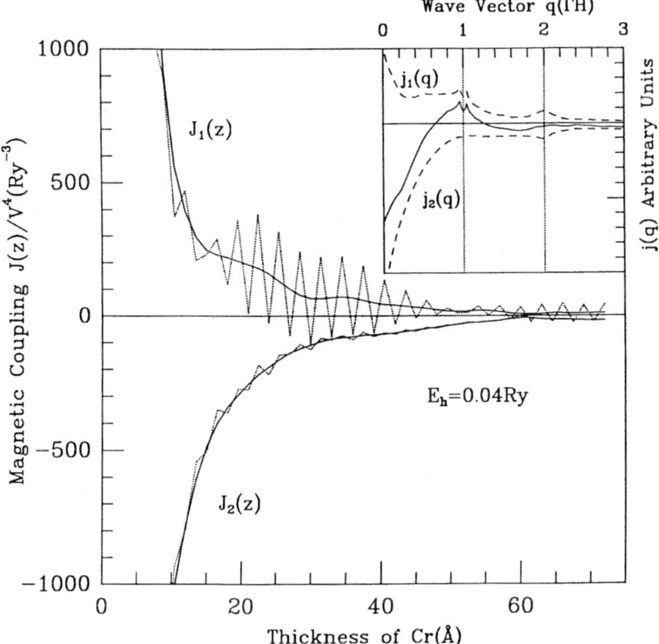

Fig. 2.14. The contributions to the exchange coupling for the model of [2.24] arising from the RKKY-like terms, J_1, and superexchange terms, J_2. The curves are shown for two different values of roughness defined by the probability, p, of finding an atom at $\pm a/2$ (where a is the lattice spacing) away from the nominal position of the interface layer: flat interfaces, $p = 0$, (dotted line) and rough interfaces, $p = 1/4$, (solid line)

interaction, parameterized by V, is first used to determine the hybridization broadening for the localized state of a single magnetic "impurity plane" embedded in a paramagnetic medium. This determines the occupation and magnetic polarization of this virtual bound state in terms of V. Then the same V is employed to produce coupling between two magnetic layers. Thus the magnitude and phase of this coupling can be related to the magnetization and occupation of the two-dimensional virtual bound state, via V, if these are known or can be estimated. The oscillation period for large PM interlayer thickness is given by the RKKY part of the interaction, deriving from those wave vectors for which the measure of the Fermi surface of the PM is stationary. (The relative magnitudes of the superexchange and RKKY contributions to the exchange coupling are, one assumes, determined by the initial, unhybridized position of the discrete states.)

Note that all three of the calculations for transition metal multilayers based on *Anderson*-like Hamiltonians [2.24, 31, 32] adopt the point of view that the FM spins act like impurity spins embedded in a paramagnetic medium, and the calculations are carried out using perturbation theory. Because of this the strength of the magnetic perturbation comes only into the coefficient of the range function and, unlike in the non-perturbation calculations of Sect. 2.1.2, does not affect the shape of the range function. For the first two calculations the magnitude of the coupling cannot be derived from the model, as it depends on the unknown mixing parameter, V, which represents the strength of the s–d contact interaction. In the third calculation by *Bruno* the magnitude of the exchange coupling is related to the magnetization and occupation of the virtual bound state of the magnetic layer through their common dependence on the mixing parameter.

A model which differs somewhat in philosophy, but which also yields a non-oscillatory superexchange type coupling has been presented by *Garcia* and *Hernando* [2.34]. In this model the discrete states occur in the PM spacer layer which is assumed to be a quantum well. The FM layers polarize the electrons in the spacer layer via a decaying proximity field which penetrates into the spacer layer with an exponential decay. For PM electron densities appropriate for Cu and Cr (and proximity fields reasonable for Fe) this calculation gives an exchange coupling as depicted in Fig. 2.15. Oscillations are produced when

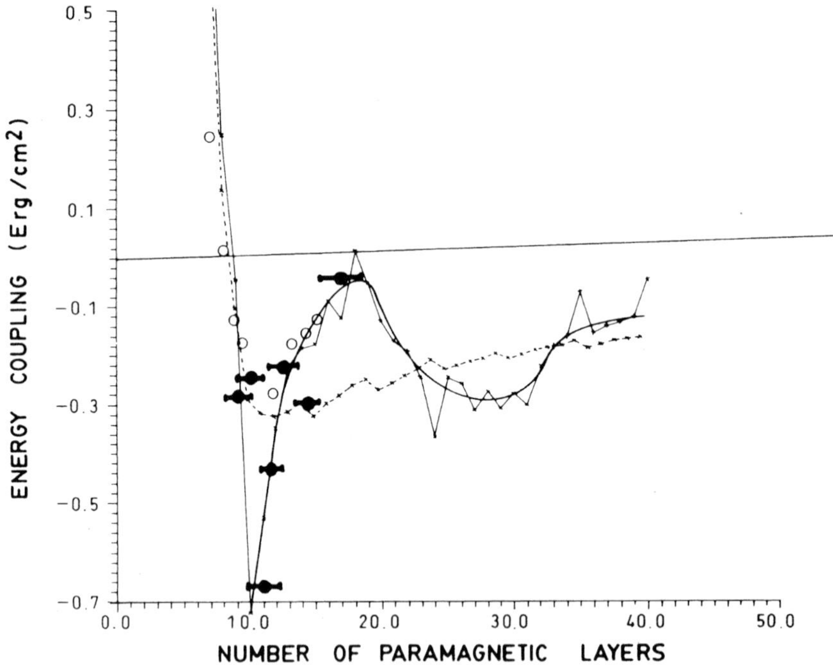

Fig. 2.15. Exchange coupling energy as a function of the number of paramagnet layers for ferromagnet/paramagnet/ferromagnet structure, as modeled in [2.34]. Curves show two different degrees of roughness obtained by modulating the proximity field due to the ferromagnets by a function $D_Q \cos(Qd)$, d the paramagnet thickness: for $Q = 0.02\pi/d$ and $D_Q = 0.85$, and for $Q = 0.2\pi/d$ and $D_Q = 0.85$

roughness is introduced into this model, as shown by the dashed lines in Fig. 2.15.

A word about roughness: Adding roughness at the interface can have dramatic affects on the calculated exchange coupling: it can smooth out oscillations with small periods (less than the roughness "period"), or it can produce oscillations in otherwise non-oscillating range functions. Furthermore, as in any random system, the number of parameters involved in describing the roughness may range from a small number to infinity. The recent observation of coincident short-period and long-period oscillations in Fe/Cr/Fe with very gradually increasing thickness wedges of Cr may indicate that roughness or, more likely, the irregular terrace structure of non-wedge samples masks short-period oscillations. To our knowledge no study of the effect of controlled roughness on the exchange coupling has been reported. The effects of roughness on the calculations contained in [2.24, 34] is illustrated in Figs. 14, 15, respectively.

One theoretical treatment of exchange coupling for which roughness plays an essential role is the theory of biquadratic coupling proposed by *Slonczewski* [2.35] to explain the preferred 90° alignment of the magnetizations in adjacent FM layers observed in several systems. This brings us to the general topic of non-Heisenberg exchange, which we will now discuss.

2.1.5 Non-Heisenberg Exchange

The recently observed biquadratic exchange coupling has been treated theoretically by *Slonczewski* [2.35] who attributes the effect to a fluctuation mechanism leading to a frustration of the bilinear (Heisenberg) exchange coupling. The phenomenological expressions for macroscopic coupling are given by

$$E = A_{12}(1 - \boldsymbol{m}_1 \cdot \boldsymbol{m}_2) + 2B_{12}(1 - (\boldsymbol{m}_1 \cdot \boldsymbol{m}_2)^2), \tag{2.31}$$

where A_{12} is the bilinear coefficient and B_{12} is the biquadratic coefficient. In general, the exchange energy associated with coupling between two localized spin operators can be expanded in the form $\sum J_n (S_1 \cdot S_2)^n$. Thus a biquadratic term, $J_2(S_1 \cdot S_2)^2$ is allowed and may arise from an intrinsic mechanism as discussed below. However the original observations of 90° alignment in Fe/Cr occurred only for Cr thicknesses where the long-period oscillating bilinear coupling was close to zero. The Fe/Cr system also exhibits evidence of short-period (two monolayer) oscillations. These observations prompted *Slonczewski* to consider an extrinsic mechanism for the biquadratic coupling. We summarize his calculation for a fluctuation model as follows.

Terraces (with $\sim$one monolayer steps) in the FM layers cause fluctuations in the intrinsic bilinear coupling. The exchange stiffness of the FM resists the torques due to the fluctuations, and the system compromises by lowering the energy through the formation of static waves of magnetization. As a simple example consider two FM layers, each t thick, with exchange stiffness A. They are coupled by an oscillating exchange which varies due to the existence of terraces of length L, assumed for simplicity in one planar direction only. If we consider only a single Fourier component of the terrace-induced step-wise lateral variation

$$J(x, y) = J_k \sin kx. \tag{2.32}$$

The energy is written as

$$W_k = \frac{1}{2L} \int dx \{ -J_k \sin kx \cos[\theta(x, 0) - \theta'(x, 0)] \}$$

$$+ A \int_0^t dz(\theta_x^2 + \theta_z^2) + A \int_0^t dz'(\theta_x'^2 + \theta_z'^2), \tag{2.33}$$

where $\theta(x, y, z)$ and $\theta'(x', y', z)$ are the angles of magnetization in the two films. If J is small the static equilibrium solution is given in terms of $\bar{\theta}$, the average value of θ by

$$\theta = \bar{\theta} - J_k \frac{\sin(\bar{\theta} - \bar{\theta}')\sin kx \cosh k(t - z)}{2Ak \sinh kt} \tag{2.34}$$

and similarly for θ'. This solution is a static spin wave in the plane with period π/k and with exponential dependence on z in the direction perpendicular to the plane. A superposition of "wavelets" of this form in both the x and y directions will give the appropriate solution for realistic cases of roughness. Substituting such a superposition solution into the expression for energy gives

$$W_{\min} = -\frac{1}{4} \sum_{k \neq 0} \left(\frac{J_k^2}{k} \right) \frac{\coth kt}{A} \sin^2(\bar{\theta} - \bar{\theta}'). \tag{2.35}$$

Since $\sin^2(\bar{\theta} - \bar{\theta}') = 1 - (\boldsymbol{m}_1 \cdot \boldsymbol{m}_2)^2$ this gives $B_{12} < 0$ in all cases. The coefficient B_{12} can be evaluated for particular forms of roughness by specifying the Fourier coefficients, J_k.

The possibility of an intrinsic mechanism for not only biquadratic exchange but a general non-Heisenberg expansion of the form $\sum J_n(S_1 \cdot S_2)^n$ is apparent in the expression for the exchange coupling of the free-electron model, discussed in Sect. 2.1.3. In the calculation of the torque from the spin-current the expression for J (2.23) contains a denominator with a complicated dependence on ϕ_{12}, which reduces to a Heisenberg bilinear form only near antiferromagnetic alignment. In a more general treatment [2.15], (2.23) can be expanded as a sum in $\cos \phi_{12}$ as follows

$$J(\phi_{12}) = \frac{A\hbar^2 k_0^4}{4\pi^2 m} \text{Im} \sum_{l=0}^{\infty} \cos^l \phi_{12} \sum_{n=l}^{\infty} \left(\frac{1}{2} \right)^n \binom{n}{l}$$

$$\times \int_{z_F}^{\infty} dz\, z(z^2 - z_F^2)(z - \sqrt{1 + z^2})^{4(n+1)} e^{i2(n+1)k_0 dz}, \tag{2.36}$$

where $\binom{n}{l}$ are the binomial coefficients. This equation defines J^n as

$$J(\phi_{12}) = \sum_{n=0}^{\infty} J^n \cos^n \phi_{12} . \tag{2.37}$$

J^1 can be identified with A_{12} and J^2 with B_{12}. The first few terms dominate the expansion (i.e. the coupling is nearly Heisenberg-like) for weak coupling (z_F large) since $(z - \sqrt{1 + z^2}) \to 0$ as $z \to$ infinity, with higher order terms becoming more important as the coupling increases. A comparison of B_{12} and A_{12} for

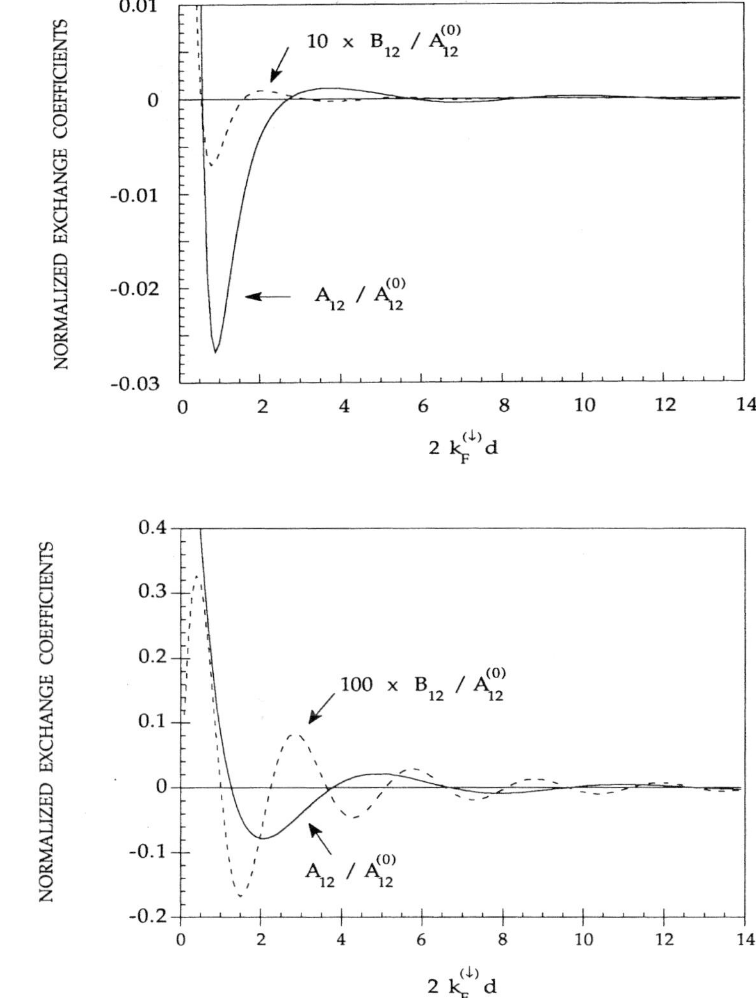

Fig. 2.16. Relative values of the bilinear (A_{12}) and biquadratic (B_{12}) terms in the exchange coupling for the free-electron model of Sect. 2.1.3 [2.15]: (**a**) for strong coupling $k_0/k_F = 5$ with B_{12} shown $\times 10$, and (**b**) for moderate coupling $k_0/k_F = 1.0$, with B_{12} shown $\times 100$

both strong and weak coupling is shown in Fig. 2.16. Note that B_{12} will produce $90°$ relative alignment only when it is negative and larger in magnitude than $1/4|A_{12}|$. Since the oscillations in A_{12} and B_{12} are incommensurate, and the incommensuration changes with coupling strength, the satisfaction of these conditions can only be predicted by actual calculation. These conditions are satisfied near several of the nodes of A_{12} for the cases shown in Fig. 2.16. A more complete discussion is given in [2.15].

Four other calculations of the intrinsic non-Heisenberg contributions to the exchange coupling have recently been published. *Edwards* et al. [2.36] have generalized the calculations of [2.18] using the spin-current approach, with the d holes treated as a gas, with results similar to those discussed above. They have also generalized the intrinsic temperature dependence of (2.38) to arbitrary ϕ_{12}. *Slonczewski* [2.37] has carried out a spin-current calculation of the ϕ_{12}-dependent exchange coupling for a general (unmatched) free-electron single-band model, focusing on the effect of the PM Fermi wave vector. *Barnas* and *Grunberg* [2.38] have generalized the calculations of [2.16] to obtain energy differences for parallel, antiparallel and perpendicular alignment of magnetizations, yielding bilinear and biquadratic coupling terms. *Bruno* [2.39] has derived a model-independent formulation which uses the force theorem to obtain the exchange coupling from one-electron energies which are expressed in terms of ϕ_{12}-dependent reflection coefficients of the wavefunctions at the interfaces. It would be interesting to obtain an experimental test of the relative contributions of intrinsic and extrinsic mechanisms to the biquadratic coupling by investigating systems with controlled amounts of roughness.

2.1.6 Band Structure Results

Given the concerns raised in earlier sections about the difficulty of including the non-perturbation nature of the interactions, as well as treating wave functions that reflect all of the lattice symmetries, it might seem most appropriate to obtain the exchange coupling directly from total energy band structure calculations. It is especially attractive that such calculations treat the sp and d electrons on an equal basis and the potentials are self-consistent. *Ab initio* calculations based on the local spin density approximation offer the most accuracy and have been shown to give results in reasonable agreement with experiment for bulk magnetic properties. Calculations for thin films of transition metals show that the electronic screening lengths are sufficiently short that the second and in some cases the first atomic plane below a surface or interface already has very nearly bulk properties. This means that the bulk moment of the FM layers will be maintained through most of each layer as long as the layer thickness exceeds five atomic planes. The magnetic moment of the atomic planes at the interface is determined both by hybridization with the PM (as in the *Anderson* model) and by interactions with the inner atomic planes of the FM. In this respect the *ab initio* calculations give results in accord with the assumptions of our simple free-electron model, but here the electronic and spin densities (and potentials) are iterated to self-consistency. All of the length scales relevant to determining the oscillations in the coupling function (Fermi wave vector(s), exchange splitting(s), and lattice periodicities) are included naturally in first-principles calculations – the first two emerge from the conditions of particle conservation and energy minimization, and the third is specified by the atomic positions used for the calculation. The usual procedure is to assume an epitaxial relationship (lattice matching) between the two components (FM and PM) of the multilayer, so that the lattice sites in the plane are in registry and close to the individual bulk lattice constants. A layered system may then be treated as an anisotropic compound with unit cell elongated in the direction normal to the layers. Thus the periodicity in the layer planes is the same or closely related to that of the bulk constituents and the periodicity normal to the layers reflects the periodicity of the multilayer. The spacings between individual atomic planes may be specified separately for each atomic plane but are also usually set to bulk spacing within the FM and PM layers and to some average of bulk values at the interface. Optimum relaxations of the atomic layer spacings normal to the plane can be determined from total energy calculations, and in some cases are known to affect the magnetic properties of the interface, but such calculations are very time consuming. In summary the major limitations of *ab initio* calculations of layer exchange couplings are not the physical approximations, of which there are very few, but the limitations imposed by computer size and speed. In practice, unit cells are limited to about 10 atoms, which means a simple lattice-matched multilayer can have a total of m FM atomic planes and n PM atomic planes with $m + n < \sim 10$. There are limitations on mesh spacings and basis sets, as well, which will be discussed later.

The limit on unit cell sizes in *ab initio* calculations indicates that simpler band structure calculations might be useful for treating larger repeat units. Accordingly we begin by discussing real space tight-binding calculations reported by *Stoeffler* and *Gautier* [2.40]. They considered Fe/Cr and Fe/V bcc (0 0 1), Co/Pd fcc (1 1 1) and Co/Ru hcp (0 0 0 1) structures. The Hamiltonian is written in terms of Slater–Koster parameters between first neighbors (for all three symmetries) and second neighbors (bcc only), with A–B interaction parameters obtained from the means of the A–A and B–B parameters. Self-consistency in the charge transfers and magnetic moments was neglected after test calculations showed the effect to be small. They report results for interlayer exchange couplings up to six atomic planes of the PM spacer for Co/Ru, Fe/V, and Co/Pd, and up to 12 atomic planes of Cr in Fe/Cr.

For both Co/Ru and Fe/V Friedel oscillations in the coupling with a period of approximately two to three atomic planes are obtained. For Co/Ru the coupling at one atomic plane of Ru is antiferromagnetic and for Fe/V the coupling at one atomic plane of V is ferromagnetic. Beyond about five atomic planes of PM the magnitude of the coupling in both cases was less than the estimated errors in the calculation. Both the Co and Fe moments are somewhat reduced at the interface and slightly enhanced within the FM layers. The Ru spacer layers are essentially non-magnetic but the V interface layers have small moments and couple antiferromagnetically to the Fe.

In Co/Pd when a susceptibility enhancement for Pd is incorporated into the calculation the Pd is slightly ferromagnetic and couples ferromagnetically to Co, and the Co interface moments are not reduced. The Co–Co layer coupling is ferromagnetic for all Pd thicknesses with only a slight oscillatory behavior. Friedel oscillations are recovered when the susceptibility enhancement is removed from the calculation.

For Fe/Cr the Cr ions clearly prefer to align antiferromagnetically in the layer normal direction (in-plane antiferromagnetic Cr spin structures were not considered) and have an even stronger antiferromagnetic interaction with the neighboring Fe atoms. The Cr antiferromagnetism is frustrated to some degree for odd numbers of Cr atomic planes when the Fe layers are aligned antiferromagnetically, and for even numbers of Cr atomic planes when the Fe layers are aligned ferromagnetically (Fig. 2.17). This frustration leads to a rigidly oscillating interlayer exchange coupling with period two atomic planes and a very slow decay in magnitude out to the maximum thickness considered of 12 atomic planes. When the sp electrons are included along with the d electrons the conditions for frustrating the Cr antiferromagnetism and thus the two atomic

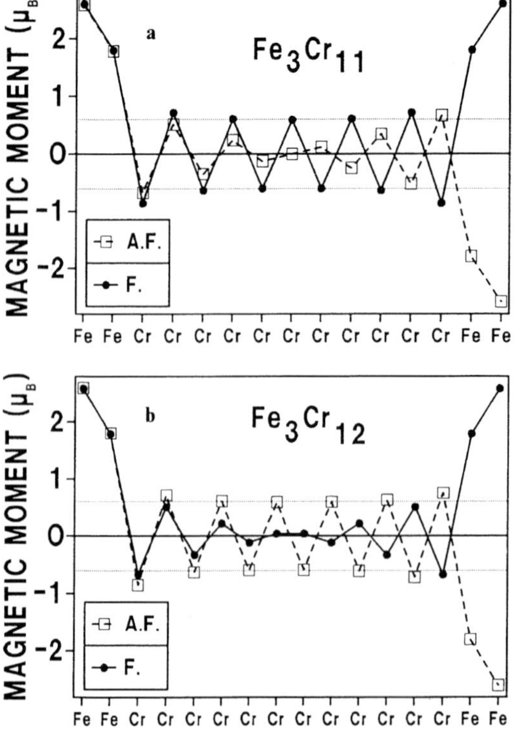

Fig. 2.17. Magnetic moments in individual atomic planes for (a) Fe_3Cr_{11} and (b) Fe_3Cr_{12}, as calculated in [2.40]. Note the frustration of the Cr antiferromagnetism near the center of the Cr layer for antiferromagnetic Fe moment alignment in (a) and for ferromagnetic Fe moment alignment in (b)

plane oscillation period in the exchange coupling remain. However the magnitude of J is increased for small Cr separations and the coupling decays much more rapidly, reflecting perhaps the RKKY-like contribution of the sp electrons. The possibility of reducing or eliminating the frustration of Cr antiferromagnetism by allowing tilting of the moments was also investigated, for odd numbers of Cr layers, but the resulting configurations were less stable than the (frustrated) colinear arrangements.

Hasegawa [2.41] has obtained similar results in a tight-binding calculation of Fe/Cr (0 0 1) augmented with calculations of bcc Co/Cr (0 0 1) and bcc Ni/Cr (0 0 1) for comparison, showing interesting trends. The exchange interactions between interfacial atomic planes are antiferromagnetic for Fe/Cr, ferromagnetic for Ni/Cr and give stable states for either ferromagnetic or antiferromagnetic alignment for Co/Cr. Accordingly the conditions for frustration of Cr antiferromagnetism are opposite in Fe/Cr and Ni/Cr systems, with the Cr moments generally enhanced when the Cr antiferromagnetism is not frustrated and reduced otherwise. The individual Ni atomic planes are not ferromagnetic and have very low moments. A major difference for Fe/Cr as compared to the results of [2.40] is that calculation of the exchange interaction between the FM layers gives antiferromagnetic coupling for all Cr thicknesses from two atomic layers to six atomic planes. This difference may be due to the fact that *Hasegawa* iterated local moments and electron occupancies to some level of self-consistency, and will be further discussed below.

We turn now to the self-consistent *ab initio* calculations for comparison. We are reminded that these calculations based on density functional theory allow an exact physical identification of the ground state energy of the spin-resolved electron densities. The individual eigenvalues (band structures) do not have a precise physical significance as this is not a one-electron problem. Exchange coupling is calculated in these calculations by comparing two different magnetically constrained ground state energies. Thus whichever turns out to be the least stable alignment of the layer magnetizations must be considered to be an excited magnetic state. These considerations may affect the mathematical precision of the calculated exchange coupling energies, but the trends obtained should be reliable.

An early *ab initio* (local spin-density functional) calculation for Fe/Cr (0 0 1) was reported by *Levy* et al. [2.42] using the Augmented Spherical Wave method. This method uses a linear combination of atomic orbitals basis and involves spherical approximations for the potential, but should be reasonably accurate as long as the atomic environments are not too anisotropic. These calculations exhibit similar behavior for the individual moments and the total coupling as the tight-binding calculation above. The Fe interface moments are reduced from bulk values ($1.8\mu_B$ vs. $2.15\mu_B$); the interior Fe moments are enhanced relative to bulk values ($2.5\mu_B$); the Cr tries to be antiferromagnetic, with ionic moments reduced slightly below the bulk Cr spin wave value of $0.6\mu_B$; the Cr moments again couple strongly antiferromagnetically to the Fe, thus the Cr antiferromagnetism is frustrated for odd numbers of Cr atomic planes with antiferromagnetic

Fe layer alignment and even numbers of Cr atomic planes with ferromagnetic Fe alignment, as in the tight-binding calculations. For three and four atomic planes of Cr this frustration determines the sign of the Fe layer exchange coupling to be ferromagnetic for three atomic planes and antiferromagnetic for four atomic planes. Unlike the tight-binding calculation of [2.40] for $n = 5$ (and $n = 7$), the frustration of Cr antiferromagnetism for antiferromagnetic alignment does not raise the energy greater than that for ferromagnetic alignment. A similar trend occurred in *Hasegawa's* calculation of antiferromagnetic exchange coupling for three and five Cr atomic layers. A comparison of *ab initio* and tight-binding results is shown in Table 2.1. The authors of [2.40] speculate that the differences for $n = 5$ and $n = 7$ may be due to bands near the Fermi level which are very sensitive to FM layer alignment. They may also be affected by the difficulties inherent in getting sufficient accuracy in a k-space sum to evaluate total energy. *Ab initio* calculations of semiconductors (where there is no Fermi surface) achieve acceptable precision in total energy for a very small number of k-points. For metals the energy bands (eigenvalues) are usually interpolated between values obtained for a reasonably large number of discrete k-points. These extrapolations are most reliable when the bands are smooth and pass rapidly through the Fermi level (i.e. are free-electron like), as for simple Fermi surfaces. Calculations of total energy for metals with complex Fermi surfaces should only be considered reliable when additional calculations at larger numbers of k-points show either no change or a behavior that can be reliably extrapolated to an infinite number of k-points. Failure to achieve k-point convergence has proved troublesome in calculations comparing the stability of various structural phases of transition metals, in attempts to calculate magnetic anisotropy energies, etc. *Edwards* et al. [2.18] have studied the effects of k-point convergence

Table 2.1. Total energy difference/unit cell between ferromagnetic and antiferromagnetic alignments of magnetizations in FM layers, $\Delta E = E(F) - E(AF)$

Number of PM monolayers, n	*Ab initio* Fe_3Cr_n [2.24]	Tight-binding Fe_3Cr_n-d only [2.40]	Tight-binding Fe_3Cr_n-spd [2.40]	Tight-binding $Fe_1Cr_n^*$ [2.41]
1		− 35		− 2
2		33	143	13
3	− 17	− 34	− 122	14
4	27	23	90	16
5	152	− 26	− 62	8
6	26	25	59	4
7	18	− 20	− 49	
8		19		
9		− 19		
10		20		
11		− 20		
12		21		

* Obtained from calculated results for J rather than E; note different Fe layer thickness.

using their tight-binding model Hamiltonian, by recasting the calculation to obtain the exchange coupling energy by a sum in k-space. They estimate that the total energy will not converge until calculations are performed at 10^6 points in k-space!

An *ab initio* calculation which explicity considers this problem of k-point conversion is discussed by *Herman* et al. [2.43]. They obtain oscillatory exchange couplings for bcc Fe/Cu and fcc Co/Cu multilayers as a function of Cu thickness. Figure 2.18 shows that their total energy differences for Co/Cu (0 0 1) for 864, 1536, and 2904 k-points, clearly have not yet reached convergence. Viewed as preliminary results, however, some interesting trends emerge from these calculations. Differences in coupling obtained for three different orientations of fcc Co/Cu ([0 0 1] and [1 1 1] from 864 k-points and [1 1 0] from 1944 k-points) are demonstrated in Fig. 2.18. Ignoring the curves drawn as guides to the eye, the [1 1 1] result seems to show evidence of a long-period directly and the [0 0 1] and [1 1 0] indirectly in that they do not oscillate symmetrically around zero.

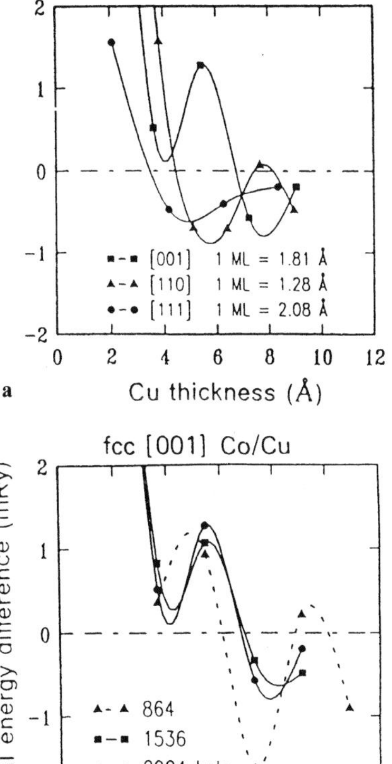

Fig. 2.18. Calculated *ab initio* exchange coupling from [2.43] for Co/Cu superlattices vs. Cu layer thickness for (**a**) various numbers of k-points, and (**b**) various crystallographic orientations

In concluding this section it appears that while *ab initio* calculations can provide some insight into individual atomic plane moments and exchange interactions, and perhaps hint at the properties of the layer exchange coupling, the problems inherent in calculations for large unit cells for very large numbers of k-points do not justify major efforts in brute force total energy calculations at this time.

2.1.7 Temperature Dependence of Exchange Coupling

The observed temperature dependence of the exchange coupling may be produced by two very different mechanisms. The first comes from the intrinsic temperature dependence of the electronic structure (states, wave functions, occupancies) of the multilayer. Temperatures below room temperature are small on the scale of electronic energies. However, temperatures can affect the occupancies of states near the Fermi level and thus reduce those properties which derive from the sharp discontinuity there at $T = 0$. A second and distinct mechanism is the temperature disordering of the magnetic moments in the ferromagnets, especially those at the interfaces. This is independent of which specific model is adopted as the origin of the exchange coupling, and the temperature dependence which derives from the mechanism is in some sense an artifact of obtaining exchange coupling from magnetization measurements. We discuss each of these mechanisms in turn.

2.1.7.1 Intrinsic Temperature Dependence

In the RKKY model based on free-electrons, the singularity in $\chi(q)$ at $q = 2k_\mathrm{F}$ produces the oscillations in the Fourier transform of χ which enter into J. Finite temperatures (or finite relaxation times for the conduction electrons) smooth out the step function in the occupied density of states according to the Fermi distribution, and this correspondingly smooths out the oscillations in J. The amplitude of the oscillations is thus expected to decrease with T with the period unchanged. "Aliased" long period oscillations arising from interference between RKKY oscillations and those due to the lattice periodicity are consequently reduced in amplitude as well. Similar effects should be observed in models with more general band structures or susceptibilities, whenever the oscillations arise because of the existence of a (sharp) Fermi surface. *Edwards* et al. [2.18] generalized their tight-binding model to an arbitrary band (retaining the assumption of an infinite U parameter) and obtained a general asymptotic finite temperature expression for the exchange coupling:

$$J(N-1) = \frac{1}{4\pi Na}\,\mathrm{Re}\sum_{s=1}^{\infty}\frac{\sigma'}{s^2}\left|\frac{\partial k_z^2}{\partial k_x^2}\frac{\partial k_z^2}{\partial k_y^2}\right|^{-1/2}\frac{\exp(2isNak_z^0(\mu))}{T^{-1}\sinh(2\pi sNaT\partial k_z/\partial\varepsilon)}. \tag{2.38}$$

N is the number of atomic layers in the PM, a the lattice constant, σ' is a phase factor, z is the layer normal direction and $k_z^0(\mu)$ is the caliper measurement of the Fermi surface in the direction perpendicular to the layers. The temperature dependence of J is determined by the velocity of carriers at the stationary points on the Fermi surface $(\partial\varepsilon/\partial k_z)$ which goes to zero near a Brillouin zone boundary. This implies that, for this model, long-period oscillations (which arise from k_F near a zone boundary) go together with strong temperature dependence. For the parameters of their model they predict a temperature dependence on the scale of 100 K.

2.1.7.2 Temperature Dependence Deriving from the Disordering of the Ferromagnet Moments

In most of the transition metal multilayer structures studied the FM layers are reasonably thick and have bulk Curie temperatures far above room temperature. These layers do not exhibit a large decrease in magnetization for temperatures less than room temperature. However the antiferromagnetic exchange coupling, usually measured by the critical field needed to ferromagnetically align the coupled layers, decreases rapidly and approximately linearly with T. Most of the theories of exchange coupling discussed above assume that the coupling involves only those few atomic layers in the FM which are nearest the interfaces. Experiments doping the interfaces with other ferromagnetic ions seem to confirm this view to some degree. Thus a picture of spin waves as producing the temperature dependence should reflect the two-dimensional nature of the structure. Here we treat only the case of an infinite superlattice. A treatment of a FM/PM/FM sandwich is given along with further details of the superlattice calculation in [2.44]. In order to derive the spin-wave-driven temperature dependence of the coupling it is necessary to relate the critical field, H_c, observed at finite T, to the intrinsic (assumed non T-dependent) exchange coupling. We assume a Heisenberg model for all exchange interactions between spins in the FMs. This exchange is given by *intra*layer exchange J. We subsume the PM layer properties into an antiferromagnetic *inter*layer exchange coupling J_A (assumed $<J$) between the interfacial monolayers of two adjacent FM layers. We first need to relate the critical field for parallel layer alignment, H_c, to J_A at $T = 0$. We minimize the free energy and linearize for small values of the angles, θ and θ', made by the magnetizations in the two FMs away from the applied field. This gives a self-consistency condition for H_c'. For thick FM layers at $T = 0$

$$H_c' = 2J_A^2/Jg\mu_B. \tag{2.39}$$

Each FM layer can be divided into "bulk" and surface regions with only the surface regions participating in the spin waves. For thin FM layers (defined by $J_A N/J \ll 1$), where all the spins participate in the surface spin waves, H_c goes to the limit

$$H_c' = 4J_A/Ng\mu_B. \tag{2.40}$$

We investigate the temperature dependence for the thin FM layer case. The expression (2.40) for H'_c can be extended to finite temperatures by inclusion of spin correlation functions defined by

$$H'_c = \frac{4J_A}{N} \frac{\sum_\alpha \left\langle \sum_{ij} S^\alpha_{iNm} S^\alpha_{j1m+1} \right\rangle}{\left\langle \sum_i g\mu_B S^z_i \right\rangle} \equiv \frac{4J_A \hat{C}(T)}{N\hat{M}(T)}. \tag{2.41}$$

S is a local spin with vector components specified by α, i and j denote sites in the x–y plane of the layers, N and 1 denote the last and first atomic layers (the interfacial atomic layers) in the adjacent mth and $(m+1)$th FM layers, respectively. Thus, the temperature dependence will derive from the interfacial correlation function (i.e. the correlation between spins in the interface atomic layers of adjacent FM layers), $\hat{C}(T)$, and the magnetization, $\hat{M}(T)$, calculated for spin eigenfunctions which minimize the free energy. The quasi-three-dimensional nature of the superlattice structure insures the stability of the ferromagnetic state within the FM layers for $H < H'_c$, even in the absence of anisotropy or dipolar forces. Only the low frequency spin wave modes make significant contributions to the temperature dependence. At very low temperatures both $\hat{C}$ and $\hat{M}$, and thus H'_c, as well, vary like $T^{3/2}$. For physically reasonable values of J and J_A we may assume $J_A \ll k_B T \ll J$ and

$$\hat{C}(T) = S^2 - \frac{k_B T S a^2}{\pi D N} \ln \frac{N k_B T}{2.55 J_A} \tag{2.42a}$$

and

$$\hat{M}(T) = g\mu_B \left[S - \frac{k_B T a^2}{2\pi D N} \ln \frac{N k_B T}{2 J_A} \right], \tag{2.42b}$$

where D is the bulk spin-wave stiffness reduced by a surface term. These expressions yield for H_c:

$$H'_c(T) = H'_c(0) \left[1 - \frac{k_B T}{2\pi D N} \ln \frac{N k_B T}{3.75 J_A} \right],$$

$$\frac{H'_c(0) - H'_c(T)}{H'_c(0)} = \frac{k_B T a^2}{8\pi^2 D N} \ln \frac{4.45 N k_B T}{J_A}. \tag{2.43}$$

Thus $H'_c(0) - H'_c(T)$ varies like $T \ln T$ and also depends on the number of atomic layers N in the FM layer, the relative change in H'_c with T being greater for thinner FM layers. The T dependence of H'_c for small values of N is shown in Fig. 2.19. A similar derivation of the temperature dependence of the exchange coupling, for the case where the FM layers are thin enough to be approximated as single atomic planes, was obtained for *both* antiferromagnetic and ferromagnetic coupling by *Qiu* et al. [2.45] with also a $T \ln T$ behavior. For sandwiches the T dependence of H'_c is more nearly linear. These linear and quasilinear

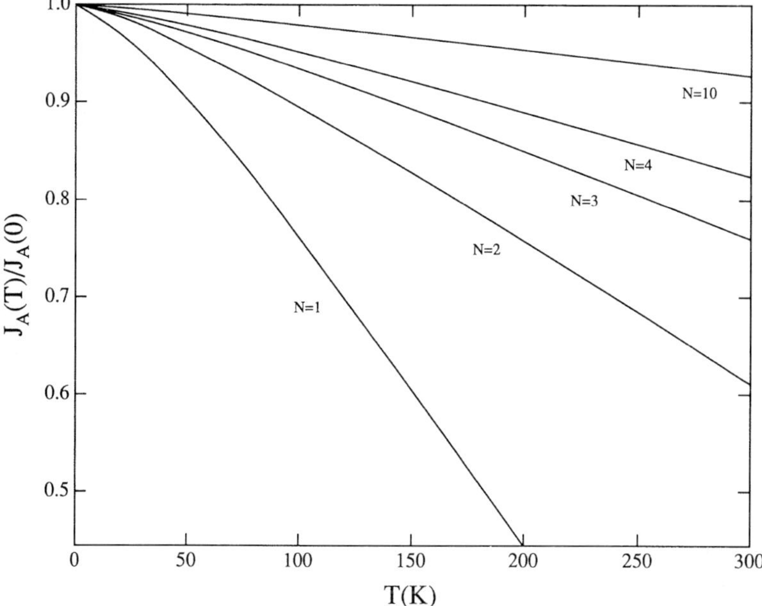

Fig. 2.19. Calculated temperature dependence of the antiferromagnetic exchange coupling (normalized to its value at $T = 0$) of a ferromagnet/paramagnet superlattice for various thicknesses (N atomic layers) of the ferromagnet layers [2.44]

temperature dependences seem to be in agreement with observations and have approximately the right order of magnitude. However both this extrinsic spin wave mechanism and the intrinsic (Fermi surface) mechanism discussed earlier for temperature dependence may be active at the same temperatures, and both act in the same way, to reduce the amplitude of the oscillating coupling.

2.1.8 Conclusions

It is clear from the preceding sections that transition metal multilayers with exchange-coupled magnetic layers are a fascinating testing ground for theories based on the concepts of itinerant magnetism. As more experimental systems are studied and the effects of physical structure (lattice structure and orientation, strain, roughness, etc.) are sorted out from the effects of chemistry and electronic structure, more definitive comparisons between theory and data will be possible. We have not attempted to survey the various theories of magneto-transport in these systems, which are covered elsewhere in this volume, but surely the same itinerant electrons assumed to mediate the exchange coupling must play a role in the transport. An understanding of the relationship between these two phenomena, exchange coupling and magneto-transport, is essential to optimizing materials for technological applications.

2.2 Interlayer Coupling and Magnetoresistance in Multilayers

A. FERT and P. BRUNO

In metallic systems, exchange interactions are propagated by itinerant electrons and thus can be transmitted over relatively long distances. It follows that exchange interactions can couple magnetic layers through non-magnetic metallic layers. Although the possibility of exchange coupling across non-magnetic layers has been considered for a long time, it was only clearly identified in 1986 and characterized in rare earth based multilayers [2.46] and Fe/Cr structures [2.47, 48]. In 1990 the interest of the interlayer exchange was increased by the discovery of unexpected long range oscillations in the variation of the exchange constant with the thickness of the spacer layers [2.49] and a large number of systems have been investigated during the following years. The interest of the exchange coupled multilayers has also been enhanced by the discovery of the giant magnetoresistance. Prior to the discovery there were reports of unusual magnetoresistive effects in layered structures [2.50], and in 1988 giant magnetoresistance effects were first clearly observed and characterized in antiferromagnetically coupled Fe/Cr systems [2.51, 52]. The promises of these magnetoresistive effects for applications (magnetic sensors) have triggered a large number of experimental and theoretical works.

Exchange and magnetoresistance properties of multilayers are also presented in Sect. 2.4 by *Parkin*, and the theoretical models of the interlayer exchange are reviewed in Sect. 2.1 by *Hathaway*. Here we review experimental results on interlayer exchange in Sect. 2.2.1. Section 2.2.2 is devoted to the theory of exchange and to the interpretation of experimental data, with special attention to the simple case of exchange through noble metals. Sections 2.2.3–5 are devoted to the magnetoresistance properties and focused on experiments in exchange coupled structures. We also review the theoretical models, and we discuss the current understanding of the mechanisms of magnetoresistance.

2.2.1 Interlayer Coupling. Review of Experiments

Interlayer coupling between ferromagnetic transition-metal ultrathin films has been observed for a wide range of spacer materials: (i) antiferromagnetic transition metals (Cr, Mn) [2.47–49, 51, 53–58], (ii) non-magnetic transition metals (Ru, Ir, Mo, Pd, etc.) [2.49, 59–62], (iii) and noble metals (Cu, Ag, Au) [2.63–79]. Oscillatory behavior has been reported in almost all cases.

The exchange coupling per unit area between two ferromagnetic layers F1 and F2 is commonly expressed as

$$E_{1,2} = I_{1,2} \frac{M_1 \cdot M_2}{M_1 M_2} = I_{1,2} \cos \phi_{12}, \tag{2.44}$$

where M_1 and M_2 are, respectively, the magnetizations of F1 and F2, ϕ_{12} the angle between M_1 and M_2, and $I_{1,2}$ the coupling constant per unit area. With the above definition $I_{1,2}$ has the dimension of an energy/surface and is positive (negative) for ferromagnetic (antiferromagnetic) coupling, but other conventions for the sign and/or dimension of $I_{1,2}$ are frequently found in the literature.

Recently, a coupling of the form

$$E_{1,2} = B_{1,2}\cos^2\phi_{12}, \quad \text{with } B_{1,2} > 0, \tag{2.45}$$

has been observed [2.56, 57]. This kind of coupling favors a perpendicular alignment of M_1 and M_2 with respect to each other, and is referred to as *biquadratic* or $90°$-coupling.

2.2.1.1 Experimental Techniques

A great variety of experimental techniques have been used to study the inter-layer exchange coupling, including (i) magnetometry, (ii) RF techniques, (iii) neutron scattering, and (iv) magnetic domain microscopy. These are dis-cussed more fully below.

(i) The simplest – and most widely used – method is to measure mag-netoresistance, magnetization or magneto-optical Kerr effect (MOKE) loops [2.49, 51, 53, 55, 58–61, 63, 68–71, 73–79]. The principle is to use an external field to bring the system from antiparallel to parallel alignment. Thus this technique works only for antiferromagnetic coupling; however, *Parkin* and *Mauri* [2.59] and *Fert* et al. [2.123] used a clever trick which allowed them to measure ferromagnetic coupling. In the absence of magnetic anisotropy, the interpretation of the results is straightforward, and the coupling strength is directly related to the saturation field. If anisotropy is present, or if the magnet-ization cannot be assumed uniform throughout the layers, a more sophisticated analysis is needed [2.80]. In contrast to magnetometry and magnetoresistance, which probe the sample as a whole, the MOKE can be used to *locally* probe a small part of the sample; this feature allows investigation of samples with a wedge-shaped spacer, and thus continuous variation of the spacer thickness. This is a great advantage, because the identification of coupling oscillations requires a large number of samples of different spacer thicknesses to be meas-ured. Indeed, the combined MOKE-wedge method enabled coupling oscilla-tions to be measured with unprecedented resolution [2.55, 57, 58, 76, 77]. This is exemplified in Fig. 2.20, which shows the results of *Purcell* et al. [2.55] for the Fe/Cr/Fe(0 0 1) system.

(ii) Ferromagnetic resonance (FMR) and Brillouin light scattering (BLS) have been used in a number of cases [2.47, 57, 63, 65, 67, 73]. They essentially rely on measurements of spin wave frequencies for the optical and acoustic modes. The analysis allowing coupling strengths to be obtained from FMR and BLS has been described by various authors [2.67, 81]. An important difference between FMR and BLS is that BLS is *local* and can thus be used in combination with a wedge-shaped spacer layer [2.57].

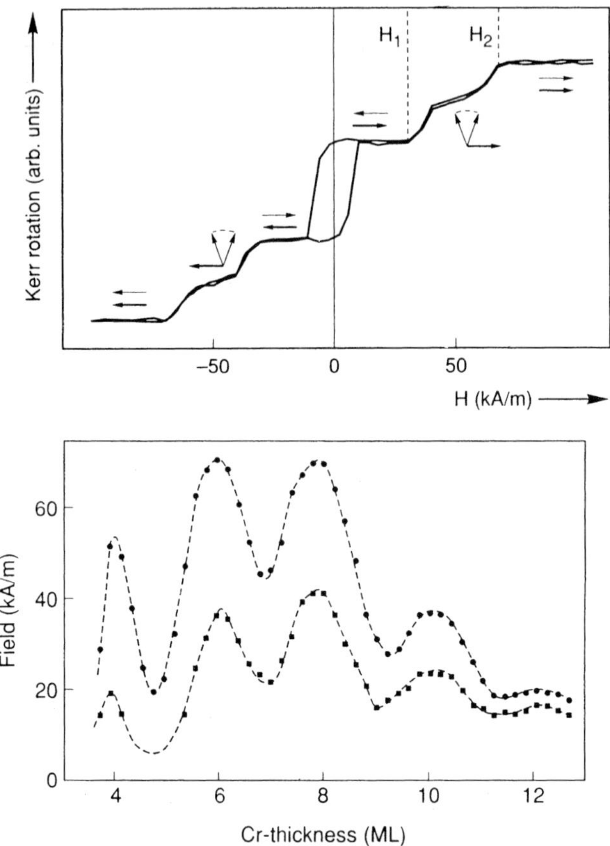

Fig. 2.20. Top: longitudinal Kerr hysteresis loop measured from a Au(2.0 nm)/Fe(5.0 nm)/
Cr(0–1.8 nm wedge)/Fe(0 0 1)-whisker sample; the thickness of Cr was about 6 ML at this position
of the laser spot; the plane of incidence of the light was parallel to the long axis of the whisker; the
thin (thick) arrow indicates the direction of the Fe overlayer (whisker) magnetization. Bottom: the Cr
thickness dependence of the critical fields H_1 (lower curve) and H_2 (upper curve) at which the
overlayer magnetization rotates with respect to the whisker magnetization; the large oscillations in
H_1 and H_2 have a period of two Cr monolayers and correspond to oscillations in the interlayer
coupling; the Cr was deposited at 150 °C. From the work of *Purcell* et al. [2.55]

(iii) In the presence of antiferromagnetic coupling in magnetic multilayers,
the magnetic unit cell is twice the size of the chemical unit cell. Neutron
scattering has been used to evidence this period doubling [2.53, 62, 64, 72]. The
results of *Rodmacq* et al. [2.72] for a Ni/Ag(1 1 1) superlattice are shown in
Fig. 2.21.

(iv) Magnetic domain imaging, when combined with a wedge-shaped spacer
layer, allows one to visualize the interlayer coupling oscillations in a very
spectacular fashion [2.54, 56]. Various techniques such as scanning electron

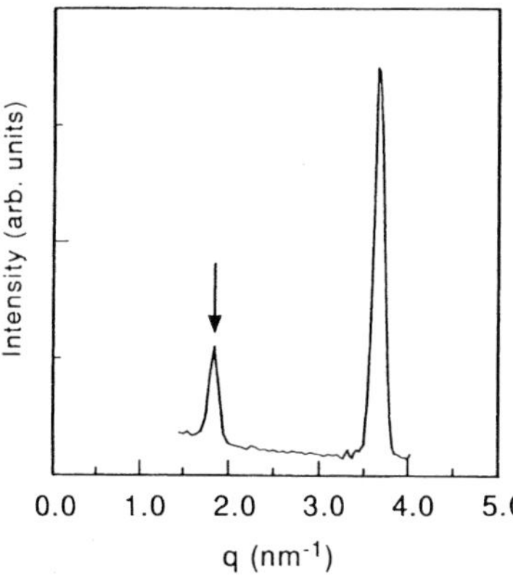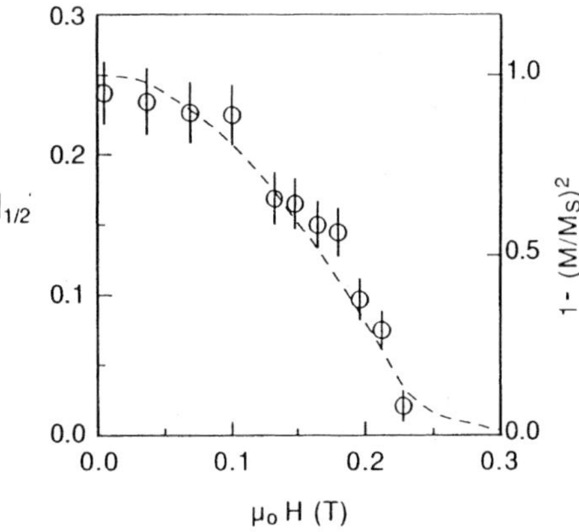

Fig. 2.21. Left panel: Low-angle neutron diffraction diagram of a Ni(0.7 nm)/Ag(1.1 nm) multilayer at 4.2 K; the arrow indicates the period-doubling peak due to the antiparallel alignement; neutron wavelength: 0.236 nm. Right panel: Variation with applied magnetic field of the intensity of the antiferromagnetic period-doubling peak (circles); the dashed line gives the variation of the quantity $1 - (M/M_s)^2$, where M/M_s is the reduced magnetization. From the work of *Rodmacq* et al. [2.72]

microscopy with polarization analysis (SEMPA) or Kerr microscopy have been used to image the magnetic domains. This method does not allow direct measurement of the coupling strength, but merely its sign: the oscillation periods are obtained directly. Actually, this method gave the first evidence of a short-period coupling for the Fe/Cr/Fe(0 0 1) system [2.54].

2.2.1.2 Cr and Mn Spacer

Antiferromagnetic coupling between Fe(0 0 1) layers separated by a Cr(0 0 1) spacer was first reported by *Grünberg* et al. [2.47]. This observation has been confirmed by *Carbone* and *Alvarado* [2.48] and by *Baibich* et al. [2.51]. No coupling oscillations were observed in these early experiments; rather, the coupling decreases monotonically with increasing spacer thickness; the coupling strength for 6 ML of Cr is 0.4 mJ m^{-2} [2.51].

The observation by *Parkin* et al. [2.49] of coupling oscillations in sputtered Fe/Cr and Co/Cr multilayers was then a major breakthrough. The oscillation period is 12.5 ML and the coupling strength is 0.66 mJ m^{-2} for 5 ML of Cr (first antiferromagnetic maximum) in Fe/Cr [2.53]. Similar results for epitaxial Fe/Cr/Fe(0 0 1) sandwiches have been obtained by *Demokritov* et al. [2.57], with a period of 14 ML and a coupling strength of 1.3 mJ m^{-2} for 3.8 ML of Cr.

Another important step was the almost simultaneous observation by *Unguris* et al. [2.44], *Purcell* et al. [2.45], and *Demokritov* et al. [2.57] of coupling oscillations with a period of 2 ML in epitaxial Fe/Cr/Fe(0 0 1). This has been made possible by preparing the Cr spacer layer at higher temperature

($\approx 150\,°\text{C}$), thus achieving almost perfectly flat interfaces. This short-period oscillation is superimposed with the previously reported long-period oscillation. The coupling strength reported by *Purcell* et al. [2.55] is $I_{1,2} = 0.6\,\text{mJ m}^{-2}$ for 8 ML of Cr. Their results are shown on Fig. 2.20. Oscillations with a period of 2 ML have also been observed for a $\text{Mn}(0\,0\,1)$ spacer by *Purcell* et al. [2.58]; the coupling strength in $\text{Fe}/\text{Mn}/\text{Fe}(0\,0\,1)$ is $0.14\,\text{mJ m}^{-2}$ for 8 ML Mn, i.e. appreciably smaller than for $\text{Fe}/\text{Cr}/\text{Fe}(0\,0\,1)$.

By examining the domain structure in a $\text{Fe}/\text{Cr}/\text{Fe}(0\,0\,1)$ sandwich with a wedge-shaped Cr spacer, *Rührig* et al. [2.56] found that, at the interface between the regions of ferromagnetic and antiferromagnetic coupling, the moments of the two Fe layers adopt a $90°$ configuration. This interpretation is also supported by the analysis of magnetization curves, as measured by MOKE. This $90°$ configuration is not expected from a coupling interaction of the form (2.44), so they postulated that it arises from a *biquadratic coupling* of the form (2.45). They evaluated the strength of the biquadratic coupling to be $B_{1,2} \approx 0.15\,\text{mJ m}^{-2}$ for a Cr thickness of 3.5 ML; this is typically one order of magnitude smaller than the usual quadratic coupling (2.44); so that the biquadratic coupling manifests itself only as the quadratic coupling crosses zero.

2.2.1.3 Non-Magnetic Transition Metal Spacer

Oscillatory coupling between Co layers across a Ru spacer was first reported by *Parkin* et al. [2.49]. The oscillation period is of 5.6 ML, and the coupling strength for 1.4 ML is of $5\,\text{mJ m}^{-2}$. This coupling strength is much larger than the ones observed in other systems. *Brubaker* et al. [2.60] have observed oscillatory coupling in $\text{Fe}/\text{Mo}(1\,1\,0)$ superlattices, with a period of 4.9 ML and a coupling strength of $0.3\,\text{mJ m}^{-2}$ for a Mo thickness of 4.9 ML.

Among the transition metals, Pd is of particular interest because of its strong Stoner-enhanced susceptibility, which makes it nearly ferromagnetic; thus ferromagnetic coupling accross Pd spacers may be expected for low Pd thickness. Indeed, this has been verified experimentally by *Celinski* et al. [2.63], who studied $\text{Fe}/\text{Pd}/\text{Fe}(0\,0\,1)$ sandwiches. Their results for the interlayer coupling can be interpreted as the superposition of (i) a ferromagnetic coupling (strength: $-0.3\,\text{mJ m}^{-2}$ for 5 ML) which decreases monotonically and disappears at approximately 10 ML, and (ii) an oscillatory coupling with a period of 4 ML (strength: below $0.05\,\text{mJ m}^{-2}$).

A systematic study of interlayer coupling in sputtered Co-based multilayers with 3d, 4d and 5d transition metal spacers was carried out by *Parkin* [2.61]. He found an oscillatory coupling in almost all cases, with an oscillation period of 5–6 ML (except for Cr which gives a period of 12.5 ML). The coupling strength presents important variations as a function of spacer material.

2.2.1.4 Noble Metal Spacer

Antiferromagnetic coupling across a Cu(0 0 1) spacer has been observed by *Cebollada* et al. [2.64] in epitaxial Co/Cu(0 0 1) superlattices, for a Cu thickness of 5 ML; and further studies on the same system by *de Miguel* et al. [2.69] have revealed an oscillatory behavior of the coupling, with a period of 6 ML of Cu. Some hints of oscillatory behavior were also reported by *Pescia* et al. [2.66], in Co/Cu/Co(0 0 1) films. *Bennett* et al. [2.68] have observed clear coupling oscillations in γ-Fe/Cu/γ-Fe(0 0 1) sandwiches. They found an oscillation period of 7.5 ML, and a coupling strength of 0.3 mJ m^{-2} for 7.5 ML of Cu. *Heinrich* et al. [2.73] have studied the interlayer coupling in Co/Cu/Co(0 0 1) systems. For Cu thicknesses of 6 and 10 ML, they found an antiferromagnetic coupling of 0.05 mJ m^{-2}; in addition, they observe a biquadratic coupling, with $B_{1,2} \approx 0.015$ mJ m^{-2}.

Very recently, *Johnson* et al. [2.77] have performed a very detailed study of the Cu thickness dependence of the interlayer coupling in an epitaxial Co/Cu/Co(0 0 1) sandwich grown onto a Cu(0 0 1) single crystal, with a wedge-shaped Cu spacer. Their results show that the interlayer coupling consists of a short-period oscillatory component, superimposed with a long-period component; the periods are respectively 2.6 and 8.0 ML. The strength of the coupling is of 0.4 mJ m^{-2} for 6.7 ML of Cu.

The problem of oscillatory coupling across Cu(1 1 1) spacers is a controversial matter. *Mosca* et al. [2.71] and *Parkin* et al. [2.70] have observed coupling oscillations in sputtered Co/Cu multilayers with a predominent (1 1 1) texture. They obtained an oscillation period of 5–6 ML, and a coupling strength of 0.15–0.3 mJ m^{-2} for 4.5 ML of Cu. *Pétroff* et al. [2.75] have observed an oscillatory coupling with a period of 6 ML in sputtered Fe/Cu multilayers, with a predominent (1 1 1) orientation of Cu. *Egelhoff* and *Kief* [2.78] then attempted to reproduce these results for epitaxial Co/Cu(1 1 1) superlattices and failed to observe any coupling; thus they argued that the previously observed coupling oscillations for sputtered multilayers were a spurious effect, which they attributed to the presence of (0 0 1)-oriented grains. This interpretation has been refuted by four different groups [2.79] who found clear evidence of antiferromagnetic coupling in epitaxial Co/Cu systems. A detailed study of the Cu thickness dependence to confirm the oscillatory behavior is still lacking.

When grown onto a (0 0 1) α-Fe surface, Cu adopts a metastable bcc structure. *Heinrich* et al. [2.65] and *Cochran* et al. [2.67] found that the interlayer coupling in Fe/bcc-Cu/Fe(0 0 1) sandwiches is strongly ferromagnetic for Cu thicknesses lower than 9–10 ML, and becomes antiferromagnetic at larger thicknesses; for 12 ML of Cu they obtained a coupling strength of about 0.2 mJ m^{-2}. These findings have been confirmed and refined by *Johnson* et al. [2.77] who used a wedge-shaped spacer of bcc-Cu(0 0 1): they obtain a ferromagnetic coupling below 10 ML of Cu, and an oscillatory antiferromagnetic coupling (period: 2 ML) above 10 ML; the coupling strength for 12 ML of Cu is 0.1 mJ m^{-2}.

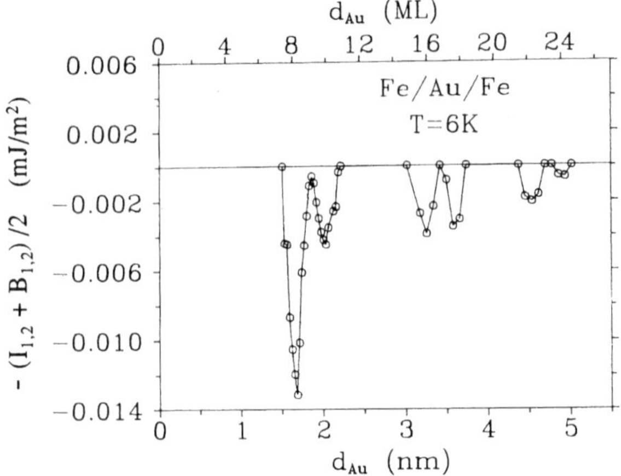

Fig. 2.22. Interlayer coupling in a Fe/Au/Fe(0 0 1) sample with a wedge-shaped Au(0 0 1) spacer, as a function of the Au thickness d_{Au}. Only the antiferromagnetic part of the coupling oscillations could be measured. From the work of *Fuss* et al. [2.76]

The interlayer coupling in sputtered Ni/Ag(1 1 1) multilayers have been studied by *Rodmacq* et al. [2.72] and *dos Santos* et al. [2.74]. They observed an antiferromagnetic coupling (see Fig. 2.21) which presents a marked peak for Ag thicknesses around 5 ML; there is no clear evidence for further oscillations. The coupling strength for 5 ML of Ag is $0.02\,\mathrm{mJ\,m^{-2}}$.

The interlayer coupling across a Au(0 0 1) spacer has been studied by *Fuss* et al. [2.76]. They have used a Fe/Au/Fe(0 0 1) sandwich, with a wedge-shaped Au layer grown onto a GaAs(0 0 1) single crystal. Their results are shown in Fig. 2.22: short-period oscillations (period: 2 ML) superimposed with long-period oscillations (period: 7–8 ML) are clearly identified. The coupling strength is much smaller than for Cu spacers.

2.2.2 Interlayer Exchange Coupling. Theoretical Models

The coupling interactions that are observed in the experiments discussed above are too large to be ascribed to magnetic dipolar interactions; thus one has to consider some indirect exchange mechanisms. There are basically two strategies that have been used to theoretically study the interlayer exchange coupling: (i) total energy calculations, and (ii) perturbative models. A survey of the principles and results of these two different approaches are given below.

2.2.2.1 Total Energy Calculations

The first method is in principle straightforward: it consists in calculating the coupling as the energy difference between the states with antiparallel and parallel magnetization alignments. Such calculations have been performed either within a tight-binding scheme [2.82], or from first principles [2.83, 84].

In practice, this kind of calculation is very difficult, essentially because the energy difference is several orders of magnitude smaller than the total energy itself. Thus, one has to pay close attention to the delicate problems of convergence and numerical accuracy in order to avoid artifacts. Another important restriction is that total energy calculations are very demanding of computer time; because the unit cell must be twice the chemical unit cell and the computation time increases very rapidly with the size of the unit cell, such calculations have been restricted so far to fairly low spacer thicknesses. Thus (at least with present-day computers) this method is probably not capable of determining long-period coupling oscillations.

Most calculations concern $Fe/Cr(0\,0\,1)$ superlattices, for which there is a good lattice matching and a wealth of experimental results. *Ounadjela* et al. [2.83] have performed first-principles calculations of the interlayer coupling in $Fe/Cr(0\,0\,1)$ superlattices with 3 ML of Fe and 3–7 ML of Cr. They used the local spin-density functional (LSDF) formalism and the augmented spherical-wave (ASW) method. They find that the interlayer coupling is ferromagnetic for 3 ML of Cr, and antiferromagnetic for Cr thicknesses between 4 and 7 ML. They obtain no indication of the oscillations with a period of 2 ML that have been observed in the most refined experiments [2.54, 55, 57]. Another point of discrepancy with experiment is the strength of the coupling $I_{1,2}$: they obtain values ranging between 20 and 150 mJ m^{-2}, whereas experiments for similar Cr thicknesses yield $I_{1,2} \leq 1$ mJ m^{-2}.

Herman et al. [2.84] have performed a very similar study of the $Fe/Cr(0\,0\,1)$ system: they performed LSDF–ASW calculations with 2 ML of Fe and 2–5 ML of Cr. In contrast to *Ounadjela* et al., they find that the coupling is antiferromagnetic for 2 and 4 ML of Cr, and ferromagnetic for 3 and 5 ML of Cr. This is consistent with the 2 ML-period oscillatory behavior observed experimentally [2.54, 55, 57], and with an antiferromagnetic ordering of the Cr spacer layer. The calculated coupling strengths are of the order of 60 mJ m^{-2}, i.e. still much larger than the experimental ones.

Tight-binding calculations for the $Fe/Cr(0\,0\,1)$ system have been performed by *Stoeffler* and *Gautier* [2.82]. Since tight-binding calculations are less demanding than first-principles calculations, they were able to investigate Cr thicknesses up to 15 ML. Like *Herman* et al., they find that the coupling is antiferromagnetic (ferromagnetic) for an even (odd) number of Cr atomic layers, which is in agreement with experiment [2.54, 55, 57]. They interpret this 2 ML-period oscillatory coupling as the energy associated with a magnetic defect appearing in the Cr layer, when the natural antiferromagnetic ordering of the latter is frustrated by an unfavorable alignment of the magnetizations of the Fe layers. Nevertheless, they are not able to find the long-period (12 ML) oscillations observed experimentally. The calculated coupling strength is of the order of 25 mJ m^{-2}; this result is closer to experiment than that determined by *Herman* et al., but still one order of magnitude larger. Their use of a direct-space method which does not require translational invariance (in contrast to other authors) enabled *Stoeffler* and *Gautier* [2.82] to introduce into their calculations some interdiffusion at the interface between Fe and Cr; they found that the coupling strength is thereby strongly reduced and that the oscillations are changed.

We note in passing that in all of the calculations described here, the intra-atomic exchange in Cr (which is responsible for the spin-density-wave antiferromagnetic ordering of bulk Cr) is a key ingredient for obtaining the 2 ML-period oscillations.

The only calculations for noble metal spacers are due to *Herman* et al. [2.84], who performed LSDF-ASW calculations for Co/Cu in (0 0 1), (1 1 1) and (1 1 0) orientations, and Fe/bcc-Cu(0 0 1) superlattices. For Co/Cu(0 0 1), Co/Cu(1 1 0), and Fe/bcc-Cu(0 0 1), they find a coupling which oscillates with a short period (≈ 2 ML); this is consistent with the available experimental data [2.77] (except for Co/Cu(1 1 0) where there are no experimental results). For Co/Cu(1 1 1), they find no indication of such short-period oscillations; again, this is consistent with experiment, where only long-period (≈ 6 ML) oscillations were reported [2.70, 71]. The calculated coupling strengths are of the order of 5–10 mJ m^{-2}.

Despite the encouraging results, it is a common feature of all total energy calculations that the coupling strength is at least one order of magnitude too large, as compared to the experimental data. Thus the numerical accuracy of the calculations may be questioned. This view is supported by the calculations by *Herman* et al. [2.84], who find that refining the mesh for k-space integration yields smaller values for the coupling strength, thus reducing the discrepancy with experiment. Another possible source of discrepancy between theory and experiment is that the samples used in experiments depart markedly from the perfect structure which is assumed in most theoretical studies: one expects that defects would contribute to *reduce* the magnitude of the coupling. Thus, at present time, this strong discrepancy remains a serious problem that needs to be solved.

2.2.2.2 Alternative Approaches

In view of the difficulties of total energy calculations, it is tempting to attack the problem of interlayer coupling in a different way, and try to obtain the coupling *directly*, without computing the total energy. The price to pay for this is that one has to make some approximations that must be suggested a *priori* by physical intuition. This is essentially the philosophy of the *perturbative models*.

A number of different models have been proposed [2.85–88, 90–96]. They all rely on the same underlying picture for the coupling mechanism: (i) the ferromagnetic layer (say F1) interacts with the conduction electrons of the spacer, and induces a spin-polarization of the latter; this spin-polarization extends throughout the spacer, and eventually interacts with F2, thus giving rise to an effective exchange interaction between F1 and F2. The various approaches differ mostly in the modeling of the physical system, and in the simplifying approximations which are made. Nevertheless, it is a common feature of all of them that, in the limit of large spacer thickness z, the coupling oscillates periodically with an oscillation period related to some measure of the Fermi surface of the spacer metal, and with an amplitude decaying like $1/z^2$.

In the following, we shall focus on the "Ruderman–Kittel–Kasuya–Yosida" (RKKY) model [2.85–88], which is the archetype of the perturbative theories of interlayer coupling. The presentation follows that given by *Bruno* and *Chappert* [2.87]. For a discussion of other models [2.90–96], the reader is referred to Sect. 2.1 by *Hathaway*.

We consider two ferromagnetic monolayers F1 and F2 embedded in a non-magnetic metal. The distance between F1 and F2 is $z = (N + 1)d$, where d is the spacing between atomic planes and N the number of atomic planes in the spacer. For the sake of simplicity, we restrict ourselves here to magnetic layers of monatomic thickness; this restriction is not very serious, for it has been found experimentally that the coupling is roughly independent of the thickness of the magnetic layers. The magnetic layers are assumed to consist of spins S_i located at the atomic positions R_i of the host metal.

The starting point of the RKKY model is the interaction between two magnetic impurities embedded in a non-magnetic host metal, as originally formulated by *Ruderman* and *Kittel* [2.89] for the case of nuclear spins. The interaction between a spin S_i localized at R_i and a conduction electron (spin s, position r) is described by a contact potential

$$\mathscr{V}_i(r \cdot s) = A\delta(r - R_i)s \cdot S_i. \tag{2.46}$$

This contact interaction, when used for transition metal spins, is a rather crude approximation; it usually leads to an incorrect phase for the coupling oscillations, while the coupling strength is described by an adjustable parameter A. These limitations of the RKKY model should be kept in mind when comparing its predictions with experiment. By treating the contact interaction (2.46) as a perturbation to second order, *Ruderman* and *Kittel* found that it produces an effective exchange interaction between two spins S_i and S_j:

$$\mathscr{H}_{ij} = J(R_{ij})\,S_i \cdot S_j, \tag{2.47}$$

where the exchange integral is

$$J(R_{ij}) = -\frac{1}{2}\left(\frac{A}{V_0}\right)^2 \frac{V_0}{(2\pi)^3} \int d^3q\,\chi(q)\exp(iq \cdot R_{ij}), \tag{2.48}$$

V_0 is the atomic volume and

$$\chi(q) = \frac{V_0}{(2\pi)^3}\int d^3k\,\frac{f(\varepsilon_k) - f(\varepsilon_{k+q+G})}{\varepsilon_{k+q+G} - \varepsilon_k} \tag{2.49}$$

is the nonuniform susceptibility of the host material (in units of $2\mu_B^2/\text{atom}$). The exchange interaction $J(R_{ij})$ is given as the Fourier transform of the nonuniform susceptibility because the spin S_i polarizes the host metal in the same way as would a hypothetical point source of magnetic field. Once the interaction between a pair of spins is known, the interaction between the two magnetic layers F1 and F2 may be obtained by summing over all pairs of spins belonging to F1 and F2, respectively. Thus, the expression of the interlayer coupling constant is

$$I_{1,2} = \frac{d}{V_0} S^2 \sum_{j \in F2} J(\boldsymbol{R}_{Oj}),$$ (2.50)

where O labels one site of F1 taken as the origin.

For the purpose of pedagogical clarity, it is useful to first examine the RKKY model within the free-electron approximation. This makes the calculations almost analytically tractable, so that the results are physically transparent. In the following, the host material will be approximated by a free-electron gas of equivalent density; since the model is to be applied to noble metals, we consider fcc lattices with one electron per atomic cell, so that the Fermi vector is $k_F = (12\pi^3)^{1/3}/a$, where a is the lattice parameter.

As is well known, the free-electron nonuniform susceptibility presents a logarithmic singularity at $q = 2k_F$. Physically, the singularity arises from the fact that the Fermi–Dirac distribution abruptly drops to zero at k_F. This singularity of the susceptibility manifests itself in its Fourier transform, i.e. in $J(\boldsymbol{R})$, by oscillations of period $\Lambda = \lambda_F/2$ ($\lambda_F \equiv 2\pi/k_F$ is the Fermi wavelength). This behavior, known as the "Gibbs phenomenon", is frequently found in physics. The expression of the exchange interaction has been given by *Ruderman* and *Kittel* [2.89]:

$$J(\boldsymbol{R}) = \frac{4A^2 m k_F^4}{(2\pi)^3 \hbar^2} F(2k_F R),$$ (2.51)

with

$$F(x) = \frac{x \cos x - \sin x}{x^4} \approx \frac{\cos x}{x^3} \quad \text{for } x \to +\infty.$$ (2.52)

The interlayer coupling within the free-electron approximation is obtained by performing the summation (2.50) with the above expression.

The problem can be further simplified by replacing the actual ferromagnetic layers by a continuous uniform spin distribution of equivalent areal density; i.e., we perform in (2.50) the substitution

$$\sum_{F2} \to \frac{d}{V_0} \int_{F2} \mathrm{d}^2 \boldsymbol{R}_{\parallel}.$$ (2.53)

In the above equation, $\boldsymbol{R}_{\parallel}$ is the in-plane projection of $\boldsymbol{R}_{Oj}$. This approximation amounts to discarding the crystalline character of the magnetic layers. The interlayer coupling is then given by

$$I_{1,2}(z) \approx -I_0 \frac{d^2}{z^2} \sin(2k_F z) \quad \text{for } z \to \infty,$$ (2.54)

with

$$I_0 = \left(\frac{A}{V_0}\right)^2 S^2 \frac{m}{16\pi^2\hbar^2}. \tag{2.55}$$

The interlayer coupling oscillates with a period $\Lambda = \lambda_F/2$ and decreases as z^{-2}; equation (2.54) was first derived by *Yafet* [2.85]. This result is not very satisfying because, (i) for usual electronic densities, $\lambda_F/2 \approx 1$ ML is too short as compared to experimentally measured periods, and (ii) it does not allow multiperiodic oscillations.

Actually, as pointed out by *Chappert* and *Renard* [2.86], *Coehoorn* [2.88], and *Deaven* et al. [2.93], the above result can be reconciled with the experimental observation of long oscillation periods. The argument is that the spacer thickness z does not vary continuously; rather it can assume only integer multiples of d: $z = (N + 1)d$. Because of this discrete sampling, one obtains an *effective period* which may be much larger than $\lambda_F/2$; this effect is called *aliasing*. The *effective period* Λ is given by

$$\frac{2\pi}{\Lambda} = \left|2k_F - n\frac{2\pi}{d}\right|, \tag{2.56}$$

where n is an integer chosen such that $\Lambda > 2d$.

The free-electron RKKY theory of interlayer coupling can also be reconciled with the observation of multiperiodic oscillations, if we reconsider the continuous approximation (2.53). When performing the continuous integration (2.53) over F2, the integrand is a function which oscillates with a period of the order of $\lambda_F/2$; thus, if the in-plane interatomic distance b is smaller than $\lambda_F/2$, we may expect the continuous approximation (2.53) to be valid. On the other hand, if b is large as compared to $\lambda_F/2$, it is clear that approximation (2.53) must break down. In order to develop this argument more quantitatively, we explicitly perform the summation (2.50) without making the approximation (2.53):

$$I_{1,2}(z) = -\frac{1}{2}\left(\frac{A}{V_0}\right)^2 S^2 \frac{d}{(2\pi)^3} \int\limits_{-\infty}^{+\infty} dq_z \exp(iq_z z)$$

$$\times \int d^2q_\parallel\, \chi(q_\parallel, q_z) \sum_{R_\parallel \in F2} \exp(iq_\parallel \cdot R_\parallel). \tag{2.57}$$

Due to the in-plane translational invariance, the last sum in the above equation equals zero, unless $q_\parallel$ is a vector $G_\parallel$ belonging to the (two-dimensional) reciprocal lattice of F2. Thus, the expression of the coupling becomes

$$I_{1,2}(z) = -\frac{1}{2}\left(\frac{A}{V_0}\right)^2 S^2 \frac{d}{V_0}\frac{d}{2\pi} \sum_{G_\parallel} \int\limits_{-\infty}^{+\infty} dq_z \exp(i(q_z z + G_\parallel \cdot R_\parallel^0))\chi(G_\parallel, q_z), \tag{2.58}$$

where $R_\parallel^0$ is the in-plane translation needed to bring F2 into coincidence with F1. For a given vector $G_\parallel$, if the integration over q_z crosses a singularity of the susceptibility $\chi(q)$ (i.e. if $G_\parallel < 2k_F$), one obtains an oscillatory contribution to $I_{1,2}(z)$, otherwise not. Thus the number of different oscillation periods is given by the number of (non-equivalent) vectors $\boldsymbol{G}_\parallel$ such that $G_\parallel < 2k_F$. It now clearly appears that the multiperiodicity is related to the discrete atomic structure within the planes, and that the number of oscillation periods increases with decreasing in-plane atomic density. This trend is well-illustrated by the case of a fcc spacer: as shown in Fig. 2.23, the number of different oscillation periods for the (1 1 1), (0 0 1), and (1 1 0) orientations is, respectively, 1, 2, and 3.

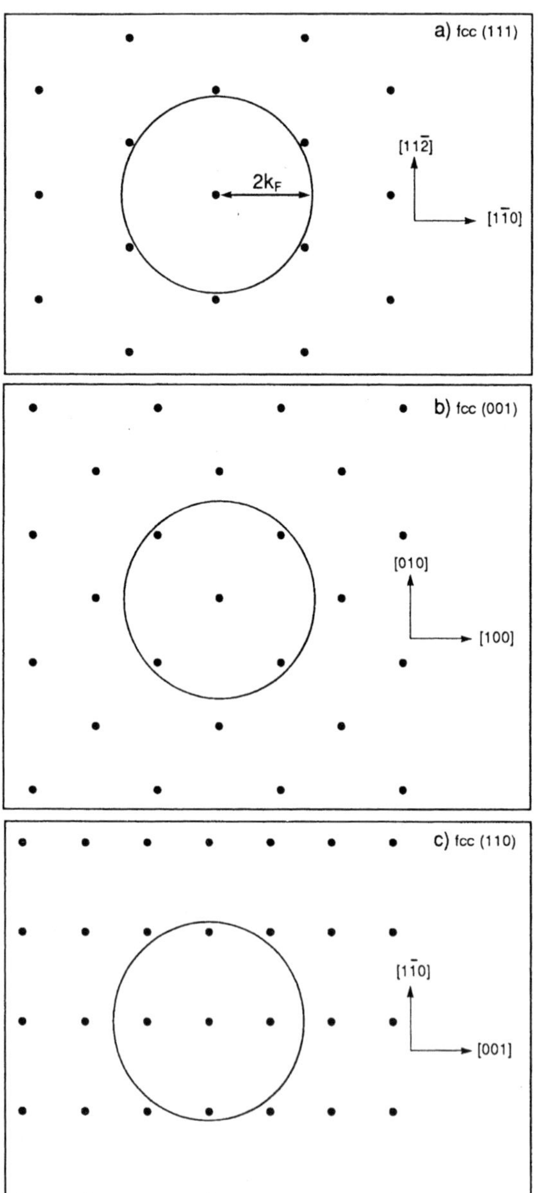

Fig. 2.23. Two-dimensional reciprocal lattice for fcc layers; (a), (b) and (c) correspond respectively to the (1 1 1), (1 0 0) and (1 1 0) orientations. The sphere of radius $2k_F$ is the locus of singularities of the susceptibility $\chi(\boldsymbol{q})$; it should not be confused with a Fermi sphere. The unit vectors have a length $2\pi/a$

The Fermi surface of real metals departs markedly from a sphere. Thus, in order to be able to make reliable quantitative predictions, one needs to release the free-electron approximation and to formulate a general RKKY theory of interlayer coupling, valid for non-spherical Fermi surfaces. This has been done by *Bruno* and *Chappert* [2.87], who showed that the oscillation periods are given by the vectors q_z^α parallel to the z direction, which span the Fermi surface and such that the corresponding Fermi velocities are antiparallel to each other (see Fig. 2.24). The period Λ_α corresponding to a given vector q_z^α is $\Lambda_\alpha = 2\pi/q_z^\alpha$ (it is always possible to chose q_z^α such that $\Lambda_\alpha > 2d$). Since the Fermi surfaces of noble metals are known experimentally with a very high accuracy from de Haas–van Alphen and cyclotron resonance measurements [2.97], they can be used to predict the oscillation periods of oscillatory coupling for noble metals. This has been done in [2.87], for Cu, Ag, and Au, in (0 0 1), (1 1 1) and (1 1 0) orientations. Figure 2.25 shows a (1 1 0)-cross section of the Fermi surface of a noble metal, and the vectors giving the oscillation periods. The number of oscillation periods is 1, 2, and 4, respectively, for the (1 1 1), (0 0 1), and (1 1 0) orientations (for the latter, in addition to the period shown in Fig. 2.25, there are three other periods

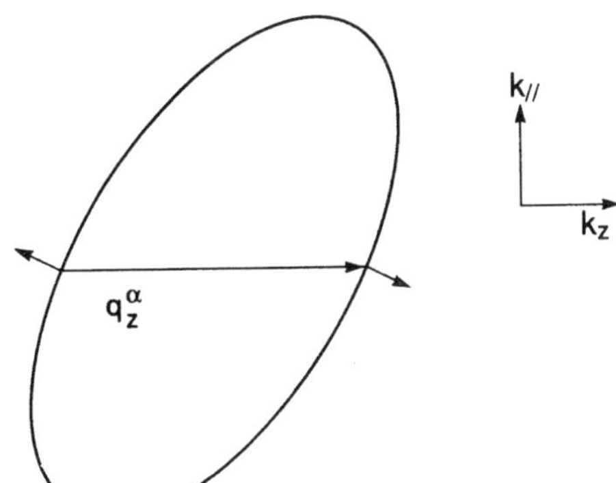

Fig. 2.24. Sketch showing the wave vector q_z^α giving the oscillation period for a non-spherical Fermi surface. The small arrows towards the exterior of the Fermi surface represent the Fermi-velocity vectors. See text for further explanations

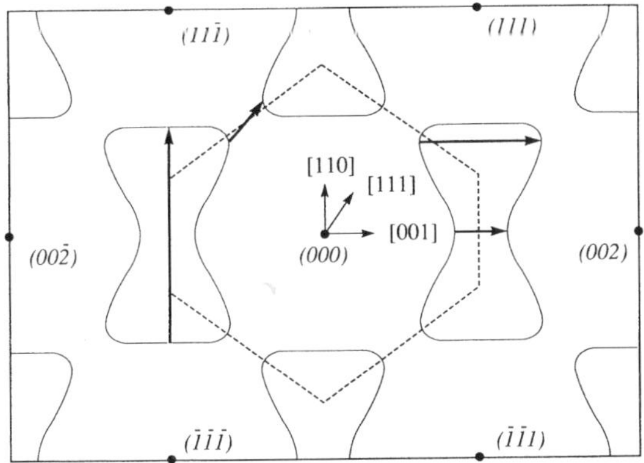

Fig. 2.25. Fermi surface of a (bulk) noble metal: (1 1 0) cross section. The bold points belong to the fcc reciprocal lattice. The first Brillouin zone is indicated by the dashed contour. The horizontal, oblique, and vertical bold arrows are the vectros giving the oscillation periods, respectively, for the (0 0 1), (1 1 1), and (1 1 0) orientations

that cannot be seen from the present cross section). Thus, the trend obtained within the free-electron approximation, stating that the number of periods increases with decreasing in-plane density, remains valid for noble metals.

The comparison between the periods observed experimentally and those predicted by the RKKY theory is shown in Table 2.2. For Cu(1 1 1), the observed period is somewhat larger than the predicted one; however, the difference is not dramatic and may be attributed to experimental uncertainties, and/or to the influence of internal strains on the Fermi surface. The (0 0 1) orientation is of particular interest because the RKKY theory predicts the coexistence of a short and a long period: this has been confirmed subsequently for Au(0 0 1) by *Fuss* et al. [2.76] (as shown in Fig. 2.22), and for Cu(0 0 1) by *Johnson* et al. [2.77]. This is a major success of the RKKY theory. Note also that the RKKY theory provides a consistent interpretation of the results of first-principles calculations by *Herman* et al. [2.84], which predict the presence of a short-period oscillation (≈ 2 ML) for Cu(0 0 1), Cu(1 1 0), and bcc Cu(0 0 1), but not for Cu(1 1 1).

These results clearly show that the RKKY theory allows prediction in an essentially correct manner of the periods of oscillatory coupling, simply by inspection of the Fermi surface of the spacer metal. However, as already mentioned above, the assumption of a contact-type interaction between the

Table 2.2. Comparison between the oscillation periods predicted by the RKKY theory [2.87] for noble metals and those observed experimentally

Theory		Experiment		Ref.
Spacer	Period(s)	System	Period(s)	
Cu(1 1 1)	$\Lambda = 4.5$ ML	Co/Cu/Co	$\Lambda \approx 6$ ML	[2.71]
		Co/Cu/Co	$\Lambda \approx 5$ ML	[2.70]
		Fe/Cu/Fe	$\Lambda \approx 6$ ML	[2.75]
Cu(0 0 1)	$\begin{cases} \Lambda_1 = 2.6 \text{ ML} \\ \Lambda_2 = 5.9 \text{ ML} \end{cases}$	Co/Cu/Co	$\Lambda \approx 6$ ML	[2.69]
		Fe/Cu/Fe	$\Lambda \approx 7.5$ ML	[2.68]
		Co/Cu/Co	$\begin{cases} \Lambda_1 \approx 2.6 \text{ ML} \\ \Lambda_2 \approx 8 \text{ ML} \end{cases}$	[2.77]
Au(0 0 1)	$\begin{cases} \Lambda_1 = 2.6 \text{ ML} \\ \Lambda_2 = 8.6 \text{ ML} \end{cases}$	Fe/Au/Fe	$\begin{cases} \Lambda_1 \approx 2 \text{ ML} \\ \Lambda_2 \approx 7\text{–}8 \text{ ML} \end{cases}$	[2.76]
Ag(0 0 1)	$\begin{cases} \Lambda_1 = 2.4 \text{ ML} \\ \Lambda_2 = 5.6 \text{ ML} \end{cases}$	Fe/Ag/Fe	$\begin{cases} \Lambda_1 \approx 2.4 \text{ ML} \\ \Lambda_2 \approx 5.6 \text{ ML} \end{cases}$	
bcc Cu(0 0 1)[a]	$\begin{cases} \Lambda_1 = 2.2 \text{ ML} \\ \Lambda_2 = 2.6 \text{ ML} \end{cases}$	Fe/Cu/Fe	$\Lambda \approx 2$ ML	[2.77]

[a] ASW calculation of the bulk Fermi surface of bcc Cu, from [2.77].

magnetic moments and the conduction electrons of the spacer is not appropriate for $3d$ transition metals. As a consequence, the RKKY model is unable to correctly describe the *strength* and the *phase* of the coupling oscillations. For this purpose, one needs to explicitly treat the hybridization between the $3d$ bands of the ferromagnetic metal and the conduction electrons of the spacer. Such studies have been done by *Wang* et al. [2.90], *Lacroix* and *Gavigan* [2.91], and *Bruno* [2.96].

2.2.2.3 Current Understanding of Interlayer Coupling

Very significant progress in this field has been made in the last two years. This is due mostly to the unprecedented improvements in sample preparation and characterization, which allowed observation of interlayer coupling oscillations with periods as low as 2 ML. In particular, in the case of noble metal spacer layers, a very good confirmation of the oscillation periods predicted by the RKKY theory has been obtained, thus demonstrating the key role played by the Fermi surface in selecting the oscillation periods.

On the other hand, our current understanding of what drives the strength of the coupling is still very preliminary. However, in view of the intense activity (both experimental and theoretical) in the field of interlayer coupling, there is no doubt that our understanding of the problem can be expected to improve significantly in the next few years.

2.2.3 Magnetoresistance: A Survey

2.2.3.1 Main Features

In Fig. 2.26, we show resistivity versus field curves for several Fe/Cr multilayers exhibiting antiferromagnetic (AF) interlayer exchange. The resistivity drops dramatically when the applied field aligns the magnetic moments of successive layers. This magnetoresistive effect, first discovered in Fe/Cr structures [2.51, 52] and labelled "giant magnetoresistance" or GMR, was subsequently found in a number of multilayer systems.

The GMR is generally ascribed to the interplay between spin dependent scattering in successive magnetic layers. As it will be discussed below, the conditions for the interplay is that the distance between the layers is relatively small in comparison with the electron mean free path (MFP). In addition there must exist some way to change the relative orientations of the magnetization in adjacent layers by applying a magnetic field. In the most classical case, an antiferromagnetic (AF) arrangement is changed into a ferromagnetic (F) one by the applied field. The AF arrangement can be provided by AF interlayer exchange, but it can also be obtained in other ways, for example by giving different coercivities to the odd and even magnetic layers, or by pinning the

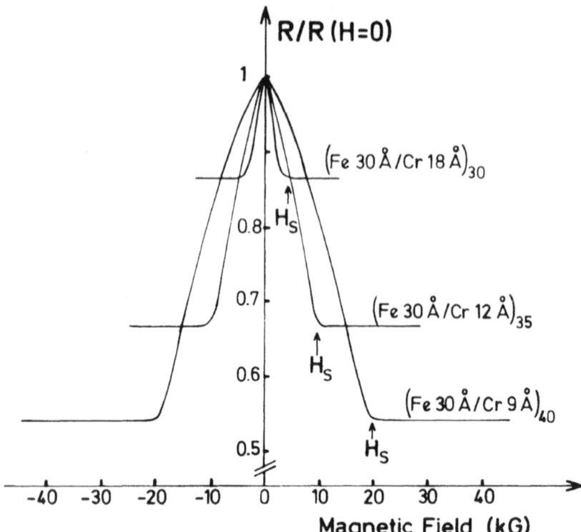

Fig. 2.26. Resistivity versus magnetic field for several antiferromagnetically coupled Fe(0 0 1)/Cr(0 0 1) superlattices. The current and the magnetic field are in the plane of the film along the [1 0 0] direction. H_s is the field needed to align the magnetic moment of all the layers and saturate the magnetization. From *Baibich* et al. [2.51]

magnetization of some layers [52, 98–100]. A non-saturated MR can also be observed if there is only a random arrangement of the magnetic moments in successive layers at low field. Here we focus mainly on the case of *exchange coupled multilayers*.

In the following paragraphs, after we have summarized the problem of spin dependent conduction in bulk ferromagnets, we describe the proposed mechanisms for the GMR and the existing theoretical models. Then we review experimental results and discuss their interpretation.

2.2.3.2 Spin Dependent Conduction in Ferromagnets

In ferromagnetic metals at low temperature, the spin-flip scattering of the conduction electron by magnons is frozen out and the spin relaxation time is much larger than the momentum relaxation time. Consequently, there is conduction in parallel by the spin↑ (majority) and spin↓ (minority) electrons [2.101–104]. If the resistivities of the spin ↑ and the spin ↓ channels are $\rho\uparrow$ and $\rho\downarrow$ respectively, the resistivity of the ferromagnet in the low temperature (LT) limit will be

$$\rho_{\mathrm{LT}} = \rho\uparrow\rho\downarrow/(\rho\uparrow + \rho\downarrow) \tag{2.59}$$

Inside each ρ_σ, we can have complications such as s- and d-bands or several types of scattering processes (by impurities, defects, surfaces, interfaces, etc...) but (2.59) still remains strictly valid. In Fe, Co, Ni and their alloys, the resistivities $\rho\uparrow$ and $\rho\downarrow$ can be very different. Schematically, the resistivity ρ_σ can be written as a function of the number n_σ, effective mass m_σ, relaxation time τ_σ and density of states at the Fermi level $n_\sigma(E_\mathrm{F})$ of the spin σ electrons in the following way

$$\rho_\sigma = m_\sigma/n_\sigma e^2 \tau_\sigma \tag{2.60}$$

with, for one type of scattering potential characterized by its matrix elements V_σ and in the Born approximation

$$\tau_\sigma^{-1} \sim |V_\sigma|^2 n_\sigma(E_F). \tag{2.61}$$

There are intrinsic origins of the spin dependence of ρ_σ that are related to the spin dependence of n_σ, m_σ or $n_\sigma(E_F)$ in the host metal. For example, in Ni and Co, as the spin $\uparrow$ d-band is below the Fermi Level, the density of spin $\uparrow$ states at the Fermi level comes from only s–p electrons and is definitely smaller than the density of spin states $\downarrow$ (s–p + d). Consequently there will be a general intrinsic tendency to have $\rho\uparrow < \rho\downarrow$ in Ni and Co-based systems. Actually, the ratio $\alpha = \rho\downarrow/\rho\uparrow$ exceeds 10 for many Ni- and Co-based alloys [2.101–104].

In addition, there are also extrinsic origins of the spin dependence of ρ_σ related to some spin dependence of the impurity or defect potential V_σ. We take the well-known example of Ni containing Cr impurities [2.101, 103]. The magnetic moment of a Cr impurity is opposite to that of the Ni atoms, which indicates that there is a strong repulsive potential for the spin $\uparrow$ electrons (more precisely, the magnetic moment change, $\Delta\mu \approx -4\mu_B$ can be accounted for by repelling 4 spin $\uparrow$ electrons from the Cr site, i.e $\Delta Z\uparrow \approx -4$ and approximately $\Delta Z\downarrow \approx 0$ if one takes into account the charge difference between Ni or Cr). The strong repulsive potential for the spin $\uparrow$ electrons at the Cr impurity sites gives rise to a strong scattering in the spin $\uparrow$ channel (with formation of a virtual bound state), in agreement with the experimental result of a coefficient $\alpha = \rho\downarrow/\rho\uparrow$ smaller than one. Both origins of the spin dependence, intrinsic and extrinsic, are generally taken into account in the theoretical models worked out for the problem [2.105]. On the experimental side the ratio α has been derived from resistivity measurements for many Ni-, Co- and Fe-based alloys and extensive tables of α values can be found in [2.103].

Departing from the low temperature limit, it is necessary to take into account the transfer of momentum between the two channels by spin-flip electron-magnon scattering. Spin $\uparrow$ (spin $\downarrow$) electrons are scattered to spin $\downarrow$ (spin $\uparrow$) states by annihilating (creating) a magnon and the transfer of momentum from the fast to the slow channel, the so-called spin-mixing effect, tends to equalize the two currents. As the shunting by the fast channel is reduced, the resistivity increases. The general expression of the resistivity is:

$$\rho = [\rho\uparrow\rho\downarrow + \rho\uparrow\downarrow(\rho\uparrow + \rho\downarrow)]/[\rho\uparrow + \rho\downarrow + 4\rho\uparrow\downarrow], \tag{2.62}$$

where $\rho\uparrow\downarrow$ is the spin-mixing resistivity term. General expressions relating $\rho\uparrow\downarrow$ to the electron–magnon scattering rates and finally to the temperature can be found in the literature [2.101, 103]. At relatively low temperatures, for $\rho\uparrow\downarrow \ll \rho\uparrow, \rho\downarrow$, (2.62) can be approximated by

$$\rho = \rho\uparrow\rho\downarrow/(\rho\uparrow + \rho\downarrow) + \left(\frac{\alpha - 1}{\alpha + 1}\right)^2 \rho\uparrow\downarrow. \tag{2.63}$$

The second term is large for $\alpha \gg 1$ or $\alpha \ll 1$, which accounts for the strongly enhanced dependence of the resistivity in a number of ferromagnetic alloys. In contrast this enhancement disappears for $\alpha \approx 1$.

At high temperature for $\rho\uparrow\downarrow \gg \rho\uparrow, \rho\downarrow$ the resistivity, (2.62) tends to

$$\rho_{HT} = \frac{\rho\uparrow + \rho\downarrow}{4}. \tag{2.64}$$

This simply expresses that, for complete spin-mixing (i.e. when the spin lifetime is shorter than the non-spin-flip relaxation time), all the electrons regardless of their spin, have the same averaged relaxation rate.

2.2.3.3 Simple Model of the Magnetoresistance in Multilayers

The ordinary interpretation of the GMR is in terms of interface or "bulk" spin dependent scattering. Bulk scattering means scattering by impurities or defects within the magnetic layers. The ordinary assumption is that the interface scattering is localized within *an infinitesimally thin layer* at the interfaces, while the bulk scattering is *uniformly distributed* within the layers. Of course, "infinitesimally thin" and "uniformly distributed" are approximate ways to describe a generally more complex distribution.

Figure 2.27 represents a schematic of the mechanism of the MR in the simplest limit, when the *MFP within the layers* (100 Å *is a typical value*) *is much larger than the thickness of the layers*. At high field, when the magnetizations of all layers are parallel, there are different scattering probabilities and therefore different resistivities, $\rho\uparrow$ and $\rho\downarrow$, for the two spin directions. The faster electrons (those with $s_z = +1/2$ in Fig. 2.27a) form a low resistivity channel throughout the sample and, as the current is shunted by this channel, the total resistivity

$$\rho_F = \rho\uparrow\rho\downarrow/(\rho\uparrow + \rho\downarrow) \tag{2.65}$$

is low. For bulk alloys, (2.65) is equivalent to (2.59).

Figure 2.27b represents the opposite case with an antiparallel arrangement of successive layers. What is the low resistivity electron species in a layer becomes the high resistivity electron species in the next. Each channel has an averaged resistivity $(\rho\uparrow + \rho\downarrow)/2$ and the final resistivity is high

$$\rho_{AF} = (\rho\uparrow + \rho\downarrow)/4. \tag{2.66}$$

The crossover from (2.65) to (2.66) is exactly equivalent to the crossover from zero to infinite spin-mixing by spin-flip scattering in bulk alloys, i.e from (2.59) to (2.64). When a "fast" electron of the magnetic layer number 1 goes into the magnetic layer number 2 where its spin direction is the "slow" one, it brings its high momentum to "slow" electrons, which is equivalent to the interchannel momentum transfer by spin-flips in bulk materials. From (2.65) and (2.66), the MR in the infinite mean free path limit described above can be simply written as:

$$R = \frac{\rho_{AF} - \rho_F}{\rho_{AF}} = \left(\frac{\rho\downarrow - \rho\uparrow}{\rho\downarrow + \rho\uparrow}\right)^2 = \left(\frac{\alpha - 1}{\alpha + 1}\right)^2, \tag{2.67}$$

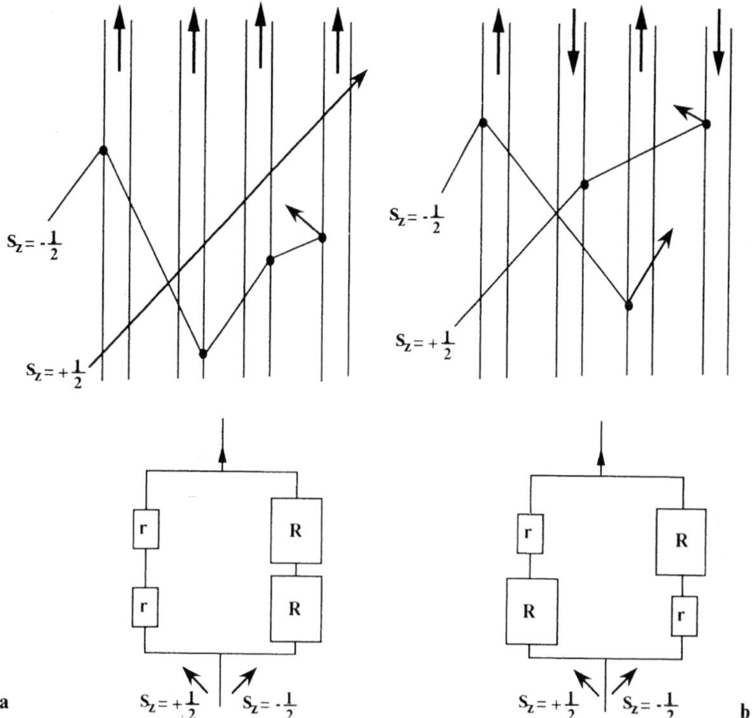

Fig. 2.27. Schematic of conduction in a magnetic multilayer for parallel magnetizations (high field) in (**a**) and for antiparallel magnetizations (zero field with antiferromagnetic exchange) in (**b**). The upper figures represent electron trajectories in a multilayer for both spin directions ($s_z = \pm \frac{1}{2}$). The scattering at the interfaces of the magnetic layers is assumed to be weaker for the majority spin direction (indicated by an arrow in each magnetic layer) and the MFP within the layers is supposed to be much larger than the individual thicknesses. The lower figures represent the equivalent resistor arrays: large and small resistors in parallel in (**a**), with a low resistance path available; large and small resistors in series in each branch (no low resistance path) in (**b**)

with

$$\alpha = \rho\!\downarrow / \rho\!\uparrow .$$

The above results are for an infinite MFP. For a finite MFP, the GMR will still exist if an electron can sample more than one magnetic layer. The MR is expected to decrease as the ratio of the MFP to the non-magnetic thickness decreases and to vanish when this ratio becomes much smaller than one. Actually, as it appears in the theoretical models (Sect. 2.2.4), the variation of the MR with the thicknesses is a little more complex. The layer thicknesses also control the ratio of interface to bulk scattering and the fraction of a magnetic layer interacting with neighboring magnetic layers. The proportion of current shunted by the over- or under-layers are also thickness dependent.

In multilayers at high enough temperature, there are two mechanisms of spin mixing: spin flip scattering and propagation through alternating magnetizations. Since channels already mixed by the former cannot be mixed much more by the latter, the spin mixing by the AF arrangement is less efficient. This gives an important contribution to the decrease of the GMR with temperature.

2.2.4 Theoretical Models of the Magnetoresistance

2.2.4.1 *Model of Camley and Barnas and Other Semiclassical Models*

The first model of the GMR was worked out by *Camley* and *Barnas* [2.106, 107] and is an extension of the Fuchs–Sondheimer semiclassical model for the conduction in thin films to the case of magnetic multilayers. *Camley* and *Barnas* assume that the electrical current is carried in parallel by the spin $\uparrow$ (majority) and spin $\downarrow$ (minority) electrons. They introduce relaxation times τ, $\tau\uparrow$ and $\tau\downarrow$, and transmission coefficient $T\uparrow$ and $T\downarrow$ to take into account the spin dependent scattering within the non-magnetic layers (τ), the magnetic layers $(\tau\uparrow, \tau\downarrow)$, and at the interfaces $(T\uparrow, T\downarrow)$. As they concentrate on the case of trilayers for which the scattering at the outer surfaces is important, they also introduce specularity factors for the reflection at the outer surfaces. The Boltzmann equation is used to describe how the perturbation of the electron distribution function by a given magnetic layer spreads and interacts with the perturbation produced by neighboring magnetic layers. If z is the axis perpendicular to the layer plane, the Boltzmann equation is written as

$$\frac{\partial g^{\uparrow(\downarrow)}}{\partial z} + \frac{g^{\uparrow(\downarrow)}}{\tau\uparrow(\downarrow)v_z} = \frac{eE}{mv_z}\frac{\partial f^{(0)}}{\partial v_x}, \qquad (2.68)$$

where

$$g^{\uparrow(\downarrow)}(\boldsymbol{v}, z) = f\uparrow(\downarrow)\,(\boldsymbol{v}, z) - f^{(0)}(\boldsymbol{v})$$

is the deviation of the electron distribution function from the equilibrium Fermi–Dirac distribution $f^{(0)}(\boldsymbol{v})$. The general solution of (2.68) is written as

$$g_{\pm}^{\uparrow(\downarrow)}(\boldsymbol{v}, z) = \frac{eE}{m}\tau\uparrow(\downarrow)\frac{\partial f^{(0)}(\boldsymbol{v})}{\partial v_x}\left[1 + F_{\pm}^{\uparrow(\downarrow)}(\boldsymbol{v})\exp\left[\frac{\mp z}{\tau\uparrow(\downarrow)|v_z|}\right]\right], \qquad (2.69)$$

where $+$ and $-$ are for $v_z > 0$ and $v_z < 0$ respectively. The transmission coefficients appear in the boundary conditions at the interfaces

$$g_{+}^{\uparrow(\downarrow)}(z = 0^{+}) = T\uparrow(\downarrow)g_{+}^{\uparrow(\downarrow)}(z = 0^{-}), \qquad (2.70)$$

$$g_{-}^{\uparrow(\downarrow)}(z = 0^{-}) = T\uparrow(\downarrow)g_{-}^{\uparrow(\downarrow)}(z = 0^{+}), \qquad (2.71)$$

for an interface at $z = 0$. Equations (2.70–71) express the fact that a proportion $T\uparrow(\downarrow)$ of the electrons are transmitted without scattering while a proportion $D\uparrow(\downarrow) = 1 - T\uparrow(\downarrow)$ are diffusely scattered (for simplicity, the model assumes that the electrons can be transmitted or scattered and cannot be specularly reflected). Boundary equations are also introduced to describe the reflection at the outer surfaces. When the magnetic moments of successive layers are not aligned, the transfer of momentum between the spin $\uparrow$ and spin $\downarrow$ channels is expressed by introducing a transmission coefficient $T\uparrow\downarrow$ at the center of the non-magnetic layers ($T\uparrow\downarrow = \cos^2\theta/2$ where θ is the angle between the magnetic moments). The functions $g_{\pm}^{\uparrow(\downarrow)}(v, z)$ can be determined from the above equations and the resistivity is then calculated for $\theta = \pi$ (AF arrangement), $\theta = 0$ (F arrangement) or any value of θ.

The model of *Camley* and *Barnas* can be applied to the case of trilayers (as in their initial publications, [2.106]) or to the standard multilayer case. The predicted main features for multilayers are:

– The MR ratio decreases as the ratio of the thickness of the nonmagnetic layers to the MFP increases. This expresses the decoupling of the spin dependent scattering in successive magnetic layers and also, in the case of interface spin dependent scattering, the decrease of the interface density (dilution effect).

– For interface spin dependent scattering, the MR ratio decreases as the ratio of the thickness of the magnetic layers to the MFP increases. This expresses that, when the thickness exceeds the larger MFP, a part of the magnetic layer becomes inactive. For bulk spin dependent scattering, the variation of the MR ratio with the thickness of the magnetic layers exhibits a maximum (as the MR is obviously zero for zero thickness, the MR first increases and then decreases).

– The MR ratio becomes independent of the number of bilayers when the total thickness of the structure is much larger than the mean free path (and also much larger than the thickness of under- or over-layers).

– The increase of the MFP as the temperature decreases gives rises to an increase of the MR at low temperatures. However, in their initial publication [2.106], *Camley* and *Barnas* had to use an unrealistically too long MFP (6000 Å at 4.2 K) to account for the increase of the MR from room temperature to the helium range. This is certainly because the model does not take into account the additional important contribution from the spin-mixing by electron–magnon collisions.

The model of *Camley* and *Barnas* has been extensively used for numerical calculations of the MR in sandwiches and multilayers [2.106–110]. Examples of interpretation of experimental data are presented in Sect. 2.2.5a. On the other hand the semiclassical approach of *Camley* and *Barnas* has also been used to derive analytical expressions of the GMR in some simple limits. We refer to the work of *Barthélémy* and *Fert* [2.111], presenting simple expressions of the GMR for the limit where the thickness of the magnetic layers is much larger than the MFP, and also the work of *Edwards* et al. [2.112] for the limit of very long MFP.

Some minor improvements have recently been introduced in the model of *Camley* and *Barnas*. *Johnson* and *Camley* [2.113] have described the interface scattering by introducing a thin interfacial layer with a shorter MFP. *Dieny* [2.109, 110] has taken into account the granular structure of sputtered samples by introducing some anisotropy of the MFP (i.e. different in-plane and perpendicular MFP). He has also emphasized the interest of analyzing the change in conductivity instead of the change in resistivity.

All of the calculations quoted above and based on the model of *Camley* and *Barnas* assume a free-electron band throughout the multilayer, thus neglecting the periodicity of the potential and taking into account only the periodic distribution of scattering centers. Several attempts have recently been made to go beyond the free-electron approach and take into consideration the effect of the multilayer potential on the electron wave function [2.114–116]. In the work of *Falicov* and *Hood* [2.115], this is done by taking into account the potential scattering by the interfaces. The extreme limit in this direction is the model of *Erlich* [2.116] who assumes that some electrons are strictly localized in quantum wells. It is too early to know whether these new models bring essential improvements to the preceding approaches.

2.2.4.2 Quantum Models

The quantum theory of transport in thin metallic films was first developed by *Tesanovic* et al. [2.117] to correct some non-physical results of the Fuchs–Sondheimer semiclassical model when the film thickness, t, becomes smaller than the electron MFP, λ. The main non-physical result of the semiclassical model is that, for $t \ll \lambda$, the resistivity varies as $t/\ln(\lambda/t)$ and therefore goes to zero as λ tends to infinity. This implies that, in the absence of bulk scattering, the scattering by the interface roughness induces no dissipation. In contrast, the prediction of quantum models varies approximately as $1/t^2$ [2.118] and a better agreement with the experimental data is obtained [2.117–119].

Quantum models of the GMR have been worked out by *Levy* et al. [2.120, 121] and *Vedyayev* et al. [2.122]. The model developed by *Levy* and coworkers is for a multilayer with an infinite number of layers and the Kubo formalism is used to treat the scattering of electron waves (free-electrons) by spin dependent potentials randomly distributed in the interface planes or within the layers (spin dependence only within the magnetic layers). The total scattering potential is written as

$$V(r,\hat{s}) = \sum_l (v_s + j_s \hat{M} \cdot \hat{\sigma}) f_l(\rho)\delta(z - z_l) + \sum_i (v_b + j_b \hat{M} \cdot \hat{\sigma}\delta(r - R_i)$$

$$+ \sum_j v_b' \delta(r - R_j), \tag{2.72}$$

where z_l is the position of the lth interface, $f(\rho = x, y)$ represents interface roughness, R_i and R_j are the positions of scattering centers (impurity or defects) in the magnetic and non-magnetic layers respectively, $\hat{M}$ is a unit vector in the direction of magnetization and $\hat{\sigma}$ represents the Pauli spin matrix. The spin asymmetry parameters are defined as $p_s = j_s/v_s$ and $p_b = j_b/v_b$ ($p =$

$(\sqrt{\alpha} - 1)/(\sqrt{\alpha} + 1)$ if α is the conventional asymmetry parameter of the two-current model for transport in ferromagnets [2.103]. In the limit of individual thicknesses much smaller than the MFP, the GMR is governed by the parameters p and, for example, one obtains

$$R = \frac{\rho_{AF} - \rho_F}{\rho_{AF}} = \frac{4p_s^2}{(1 + p_s^2)} = \left(\frac{\alpha_s - 1}{\alpha_s + 1}\right)^2 \tag{2.73}$$

for only interface scattering (compare with (2.67)) and

$$R = 4p_b^2/(1 + p_p^2 + b/a)^2 \tag{2.74}$$

for only bulk scattering (b and a are the thicknesses of the magnetic and non-magnetic layers respectively; $v_b = v_b'$ and equal concentrations of scattering centers in the magnetic and non-magnetic layers are also assumed in (2.74)).

In the general case, the behavior of the GMR as a function of the thicknesses, spin asymmetries and other parameters of the problem is described by *Zhang* et al. [2.120]. In short, for spin dependent scattering only at interfaces, the MR ratio is a continuously decreasing function of the magnetic and non-magnetic thicknesses. The shorter the MFP λ_b, the steeper is the decrease. For bulk spin dependent scattering only, the variation with the thicknesses is more complex [2.120]. Fits with experimental data are presented in Sect. 2.2.5a. Finally, a very recent improvement of the model of *Levy* and coworkers is the introduction of the periodic potential of the superlattice [2.148].

Vedyayev et al. [2.122] have worked out another quantum model, also based on the Kubo formalism. The calculation is limited to the case of bulk scattering but has the advantage of giving analytical expressions of the MR with only few approximations. It would be of great interest to compare the MR predicted with the same parameters by the quantum models of *Levy* et al. [2.120, 121] and *Vedyayev* et al. [2.122].

According to what is known for simple thin metallic films, the results of the semiclassical models deviate from those of the quantum models when the film thickness is much smaller than the electron MFP [2.117–119]. For the GMR, the problem of the deviation of the semi-classical calculations from the quantum results in the limit of small thicknesses has not been really solved. Such deviations could likely explain the less quantitative fits obtained for some systems in the semi-classical models (see Co/Cu in Sect. 2.2.5a). It would be of great interest to more carefully compare the semi-classical and quantum calculations and to determine the thickness range in which the simple semi-classical models can be used reliably.

2.2.4.3 Theoretical Models of the Spin Dependent Scattering

The models described in 2.2.4a and b assume some spin dependence for the interface and bulk scattering but are not concerned with the microscopic origin of this spin dependence. The major open problem is *the microscopic origin of the spin dependent scattering by interfaces,* i.e. the problem of the microscopic mechanism giving rise to different values of $T\uparrow$ and $T\downarrow$ (notation of the semiclassical models) or to a non-zero value of p_s (notation of the quantum model of *Levy* et al.). The spin dependence of the bulk scattering seems to be less essential in the best known systems (Fe/Cr, Co/Cu, as discussed in Sect. 2.2.5) and, anyhow, is probably related to spin dependence effects already studied in ferromagnetic metals and alloys [2.103].

Inoue and *Maekawa* [2.124] have developed a theory of the spin dependent scattering by interfaces in the two cases where the spacer metal is a non-magnetic transition metal (Cr, V, etc.) and a noble metal (Cu, Ag, etc.), respectively. For non-magnetic layers of transition metal, the calculation of *Inoue* and *Maekawa* is based on a square well picture (Kronig–Penney-like) of the multilayer potential. The potential level of each layer for a given spin direction is fixed to accomodate the right number of d electrons of this spin direction in rigid d-band picture. The calculation assumes that the scattering potentials arising from interface roughness are proportional to the square well heights, which leads to resistivities proportional to the square of these spin dependent heights. The MR is finally calculated in the simple limit when the layers are much thinner than the MFP (cf. (2.67) or (2.73)). The model predicts a large MR for Fe/Cr, in agreement with the experiments, but is less successful for Co/Cr or Co/Mn for which it would also predict a fairly large MR.

For non-magnetic layers of noble metals, *Inoue* and *Maekawa* [2.124] consider the scattering of the s electrons of the noble metal electrons by the interface magnetic atoms and reduce this problem to the classical case of scattering by magnetic impurities. The resonance of free-electrons with localized d-electrons (virtual bound state formation) is described in terms of phase shifts and, via the Friedel sum rule, the spin $\uparrow$ and spin $\downarrow$ resistivities are related to the number of d electrons for each spin direction. The model accounts for the large MR of Co/Cu [2.70, 71] but predicts a too large MR for Fe/Cu [2.75] and a too small MR for Ni/Ag [2.125]. Obviously, an impurity virtual bound state model can only be very approximate to describe the hybridization of the s- and d-band at a noble metal/transition metal interface.

The simple models developed by *Inoue* and *Maekawa* represent interesting first steps towards an accurate picture of the electron scattering by rough interfaces. However they are probably too simple for such a complex problem. It has been known for a few years that sophisticated numerical calculations are necessary to precisely predict the electronic structure properties of the interfaces and this type of calculation should now be extended in order to describe rough interfaces and their scattering effects.

2.2.5 Review and Discussion of Magnetoresistance Data

We present experimental results for the dependence of the GMR on the layer thickness, applied field, structure of the interface, doping, temperature, etc., and we use these data to discuss the mechanism of the magnetoresistance.

2.2.5.1 Dependence on Layer Thickness

Figure 2.28 plots the variation of the MR in Fe/Cr multilayers with the thickness of chromium. The experimental data was obtained by *Gijs* and *Okada* [2.126] on Fe/Cr multilayers grown by sputtering, and the calculated curve was generated by the same authors using the quantum model of *Levy* et al. [2.120]. The advantage of Fe/Cr is that the interlayer exchange is particularly strong, so that, at least in the first and second AF thickness ranges (up to about 3 nm), the resistivity change identifies with the difference between the resistivities of the AF and F configurations (strong coupling regime). In the first and second AF thickness ranges, a good agreement is obtained for the MR and also for the absolute values of the resistivity [2.126]. Between the second and third AF ranges, there is a crossover to the weak coupling regime with imperfect AF arrangement at low field and this probably explains that the experimental values of the MR are below the calculated curves at the largest thicknesses. The fit of Fig. 2.28 is obtained with the value 0.55 for the parameter p_s expressing the spin dependence of the interface scattering; this corresponds to 1/12 for the ratio α between the spin $\downarrow$ and the spin $\uparrow$ scattering rates at the interfaces. A high value of p_s is needed to account for the steep decrease of the MR with the Cr thickness. The parameter p_b for the bulk scattering in Fe is relatively small and hard to determine accurately. The main conclusion of *Gijs* and *Okada* [2.126] is that it is essential to assume a strong spin dependent interface scattering to account for the experimental data. This importance of spin dependent interface scattering for Fe/Cr and several other systems is confirmed by various data reported below.

Figure 2.29 displays experimental data and calculated curves for the variation of the MR in Co/Cu with the thickness of copper. The solid lines have been

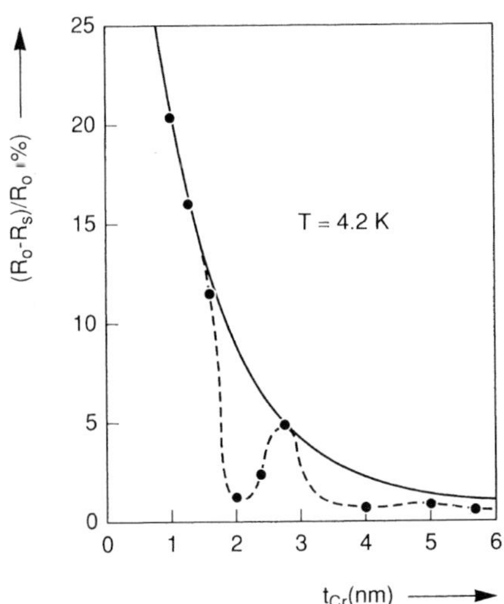

Fig. 2.28. MR ratio for Fe30 Å/Cr multilayers versus thickness of chromium. The experimental data (black dots) are obtained at 4.2 K. The solid lines are calculated within the quantum model of *Levy* et al. [2.120]. Experimental data and calculations are from *Gijs* and *Okada* [2.126]

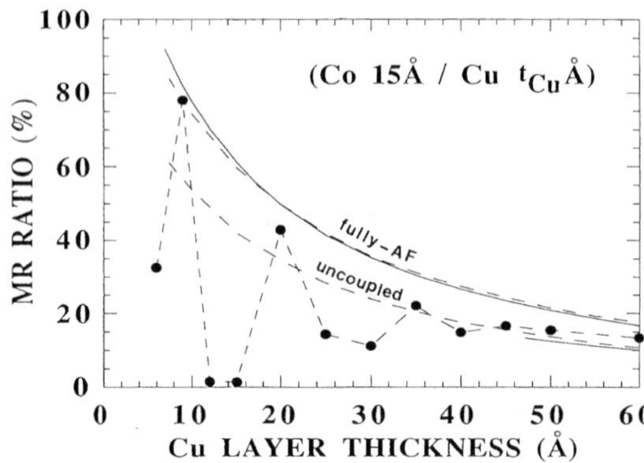

Fig. 2.29. Variation of the MR ratio in Co15 Å/Cu with the thickness of copper [2.123]. Symbols: experimental data at 4.2 K [2.71]. Dashed lines: quantum model of *Zhang* and *Levy* [2.121], in both the strong coupling and uncoupled limits. Solid lines: semi-classical model in the same limits [2.123, 127]

calculated [2.127] in the semiclassical model of *Camley* and *Barnas* [2.106, 107] and the dashed curves in the quantum model of *Zhang* et al. [2.121]. The curves calculated for the strong coupling regime, i.e. with

$$MR = (\rho_{AF} - \rho_F)/\rho_F \tag{2.75}$$

represent the variation that should be observed if the interlayer exchange was strong enough in all the AF half-periods to induce a perfect AF order at low field. The curves calculated for the uncoupled regime (random orientation of the magnetizations at low field) are relevant to interpret the experimental results for thicknesses above 45 Å approximately. *Zhang* and *Levy* [2.121] have shown that the MR of the uncoupled regime is reduced by a factor around 0.6 with respect to (2.75) and the same reduction factor has been introduced in the semiclassical calculation of Fig. 2.29. Both models are able to properly account for the decrease of the MR with the thickness of copper. The quantum model is also able to account with the same parameters for the absolute values of the resistivity (in the semiclassical fit, the absolute values are too small [2.127]). In both models the best fit is obtained with a stronger spin dependence for the interface scattering. However, owing mainly to the uncertainty on the low field random arrangement, the predominance of the interface scattering is much less definitely established for Co/Cu from the fit of Fig. 2.29 than for Fe/Cr (Fig. 2.28).

The variation of the MR with the thickness of the magnetic layers also contains potential information on the proportion of bulk and interfacial spin dependent scattering. In short, when the scattering centers are supposed to be uniformly distributed inside the magnetic layers, the number of active scattering centers increases as long as the thickness does not exceed the longer mean free path and there is a maximum of the MR ratio at a relatively large thickness. In contrast, a much smaller thickness is generally sufficient to have two well-defined interfaces and a fully developed interface scattering. *Dieny* et al. [2.139] have used the semiclassical model of *Camley* and *Barnas* [2.106] to interpret the variation of the MR in F/Cu/NiFe/FeMn spin valve structures as a function of the thickness of the ferromagnetic layer F (F = Fe, Co or NiFe, NiFe is for permalloy). Their fits for the absolute values of the conductance and conductance change between the AF and F arrangements are shown in Fig. 2.30. The best fit is obtained by assuming purely interface spin dependent scattering for Fe ($\lambda\uparrow = \lambda\downarrow = 70$ Å, $T\uparrow = 1$, $T\downarrow = 0.6$), purely bulk spin dependent scattering for NiFe ($\lambda\uparrow = 110$ Å, $\lambda\downarrow = 10$ Å, $T\uparrow = T\downarrow = 1$) and both interfacial and bulk contributions for Co ($\lambda\uparrow = 140$ Å, $\lambda\downarrow = 10$ Å, $T\uparrow = 1$, $T\downarrow = 0.2$). However, the difference in Fig. 2.30 between the behaviors for bulk and interfacial spin dependent scattering is not spectacular and it cannot be concluded that the above data on interface and bulk scattering are definitely established.

Similar analyses of the variation of the GMR with the thicknesses of the non-magnetic and magnetic layers in various systems have been recently reported. In a general conclusion, the dependence on thickness is fairly well accounted for by the theoretical models, but, with the exception of Fe/Cr, its

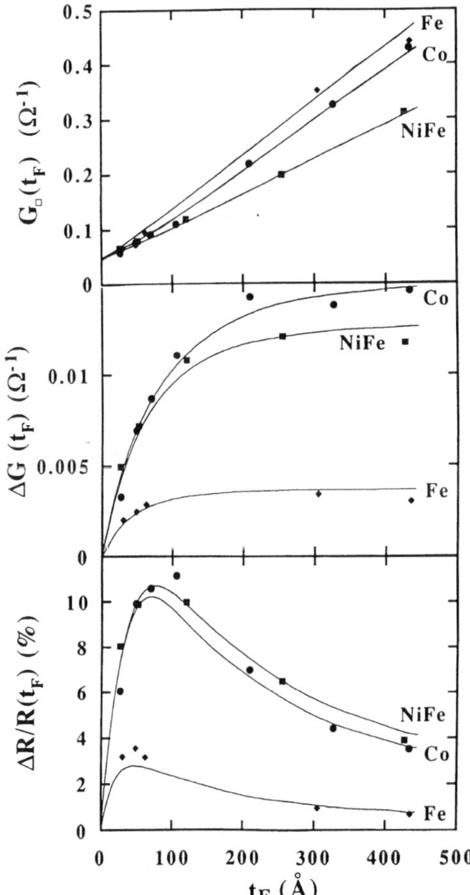

Fig. 2.30. Experimental (symbols) and calculated (solid lines) variation of the MR for M/Cu22 Å/NiFe50 Å/FeMn90 Å spin valve structures as a function of the thickness of M at 1.5 K (M = NiFe, Co or Fe, NiFe = permalloy). G is the sheet conductance, ΔG the absolute change of sheet conductance between the AF and F states, $\Delta R/R$ is the relative change of resistance. The MR is calculated in a semi-classical model with the parameters indicated in the text. From *Dieny* et al. [2.138]

analysis in terms of interface and bulk parameters is never completely conclusive. Fortunately, additional information can be obtained from samples with planar doping of the interfaces (Sect. 2.2.5b) and also from magnetoresistance measurements with currents perpendicular to the layers.

2.2.5.2 Influence of Interface Structure, Planar Doping, etc.

The resistivity is not a truly intrinsic property of the metals. It depends not only on the intrinsic band structure but also on the properties and concentrations of defects, impurities and other scattering centers (the resistivity is strictly zero in an ideally perfect crystal or superlattice). Consequently, to understand the MR of the magnetic multilayers, it is essential to identify the "defects" involved in its mechanism (scattering centers).

From the earliest experiments on Fe/Cr, it turned out that the MR was very sensitive to the growth conditions and that the scattering by interface imperfections was probably playing an important role. Also very high MR could be found in Fe/Cr samples with intentionally mixed interfaces during the growth and enhanced interface scattering (the enhancement of the interface scattering was indicated by the increase of the resistivity) [2.128]. In the following we describe several experiments aiming to correlate the MR to the properties of the interface or, more generally, to the presence of some type of scattering center.

The influence of the interface structure on the MR has been mainly studied for Fe/Cr. *Petroff* et al. [2.128] found that "roughening" the interfaces of Fe(0 0 1)/Cr(0 0 1) Molecular Beam Epitaxy (MBE) superlattices by an annealing treatment enhances the MR for annealing temperatures up to 300 °C and reduces it at higher temperatures. These experiments suggest the existence of an "optimum roughness".

Fullerton et al. [2.129] have combined low-angle x-ray diffraction, MR and magnetization experiments on Fe/Cr multilayers prepared by sputtering in various conditions. They find that the MR is higher when the interfaces are "rougher", or, more precisely, when the intensity of the low-angle peak is smaller. From magnetization measurements, they show that the higher MR is not due to a better antiferromagnetic arrangement and they conclude that the spin dependent scattering by the interfaces is enhanced by their roughness.

Obi et al. [2.130] have combined low-angle and high-angle x-ray diffraction experiments with MR and magnetization measurements for Fe/Cr multilayers prepared by sputtering under various argon pressure conditions. They find that the MR is maximum for argon pressure around 40 m Torr, when the short-range disorder derived from high-angle data is minimum. The conclusion of *Obi* et al. is that the spin dependent scattering is not due to interface structural disorder and could arise from compositional mixing or bulk scattering. We point out however that the minimum of disorder at 40 m Torr is weakly pronounced, which does not clarify the problem.

In conclusion, at the present stage and for Fe/Cr: (i) The structure of the interfaces is obviously important for the MR, which is consistent with the interpretations putting forward the role of spin dependent interfacial scattering. (ii) There is some controversy about the experimental results on the correlation between roughness and MR. (iii) One does not know the type of interface imperfection involved in the MR and the corresponding scattering process.

Experiments aiming to correlate interface structure and MR have also been performed for the Co/Cu and NiFe/Cu systems. In these systems the value of the MR is related to the structural integrity of the layers: pinholes and other types of accidental bridging can couple the magnetic layers ferromagnetically and reduce or supress the magnetoresistance. *Parkin* et al. [2.131] have discussed these effects for Co/Cu and NiFe/Cu multilayers and emphasized the importance of the buffer layer for the flatness of the layers. *Highmore* et al. [2.132] have clearly shown that, for sputtered Co/Cu, the larger MR obtained at low argon pressure are due to a greater AF alignment. In the CoFe/Cu multilayers prepared by ion beam sputtering and studied by *Saito* et al. [2.133], the strong influence of the acceleration voltage on the MR is probably due to similar effects. The origin of the ferromagnetic bridging competing with antiferromagnetic exchange in Co/Cu(1 1 1) has been discussed by *Kohlhepp* et al. [2.134].

In order to identify the interfacial scattering processes, several groups have studied multilayers in which selected elements are inserted at the interfaces between the magnetic and non-magnetic layers (planar doping). The first experiments were performed by *Gurney* et al. who inserted thin (0–4 Å) layers of Au or Ag [2.135], V, Mn, Al, Ge or Ir [2.136] between the Fe and Cr layers of Fe/Cr multilayers. Inserting Au, Ag, Al, Ge or Ir reduces strongly the MR, whereas, with V or Mn, the MR is almost as large as in a Fe/Cr sample with the same non-magnetic thickness (Fig. 2.31). This suggests that the spin scattering properties of the Fe/Cr interfaces remain for Fe/V or Fe/Mn interfaces and disappear for an interface of Fe with the other elements. *Gurney* et al. [2.136] and *Johnson* and *Camley* [2.113] have related this behavior to the different spin dependent scatterings by, respectively, Cr, V, Mn and Al, Ge, Ir, Au, Ag impurities in bulk iron [2.103].

Permalloy/Cu and Co/Cu multilayers with another element inserted at the interfaces have been studied by *Parkin* (Sect. 2.4). A few atomic layers of Co inserted between permalloy and Cu are sufficient to raise the MR ratio to the high value found in Co/Cu. This shows that the high MR of Co/Cu is mainly due to spin scattering effects at the Co/Cu interfaces. Similar effects have been observed by other groups [2.137, 138].

In Fig. 2.32 we show results obtained on Co/Cu with thin layers of Fe inserted at the interfaces. As the MR is much lower in Fe/Cu [2.75] than in Co/Cu, the interfaces between Fe and Cu are expected to exhibit much smaller spin dependent effects than those between Co and Cu. This accounts for the

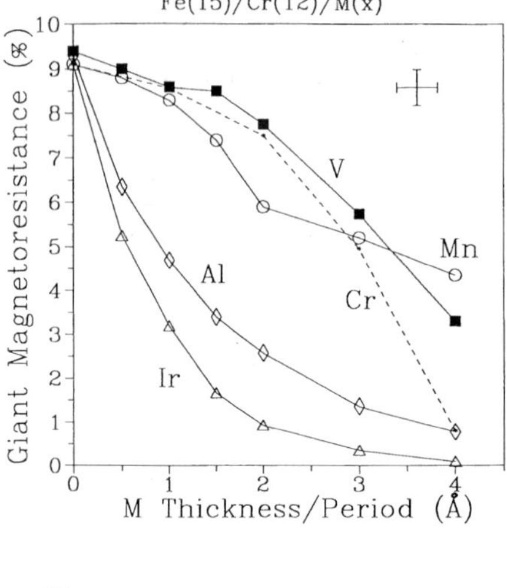

Fig. 2.31. Magnetoresistance of the multilayer structure (Fe15 Å/M/Cr12 Å/M) × 20 as a function of the thickness t of the element M for M = V, Mn, Al and Ir (solid lines), and M = Cr (dashed line). From *Baumgart* et al. [2.136]

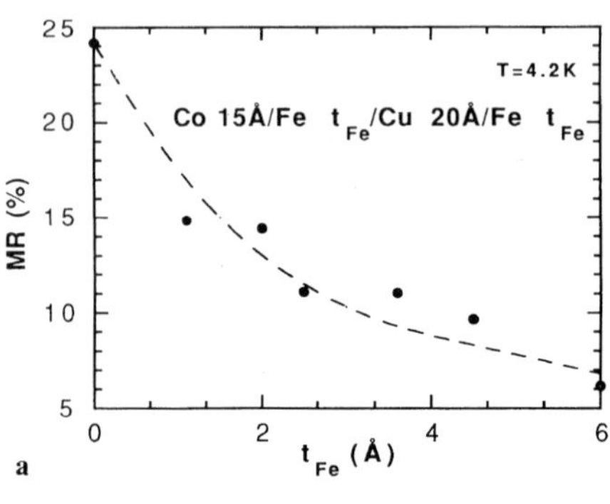

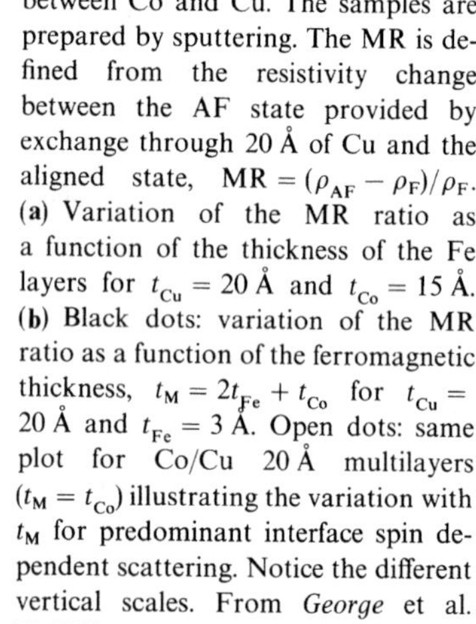

Fig. 2.32. MR of Co/Cu multilayers in which thin Fe layers have been inserted between Co and Cu. The samples are prepared by sputtering. The MR is defined from the resistivity change between the AF state provided by exchange through 20 Å of Cu and the aligned state, $MR = (\rho_{AF} - \rho_F)/\rho_F$. (a) Variation of the MR ratio as a function of the thickness of the Fe layers for $t_{Cu} = 20$ Å and $t_{Co} = 15$ Å. (b) Black dots: variation of the MR ratio as a function of the ferromagnetic thickness, $t_M = 2t_{Fe} + t_{Co}$ for $t_{Cu} = 20$ Å and $t_{Fe} = 3$ Å. Open dots: same plot for Co/Cu 20 Å multilayers ($t_M = t_{Co}$) illustrating the variation with t_M for predominant interface spin dependent scattering. Notice the different vertical scales. From *George* et al. [2.139]

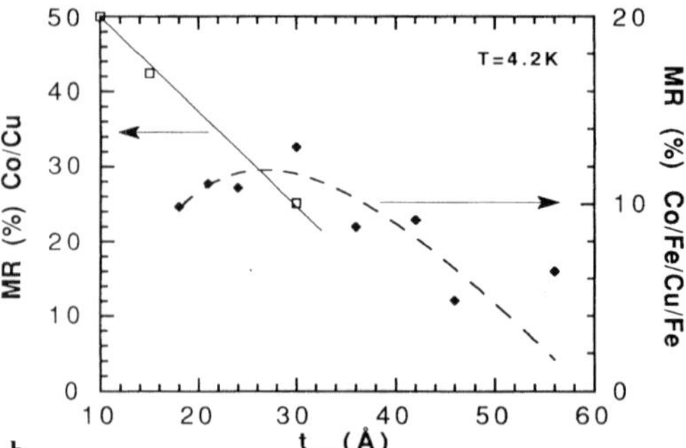

steep decrease of the MR when a few atomic layers of Fe are inserted (Fig. 2.32a). Once the contribution from the Co/Cu interface has been suppressed by the inserted Fe, the variation of the MR with the thickness of the Co layers exhibits the typical behavior expected for "bulk" spin dependent scattering, as this can be seen by comparing the experimental variation in Fig. 2.32b with the calculated curves in Fig. 2.30. This indicates that bulk spin dependent scattering also exists within the Co layers and determines the behavior of the MR when the much stronger interface spin dependent scattering has been removed. In conclusion, the high MR of Co/Cu turns out to be due to a predominant contribution from the interfaces and to a smaller but non-negligible bulk contribution. Analyses of the MR with the current perpendicular to the layers brings a more quantitative basis to this result (Sect. 2.2.5e).

2.2.5.3 Variation with Temperature

The MR ratio decreases as the temperature (T) increases, moderately in Co/Cu (by about a factor of 1.7 between 4.2 K and room temperature for Co15 Å/Cu9 Å [2.71]), more significantly in Fe-based structures (a factor of 3.1 for Fe16 Å/Cr12 Å) [2.128]. There are several effects contributing to this variation:

i) Increasing T introduces additional scattering processes (inelastic processes: phonons, magnons, etc.) and the resulting shortening of the MFP reduces the MR.

ii) The spin asymmetry factor μ of the temperature-induced scattering processes is generally different from that of the low temperature elastic processes (scattering by interfaces, defects, impurities). According to results for bulk metals, the spin asymmetry factor μ (i.e. $\mu = \rho_i \downarrow (T)/\rho_i \uparrow (T)$ is close to 1 for Fe [2.101], around 4 for Ni [2.101] and higher for Co (11 at 77 K) [2.104]. Adding spin dependent inelastic scattering (in Co layers, for example) will increase ρ and $\Delta\rho$ at the same time, whereas adding spin independent inelastic scattering (in Fe layers) will increase only ρ.

iii) The GMR is ascribed to the transfer of momentum between the spin $\uparrow$ and the spin $\downarrow$ currents (spin-mixing effects) by alternating magnetizations (Sect. 2.2.3c). At finite temperatures, there is an additional spin-mixing by electron–magnon spin-flip scattering ($\rho\downarrow\uparrow$ term, Sect. 2.2.3b). Currents significantly mixed and equalized by electron–magnon scattering cannot be mixed and equalized much more by alternating magnetizations, so that the GMR is reduced.

Only the first contribution is included in the initial model of *Camley* and *Barnas* [2.106, 107]. *Zhang* and *Levy* [2.140] have taken into account the first and second contributions and put forward the role of (non-spin-flip) scattering by interface magnons in $\rho_{i\sigma}(T)$. *Duvail* et al. [2.141] have recently extended the model of *Camley* and *Barnas* to take into account the above three contributions and interpret results on Co/Cu and Fe/Cu. At present there are too few comparisons between theory and experiments and definite conclusions cannot be established. However, it can be said that the weak temperature dependence in Co/Cu is likely due to the strong stiffness constant of Co (few magnons) and to the large value of μ quoted above. In contrast, for Fe, the stiffness constant is smaller and $\mu \approx 1$, so that a more pronounced variation with temperature is expected.

2.2.5.4 Variation with Field

The field dependence of the GMR is governed by the variation of the angle ϕ_{12} between the magnetic moments of neighboring layers [2.106]. In the ideal case with perfect AF arrangement at zero field ($\phi_{12} = 180°$), negligible anisotropy and coercivity, $\cos \phi_{12}/2$ is a linear function of H [2.147] and $\Delta\rho$ starts as $(H/H_s)^2$ (H_s is the saturation field). This leads to the rounded maximum illustrated by the MR of Co/Cr in Fig. 2.33a. In contrast, for a non-perfect AF arrangement ($\phi_{12} < 180°$ at zero field), $d\rho/d\phi_{12}$ is non-zero at zero field, which gives the sharp maximum of Fig. 2.33b for a sputtered Fe/Cu sample. Only a few systems exhibit the ideal behavior of Fig. 2.33a. These types of MR curves (more or less sharp) provide interesting information on the imperfection of the AF arrangement but can hardly be accounted for quantitatively. Only the saturation field is always of interest to determine the interlayer exchange. Additional structures in the field dependence can also be used to derive the biquadratic exchange terms and the magnetic anisotropy (in single crystal samples).

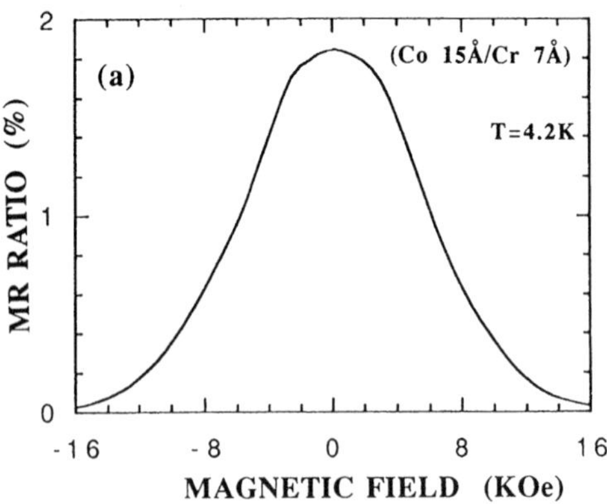

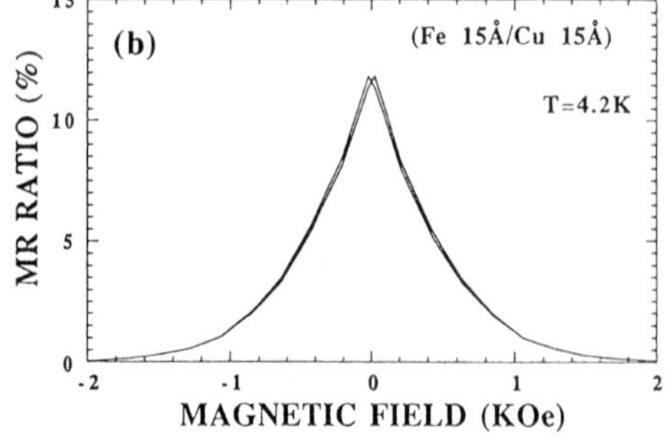

Fig. 2.33. Magnetoresistance curves for (a) Co15 Å/Cr7 Å and (b) Fe15 Å/Cu15 Å. The curves illustrate the field dependence of the MR for, almost perfect and very imperfect AF alignment, respectively

2.2.5.5 Magnetoresistance with the Current Perpendicular to the Layers

The GMR of magnetic multilayers has also been observed in the CPP (Current Perpendicular to the Planes) geometry for Ag/Co and Cu/Co structures at Michigan State University [2.142, 146]. The MR in the CPP geometry is higher than in the conventional geometry, exhibits similar oscillations as a function of the thickness of the non-magnetic layers, and decreases less rapidly at large thicknesses. In Fig. 2.34 we show an example of experimental results.

It is beyond the scope of this chapter to describe the specific problems of the CPP–MR in detail. In short, there are new fundamental problems related to the spin accumulation effects occurring when the current is perpendicular. The spin accumulation is balanced by the spin relaxation rate (spin-flip scattering), which introduces the spin diffusion length as the new scaling length of the problem. In multilayers composed with 3d transition metals or noble metals, the spin diffusion length can be estimated to be ten times (or more) longer than the momentum MFP. *Johnson* [2.143] has worked out a model of the spin accumulation by interfaces, and *Valet* and *Fert* [2.145] have adapted this model to the case of multilayers (successive interfaces). *Zhang* and *Levy* [2.144] have developed a quantum model in the limit of zero spin-flip relaxation. The experimental data on Ag/Co and Cu/Co have been recently accounted for in a simple model [2.146] that is now justified by the calculation of *Valet* and *Fert* [2.145]. The interest of the analysis of the CPP–MR data is that it leads to a straightforward and quantitative separation of the interface and bulk spin dependent scatterings. For example, in the case of the Co/Cu system, the group at Michigan State University has found that the spin asymmetry was definitely stronger for the interfaces scattering ($\alpha = 7.5$) than for the bulk scattering within the Co

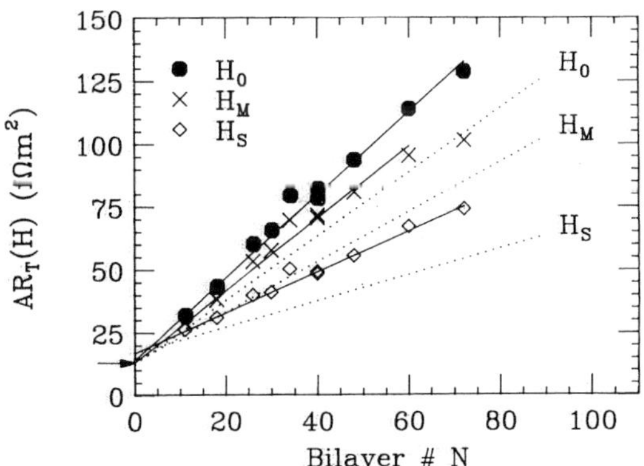

Fig. 2.34. Magnetoresistance data for sputtered Ag/Co multilayers with current perpendicular to the layers. The resistances at zero field in a virgin state, at the resistivity peak during field cycling and at the saturation field are plotted versus the number of bilayers M for samples of total thickness 720 nm $t_{Co} = 6$ nm (symbols and solid best-fit lines) and $t_{Co} = 2$ nm (dotted best-fit lines). From *S.F. Lee* et al. [2.142]

layers ($\alpha = 3$) [2.146]. These asymmetries, combined with the resistivities and interface resistances found in the analysis, indicate that, for an individual thickness of a few nm, the predominant contribution to the MR comes from the interfaces. The contribution from bulk scattering becomes significant only when the thicknesses exceed 10 nm. With the same parameters, similar proportions of the interface and bulk contributions can be anticipated for the longitudinal MR.

2.2.5.6 Current Understanding of the Giant Magnetoresistance (GMR)

We summarize the conclusions of the discussions developed in the preceding Sects. 2.2.5a–e.

Most interpretations of experimental data have been done in the framework of theoretical models based on spin dependent scattering at the interfaces or within the magnetic layers. These models account fairly well for the variation of the MR with the individual thicknesses. However, it is hard to derive very reliably all of the parameters of these models from fits of the thickness dependences. This actually depends on the system. Whereas in Fe/Cr the analysis of the thickness dependence can well show the predominance of spin dependent scattering by interfaces, similar analyses for other systems are less conclusive (Sect. 2.2.5a).

Several experiments have shown the influence of the interface structure on the GMR. It turns out that, depending on the system, the major influence is either on the scattering of the electrons or on the interlayer coupling (Sect. 2.2.5b). More quantitative correlations between interfacial spin dependent scattering and interface structure would be interesting. Doping the interfaces with additional elements is a powerful method to identify the "best" scatterers. For example, experiments with doped interfaces have clearly shown the strongly spin dependent scattering by Co/Cu interfaces (Sect. 2.2.5b). Measurements of the MR in the CPP geometry represent an interesting direction. In particular, the analysis of CPP–MR data leads to a simple and quantitative separation of the interface and bulk contributions. In the case of Co/Cu and Co/Ag, this allows the group at Michigan State University to quantify the spin asymmetries of the scattering in these systems (Sect. 2.2.5e). Although the predominant contribution from interface scattering for Fe/Cr, Co/Cu, Co/Ag seems to have been definitely demonstrated, this does not rule out a possible significant contribution from bulk scattering in other systems (for example, in permalloy based systems, as suggested by the results of *Dieny* and coworkers). The variation of the GMR with temperature is due to the development of inelastic additional scattering, with a major role of the spin-flip scattering induced by spin fluctuations. The analysis of the experimental data is complex but should clear up the problem of the progressive mixing of the spin $\uparrow$ and spin $\downarrow$ currents as one goes from zero Kelvin to T_c.

The main open question is about the microscopic origin of the spin dependent scattering by interfaces in systems such as Co/Cu or Fe/Cr. The role of resonant interface or quantum well resonant states can be put forward (it can be analogous to that of virtual bound states in bulk materials). Numerical calculations of the electronic structure and electron scattering at rough interfaces would be of great interest. Finally, although most present models take into consideration only a periodic distribution of *scattering potentials,* some attempts have recently been made to consider also the influence of the *periodic interface potential of the superlattice* (Sects. 2.2.4a, b). The interplay between scattering and periodic potentials could be an important point leading to a better understanding of the effectiveness of certain interfaces.

Acknowledgements. One of us (A.F.) acknowledges support from the European Economic Community (Esprit project BRA 6146) and the NATO (grant number 5-2-05/RE 890599).

2.3 Investigation of Exchange Coupled Magnetic Layers by Scanning Electron Microscopy with Polarization Analysis (SEMPA)

D.T. Pierce, J. Unguris, and R.J. Celotta

Artificially layered magnetic structures offer the promise of being able to tailor transport and magnetic properties to fit specific needs. For example, two ferromagnetic layers separated by a nonferromagnetic interlayer can be exchange coupled such that the magnetic moments in the two ferromagnetic layers are parallel (ferromagnetic exchange coupling) or antiparallel (antiferromagnetic exchange coupling) depending on the interlayer material and its thickness [2.149, 150]. Two magnetic layers that are antiferromagnetically coupled in the absence of an applied magnetic field can be forced into ferromagnetic alignment by the application of a sufficiently strong magnetic field. Accompanying the change from antiferromagnetic to ferromagnetic alignment is a large reduction in the electrical resistance of the multilayer structure [2.151, 152]. This "giant" magnetoresistance (discussed by *Fert* and *Bruno* in Sect 2.2 and *Parkin* in Sect. 2.4) is of interest for applications to magnetoresistive recording heads and sensors and has stimulated much of the activity in this field. The magnetoresistance and the applied field required to obtain ferromagnetic alignment of the magnetic layers can be varied widely by varying the material of the ferromagnetic layers, the interlayer material, and the interlayer thickness.

For many combinations of magnetic layer materials and nonmagnetic interlayer materials, the exchange coupling has been found to oscillate from ferromagnetic to antiferromagnetic with changing nonmagnetic layer thickness [2.149, 153]. In the case of Gd/Y multilayers [2.149], the localized rare earth magnetic moments allow the long range oscillatory exchange coupling to be adequately explained [2.154] by the Ruderman–Kittel–Kasuya–Yosida (RKKY) [2.155] interaction. When the magnetic layer is a transition metal, the less localized nature of the magnetic moments complicates the theory (Sect. 2.1).

There are many different theoretical approaches including RKKY-like models between planes of local moments [2.156–160], free electron-like particle-in-box models [2.161, 162], tight-binding models with magnetic interaction [2.163–165], and self-consistent electronic structure calculations [2.166]. There is considerable ongoing experimental effort to elucidate the nature of the magnetic coupling. The strength of the coupling can be measured by ferromagnetic resonance (FMR), Brillouin light scattering (BLS), and, in the antiferromagnetic coupling region, by the surface magneto-optic Kerr effect (SMOKE) (Sects. 3.1, 3.2, and 4.1 respectively). In favorable circumstances, oscillations in the exchange coupling over a limited range have been observed with these techniques.

Scanning Electron Microscopy with Polarization Analysis (SEMPA) has proved particularly well-suited to determining the period (or periods) of oscillation of the exchange coupling between magnetic layers. SEMPA images the magnetization of the top few layers of a magnetic material by measuring the spin polarization of secondary electrons excited from the material by the electron beam in a scanning electron microscope (SEM) as described in Sect. 2.3.1. For comparison to theory, measurements should be made of materials which are perfect crystals with sharp interfaces between the layers. This challenge to achieve perfection is very nearly met by growing epitaxial Cr films on Fe(100) single crystal whiskers, as described in Sect. 2.3.2a. The period of the oscillation in exchange coupling as a function of interlayer thickness can be determined accurately only if the interlayer film thickness can be varied nearly continuously in a reproducible manner and accurately measured. This is achieved by using spatially resolved reflection high energy electron diffraction (RHEED) to measure each atomic layer increment in the thickness of the interlayer material which is grown so the thickness varies linearly with distance along the substrate like a wedge. The observation of two different oscillation periods in the exchange coupling of Fe/Cr/Fe(100), their sensitivity to interlayer growth conditions, and the connection of the periods to Fermi surface nesting vectors are discussed in Sect. 2.3.2b. Additionally, the agreement between the two oscillation periods measured for a Ag spacer layer and the predictions based on the Ag Fermi surface are briefly discussed. Another coupling, biquadratic exchange coupling, is discussed in Sect. 2.3.2c and a brief summary is given in Sect. 2.3.2d.

2.3.1. The SEMPA Technique

2.3.1.1 Principle

The SEMPA technique (Scanning Electron Microscopy with Polarization Analysis) relies on the fact that the secondary electrons emitted from a ferromagnet have a spin polarization which reflects the net spin density in the material which in turn is related to the magnetization. The spin part of the magnetization is

$$M = - \mu_B(n_\uparrow - n_\downarrow), \tag{2.76}$$

where $n_\uparrow(n_\downarrow)$ are the number of spins per unit volume parallel (antiparallel) to the particular direction. Recall that the electron spin magnetic moment $\boldsymbol{\mu}$ (units of Bohr magneton, μ_B) and the electron spin s (units of $\hbar/2$) point in opposite directions $\boldsymbol{\mu} = - \mu_B s$. The spin part of the magnetization is a close approximation to the total magnetization in a transition metal ferromagnet in which the orbital moment is quenched by the cubic crystal field.

The spin polarization of the secondary electrons emitted from a ferromagnetic sample is a vector quantity. For the purposes of SEMPA it is adequate to consider each component of the polarization separately. The polarization along, for example, the z direction is

$$P_z = (N_\uparrow - N_\downarrow)/(N_\uparrow + N_\downarrow), \tag{2.77}$$

where $N_\uparrow(N_\downarrow)$ are the number of electrons with spins parallel (antiparallel) to the z direction. The polarization may have values $-1 \leq P \leq 1$.

The first measurements [2.167] of the energy distribution of the spin polarization of secondary electrons emitted from a ferromagnet were made on the ferromagnetic glass $Fe_{81.5}B_{14.5}Si_4$ and are shown in Fig. 2.35. The lower part of the figure shows the number N(E) of secondary electrons as a function of kinetic energy. The familiar low energy peak in the secondary electron intensity distribution is due to the secondary electron cascade process which excites many electron-hole pairs in the valence band. To the extent that the cascade electrons represent a uniform excitation of electrons from the valence band, the expected polarization is

$$P = n_B/n_v, \tag{2.78}$$

where n_v is the total number of valence electrons per atom. The number of Bohr magnetons per atom, n_B, is the difference in the number of majority (spin antiparallel to the magnetization) and minority spins per atom. This simple model predicts polarizations of 28%, 19%, and 5% for Fe, Co, and Ni, respectively. These values agree reasonably well with polarizations measured for kinetic energies between 10 and 20 eV for Fe [2.168], Co [2.168], and Ni [2.169].

The measured secondary electron polarization is not constant as a function of kinetic energy and increases below about 10 eV as seen in the top part of Fig. 2.35. From the measurements [2.168, 169] of Fe, Co, and Ni it is known that the increase at low energy is an enhancement over the predicted values, n_B/n_v. The enhancement is due to the spin dependent filtering of the low energy electrons [2.170, 171]. As the kinetic energy becomes smaller, the probability that the electron will lose energy and drop down into an unoccupied state below the vacuum level increases. In a ferromagnet, there are more minority spin unoccupied states so minority spin electrons are preferentially removed from the secondary electron distribution thereby making the polarization more positive.

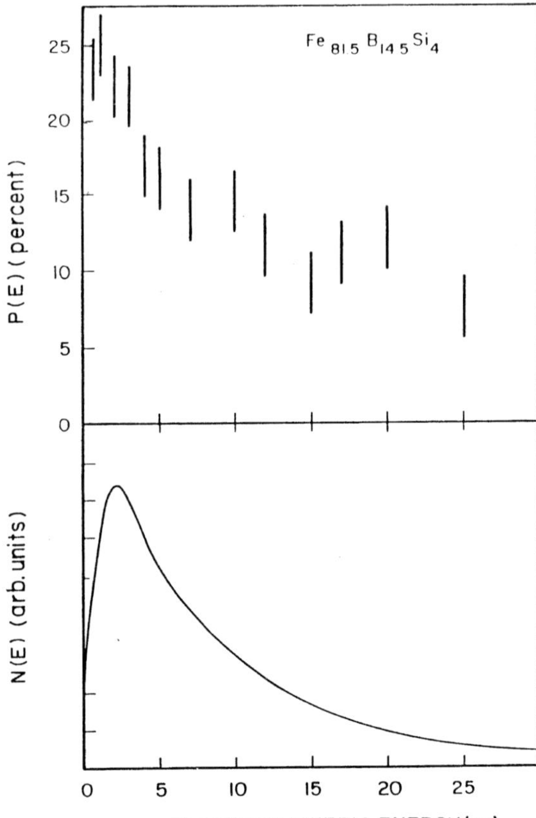

Fig. 2.35. Bottom: The energy distribution of secondary electrons emitted from the amorphous ferromagnetic Fe$_{81.5}$B$_{14.5}$Si$_4$. Top: The spin polarization of the secondary electrons as a function of kinetic energy. (From [2.167])

If the spin density which gives rise to M in (2.76) is uniformly sampled in the secondary electron distribution and the electrons are emitted without changing their polarization, then P is related to M as in (2.78). In practice, the constant of proportionality is not known precisely and will depend on the electron states sampled and on the range of kinetic energies measured, as indicated by Fig. 2.35. It may thus be different for two materials that have spin densities distributed differently in energy that therefore contribute differently to the polarization of the secondary electron distribution. Thus, the polarization of secondary electrons can be taken to be proportional to the magnetization, but it is in general not possible to deduce quantitative values for the surface magnetization or the corresponding surface magnetic moments. In other polarized electron spectroscopies, when the electrons excited to specific states are selected in an energy and angle-resolved measurement the proportionality of P and M may be even more complex (Volume I, Sects. 2.1 and 2.2).

Another important feature of the secondary electrons is that they are emitted from a region near the surface of the material. The energy of the incident electron beam is deposited in the sample to a significant depth, of the order of hundreds of nanometers for incident energies of a few keV[2.172]. However,

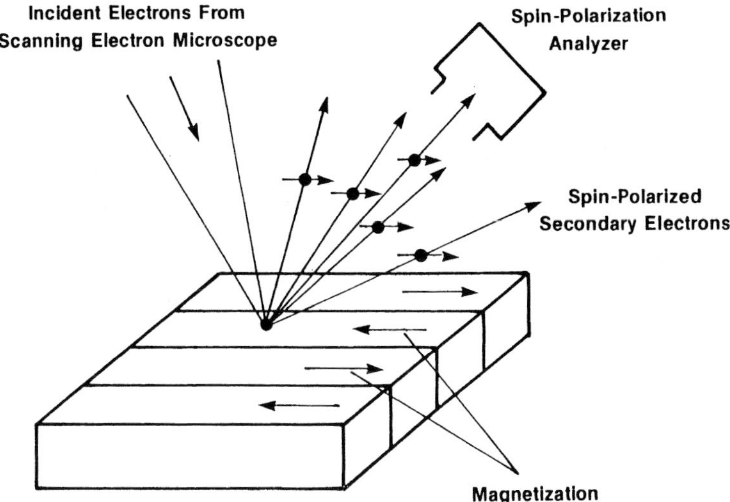

Fig. 2.36. The principle of Scanning Electron Microscopy with Polarization Analysis (SEMPA): The spin polarization of the secondary electrons generated by the scanned electron beam is measured to give an image of the specimen magnetization

only those electrons that are close enough to the surface can escape before losing sufficient energy that they fall below the vacuum level. For spin polarized secondary electrons, the $1/e$ sampling depth, or average attenuation length, is of order 1 nm, ranging from about 0.5 nm for a transition metal like Cr [2.173] to about 1.5 nm for a noble metal like Ag [2.174].

The measurements [2.167] of the secondary electron spin polarization shown in Fig. 2.35 provided a significant stimulus to the development of SEMPA. The results showed that there were many low energy electrons and that they have a sizeable polarization. Furthermore, it was clear that the spin polarization of the secondary electrons was closely related to the net spin density, and hence the magnetization, near the surface of the material. Therefore, it was suggested [2.167, 175] that by exciting the electrons with a well focused beam in a scanning electron microscope (SEM) one could obtain a spatially resolved magnetization measurement at the sample. This technique [2.176, 177], now called SEMPA, is illustrated schematically in Fig. 2.36. In the SEM, the incident beam is rastered across the sample. The spin polarization analyzer measures the polarization of the secondary electrons and the number of secondary electrons simultaneously to give a magnetization image and a topographic image of the sample.

2.3.1.2 Apparatus

The essential elements of a SEMPA apparatus are 1) a SEM column to form the focused incident electron beam, 2) an ultrahigh vacuum chamber with instrumentation for surface preparation and analysis, 3) electron spin polarization analyzers, 4) transport electron optics to collect and transport the emitted secondary electrons from the sample to the spin analyzer, and 5) a data storage and image processing system to transform raw data to magnetization images. The SEMPA apparatus [2.178, 179] used in obtaining the results described here is shown schematically in Fig. 2.37.

The choice of electron microscope and specimen chamber are constrained by the surface sensitivity of SEMPA which requires an ultrahigh vacuum surface analysis environment. The vacuum must be good enough so that background gases do not adsorb significantly on the sample surface and diminish the magnetic contrast. Conventional surface science preparation techniques like ion

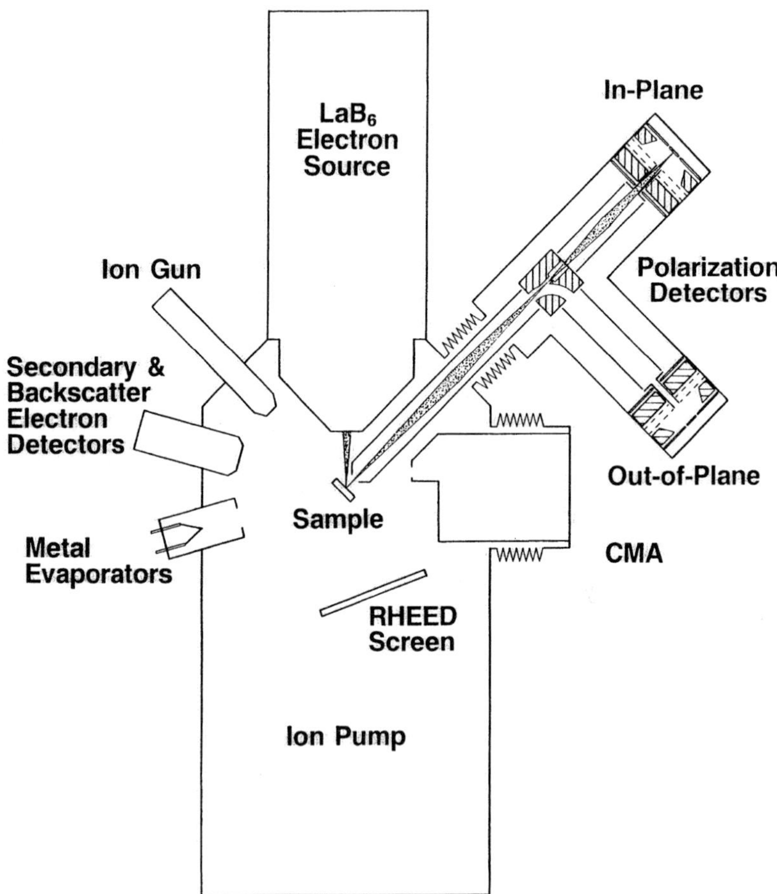

Fig. 2.37. Schematic drawing of the NIST SEMPA apparatus. The cylindrical mirror analyzer (CMA) and spin polarization analyzers are mounted on bellows. The ion gun, secondary and backscatter electron detectors, and metal evaporators are not shown in their true position

bombardment and annealing, and analysis techniques like Auger spectroscopy, to measure sample surface chemical composition, are desirable. Such requirements are met most conveniently by commercial scanning Auger microprobes to which SEMPA can be added. In our apparatus, shown in Fig. 2.37, the cylindrical mirror analyzer (CMA) used for Auger spectroscopy can be retracted to allow insertion of the SEMPA electron collection and transport optics close to the sample. The evaporation sources, which have been added, are shown schematically in Fig. 2.37 but actually are positioned near the Auger analyzer so that either SEMPA or Auger spectroscopy can be carried out during evaporation. A phosphor screen below the sample stage allows reflection high energy electron diffraction (RHEED) measurements to be carried out when the sample is tilted so the SEM beam is at grazing incidence.

Owing to the inefficiency of existing spin polarization analyzers, it is important to have a high brightness cathode in the SEM electron gun. In our apparatus, SEMPA images can be obtained in a reasonable time with a minimum beam current of 1 nA. Our apparatus has a LaB_6 thermionic emission cathode which produces the required current in a beam diameter of 50 nm. For future higher resolution measurements, an apparatus with a thermally assisted field emitter that produces 1 nA in a beam of 10 nm diameter will be used. The choice of incident electron beam energy is a tradeoff between resolution and signal intensity. With increasing beam energy, a smaller beam diameter is possible but the secondary yield decreases. Under our typical operating conditions, a 10 keV incident electron beam is used which produces a secondary yield that is approximately 20% of the incident beam current. Another tradeoff is between a shorter working distance (between the sample and the SEM objective lens) which leads to a smaller beam diameter and a longer distance which leads to a smaller magnetic field (from the objective lens) at the sample. Some minimum working distance is also necessary to extract the polarized secondary electrons. A reasonable compromise is a working distance of 10 mm and a stray field at the sample of 1 Oe ($80\ \mathrm{A\,m^{-1}}$) or less.

Transport optics must efficiently collect the spin polarized secondary electrons emitted from the sample surface and deliver them at the correct energy to the spin polarization analyzer. Additionally, the transport optics should map the spot under the scanned electron beam on the sample surface, which determines the source of polarized electrons, onto the scattering target of the spin analyzer in such a way as to minimize spurious instrumental asymmetries in the analyzer. In our apparatus, the front end of the transport optics is biased at ± 1500 V relative to the sample to collect the low energy secondary electrons. The transport optics are designed to transmit the secondary electrons with initial kinetic energies between 0 and 8 eV to the spin analyzer. The transport optics also contain descan deflectors which prevent the motion of the SEM beam on the sample, i.e. the moving secondary electron source, from being transmitted to the spin analyzer which could introduce instrumental asymmetries for large area scans. As seen in Fig. 2.37, the transport optics also contain a 90° spherical deflector to shift the beam to an orthogonal spin analyzer. Each spin analyzer can measure the two components of polarization transverse to the electron beam. Out-of-plane magnetization at the sample leads to a component of polarization along the electron beam. The spherical deflector changes the electron momentum without (at these energies) changing the electron spin direction so the orthogonal analyzer measures the component of polarization out-of-plane as well as one in-plane component. The redundant measurement of the in-plane component by both spin analyzers provides a monitor of the stability of the polarization sensitivity of each.

The spin analyzer is a key element of the SEMPA apparatus. The characteristics of the wide range of available spin analyzers have been compared [2.180, 181] and the application of several of them to SEMPA has been discussed [2.178]. A number of features are desirable in a spin analyzer for a SEMPA measurement: 1) The highest possible efficiency is required. 2) The electron optical phase space of the spin analyzer must be appropriately matched to the phase space of the secondary electrons to be analyzed. 3) The spin analyzer should accept a range of secondary electron energies to take advantage of the high polarization, high intensity peak in the secondary electron distribution (Fig. 2.35). 4) The spin analyzer should have maximum immunity to false apparatus asymmetries resulting from changes in position and angle of the electron beam entering the analyzer. 5) The spin analyzer should minimally perturb the operation of the SEM. 6) The SEMPA measurement is sufficiently demanding that the spin analyzer must be reliable and easy to use.

The spin-orbit interaction is the spin dependent interaction which is the basis of the spin sensitivity in the most commonly used spin analyzers [2.182]. When an electron scatters from the central potential of a high atomic number atom, there is an interaction between the electron's spin and its orbital angular momentum about the scattering center. This spin-orbit interaction introduces a difference in the scattering cross sections depending on whether the spin of the electron is parallel or antiparallel to the scattering plane normal. The normal to the scattering plane, n, is defined in terms of the incident and final electron wavevectors, k_i and k_f, which lie in the scattering plane as, $n = (k_i \times k_f)/|k_i \times k_f|$. The scattering cross section can be written

$$\sigma(\theta) = I(\theta)[1 + S(\theta)P \cdot n], \tag{2.79}$$

where P is the beam polarization, $I(\theta)$ is the angular distribution of back-scattered electrons, and $S(\theta)$ is the Sherman function for the spin analyzer scattering target [2.182]. The Sherman function is a measure of the strength of the spin-dependent scattering which depends on the target material and the beam energy and angle. To measure the polarization of the beam, one measures the asymmetry A_s in the number of electrons scattered to the left, N_L, and to the right, N_R, relative to the incident beam direction:

$$A_s = (N_L - N_R)/(N_L + N_R) = PS. \tag{2.80}$$

Here S is the integrated Sherman function for the range of angles collected.

For a measurement limited by counting statistics, the efficiency of a spin analyzer is described by the figure of merit

$$F_s = S^2 I/I_0, \tag{2.81}$$

where I_0 is the beam intensity incident on the spin analyzer scattering target. For typical analyzers [2.180, 181], S lies between 0.1 and 0.3, and I/I_0 is in the range 10^{-2}–10^{-4}. The figure of merit is of the order of 10^{-4} for the most efficient spin analyzers. The uncertainty δP in the polarization measurement [2.182] of a beam of N electrons is $\delta P = (NF_s)^{-1/2}$. Compared to an intensity measurement, where the relative uncertainty is $\delta N/N = N^{-1/2}$, the relative uncertainty in the polarization measurement is $\delta P/P = (P^2 N F_s)^{-1/2}$. Thus, compared to an intensity measurement, a polarization measurement with the same statistical precision will take $P^{-2} F_s^{-1}$, or at least 10^4 times, as long to make. Hence there is considerable emphasis on the efficiency of the spin analyzers in a SEMPA measurement.

No single analyzer fully meets all of the other requirements on spin analyzers in addition to efficiency. For example, the high energy (100 keV) Mott analyzer [2.182] has the highest electron optical acceptance, but because of the high voltage requirement, it is much larger and more cumbersome than low energy spin analyzers. Nevertheless, the first SEMPA measurements [2.176, 183] were made by attaching a high quality scanning electron gun to a 100 keV Mott

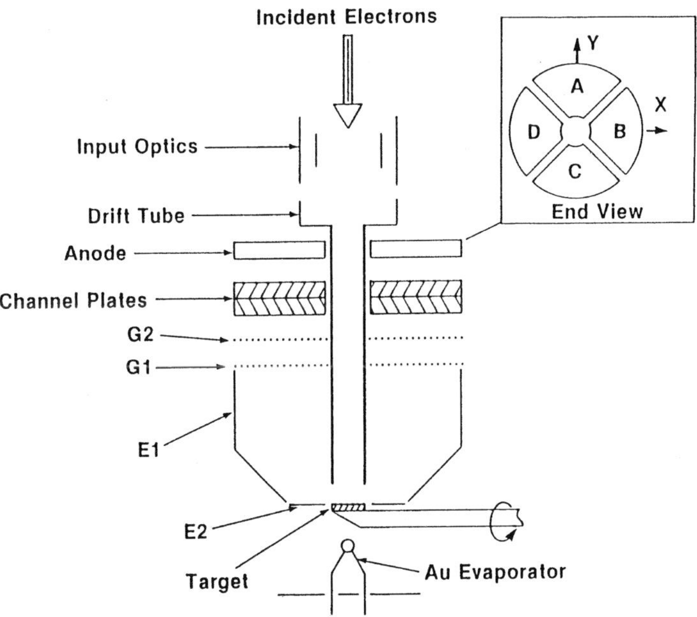

Fig. 2.38. A cross section of the low energy diffuse scattering spin analyzer used for SEMPA. The divided anode assembly is shown in the inset as viewed from the Au target

analyzer. Our approach was to develop a high efficiency, compact low energy spin analyzer which could be attached directly to an existing SEM [2.177, 180, 184]. In addition to the apparatus described here, SEMPA apparatuses currently in operation include those using [2.176, 185] high energy Mott analyzers and one using [2.186] a low energy electron diffraction (LEED) analyzer. The specifications and relative merits of the various analyzers for SEMPA have been previously discussed [2.178, 180]. Recently a high figure of merit spin analyzer based on exchange scattering has been introduced [2.187]; however, when account is taken of its acceptance phase space and limitations on the range of energies accepted by it, the overall efficiency for SEMPA is only marginally higher than the best of the other analyzers discussed. Moreover, at this stage of its development it requires a significantly greater effort to use. Here we will focus our further discussion on the low energy diffuse scattering analyzers shown in Fig. 2.37.

A schematic of the low energy diffuse scattering analyzer [2.180, 188] is shown in Fig. 2.38. In this detector a 150 eV beam is incident on an evaporated polycrystalline Au target. The electrons, which are diffusely scattered by the Au target, are deflected by the electrode E1 such that their trajectories are approximately normal to grids G1 and G2, which filter out the low energy secondaries generated at the Au target. The electrons passing the grids are multiplied by the channel plates and are detected by the anode which is divided into four equal quadrants as shown in the inset of Fig. 2.38. Two orthogonal components of the transverse polarization are measured simultaneously:

$$P_x = (1/S)(N_C - N_A)/(N_C + N_A), \tag{2.82a}$$

$$P_y = (1/S)(N_B - N_D)/(N_B + N_D), \tag{2.82b}$$

where N_i is the number of electrons counted by quadrant i. The spin analyzer is very efficient having a figure of merit of 2×10^{-4} which does not change over the 8 eV energy width of the electron beam [2.180, 188]. The electron optical acceptance is well-matched to the phase space of the electrons to be measured. The spin analyzer is reliable, compact and readily interfaced to the SEM electron optical column without destabilizing the column itself.

A common problem with any spin analyzer is the elimination of any false polarization signals due to instrumental asymmetries. Instrumental asymmetries that remain constant, such as due to a mechanical misalignment or due to a difference in electronic gain in a signal channel, can be measured and accounted for. More troubling are asymmetries resulting from actual movement of the incident beam in position or angle at the spin analyzer target. The electron trajectories at the spin analyzer can vary because of the scanning of the SEM beam on the sample, because of variations in the extraction electric field at the sample owing to the sample geometry, and because of variation in topography, work function, or stray magnetic field in the extraction region. Asymmetries due to the SEM beam scanning on the sample are minimized by the descan in the transport optics. Also, it is possible to compensate the asymmetry caused by a displacement of the electron beam at the Au target by causing that beam displacement to be accompanied by a change in angle that causes an equal and opposite asymmetry [2.188]. The input optics are designed to do just this. By these means, instrumental asymmetries are reduced to a negligible level for most SEMPA measurements. The beam polarization as defined in (2.77) contains the simultaneously measured beam intensity and thus should be insensitive to variations in intensity caused by topographic variations. In rare cases of a particularly rough sample or accentuated geometry, the trajectories are so perturbed that the topography may be visible in a polarization image. In such cases, provision has been made in the spin analyzer of Fig. 2.38 for the electron beam to scatter from a low atomic number graphite target which can be rotated into position in place of the Au target. The difference between the image taken using the graphite target and the Au target is then the true polarization.

2.3.1.3 Examples of Magnetization Images

Each spin analyzer determines the secondary electron polarization and intensity at each point as the SEM beam rasters the sample. The intensity measurement gives the topographic map as in a conventional SEM. The polarization measurement, (2.82), gives two components of the polarization. Multiplying the polarization map by -1, one obtains an image of the magnetization in direction and relative magnitude. The SEMPA image processing system has to be able to perform image processing tasks such as data storage and display, filtering and background subtraction, line scans, and other standard tasks like image rotation and expansion. Additionally, because magnetization is a vector quantity, the image processing system has to combine the components of magnetization in various ways to best display the magnetization vector field. The simplest display is of each magnetization component individually as in Fig. 2.39a and b where M_x and M_y of a region on the $(0\,0\,1)$ surface of a Fe single crystal whisker are shown. The gray scale gives the direction for each component. Thus, in Fig. 2.39a white indicates magnetization to the right and black to the left. In the M_y image of Fig. 2.39b white indicates magnetization in the upward direction. For an Fe$(0\,0\,1)$ surface there are two in plane easy axes of magnetization along $[1\,0\,0]$ type directions. The intermediate gray levels in the diamond at the center of the M_y image of Fig. 2.39b correspond to magnetization in the x direction. The difference in the gray levels shows that the magnetization axes are not exactly aligned with the axes of the spin analyzer.

The image of Fig. 2.39 is a relatively large area, low magnification (250×) scan of the Fe whisker. Typically, such an image would be acquired with a 60 nA beam that is 250 nm in diameter. The image is 256 × 192 pixels and was acquired in 4 min. The time for image acquisition varies widely depending on the purpose, that is, on the pixel density and statistics required. When selecting the desired area on the sample to image, a 64 × 48 pixel scan in 3 s makes it easy to position the sample and scan area. A typical image in routine work would be 256 × 192

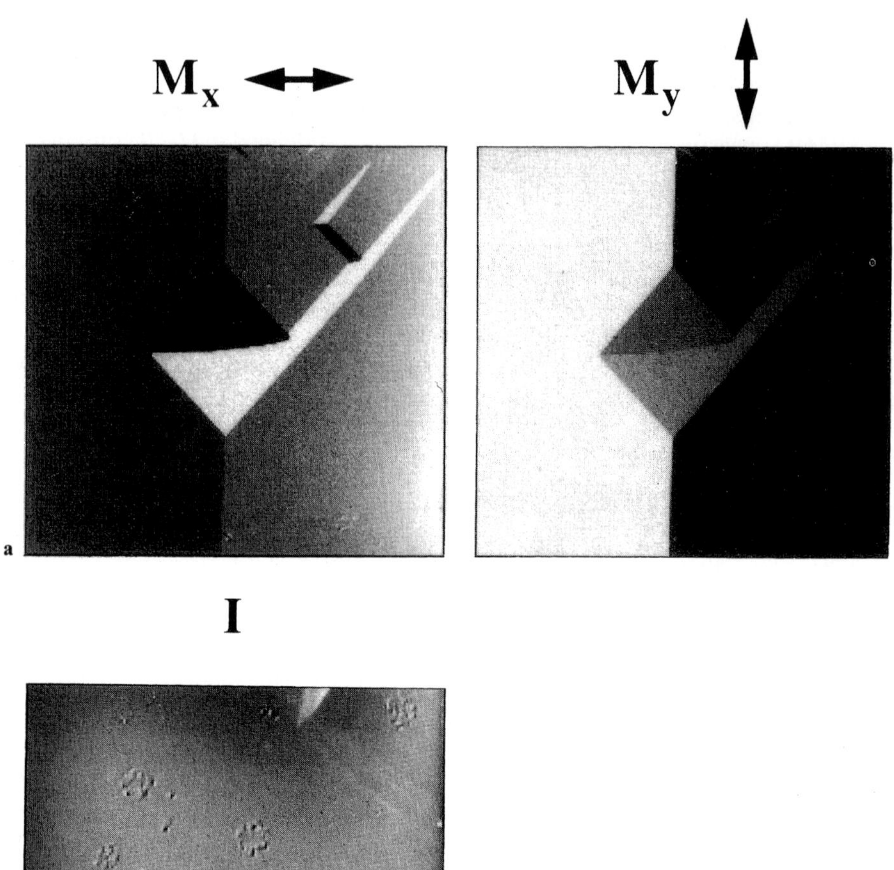

Fig. 2.39. The x and y components of the magnetization for an Fe single crystal whisker are shown for in (**a**) and (**b**) respectively. The secondary electron intensity image is shown in (**c**)

pixels and would take about 10 min. For samples with low magnetization contrast and where highest pixel resolution is required, 512×384 pixel scans lasting up to 100 min have been made. Over this long time period some drift in position, for example due to the sample stage, may be observed in high magnification images.

In a conventional SEM image, the intensity of the secondary electrons gives an image which reflects the topography of the sample. The spin analyzer directly measures the intensity which is the sum of the current to opposing quadrants, the denominators in (2.82a, b). The intensity image corresponding to M_x is shown in Fig. 2.39c; the intensity image measured with M_y is identical. Because the intensity is measured simultaneously with the polarization but independent of it, the magnetization and the topographic images can be separated allowing

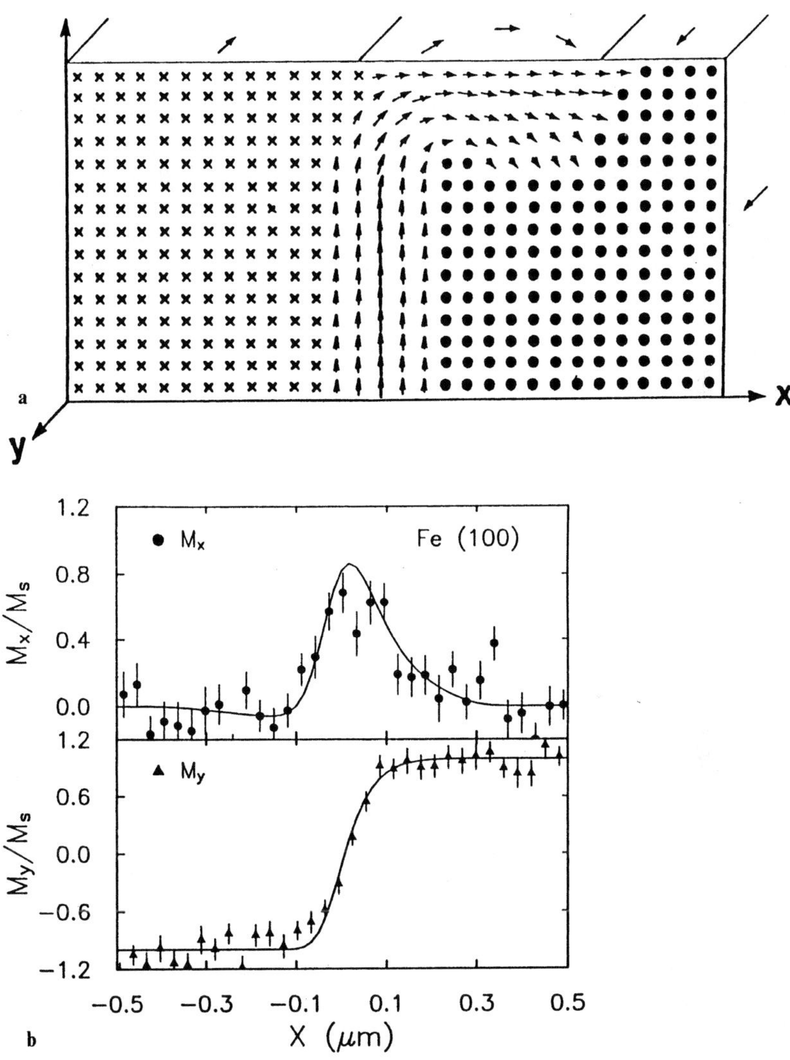

Fig. 2.40. (a) Schematic representation of calculated magnetization distribution in a cross section through the upper part of an Fe sample. The domain wall is a Bloch wall in the solid, but rotates into a Néel wall at the surface (from [2.179]). (b) Measurements of the x and y components of magnetization relative to the saturation magnetization M_s are shown in the upper and lower panels respectively versus distance in the x direction. The data were acquired with a beam current of 0.3 nA. The solid line is the result of micromagnetic theory convolved with the measured beam current (70 nm FWHM) density distribution. (From [2.191])

the investigation of the influence of the topography on the magnetic structure. For example, SEMPA has been used to observe the pinning of domain walls by point defects in a magnetic sample [2.189].

It is sometimes useful to present the magnetization information in other ways. For example, the angle between the x axis and the direction of the magnetization in the surface plane is

$$\Theta_{\text{in-plane}} = \tan^{-1}(M_y/M_x). \tag{2.83}$$

The direction of magnetization can be displayed in a color image where, through a color wheel, each color is associated with a direction. For the Fe whisker shown in Fig. 2.39 there is no out-of-plane magnetization owing to the cost in magnetostatic energy. An example of a material where SEMPA has been used to measure all three components of the magnetization is Co where there is a strong uniaxial anisotropy perpendicular to the Co(0 0 0 1) surface [2.190]. The angle of the magnetization relative to the surface is

$$\Theta_{\text{out-of-plane}} = \tan^{-1}[M_z/(M_x^2 + M_y^2)^{1/2}]. \tag{2.84}$$

A useful check on the data is to form the quantity

$$|M| = (M_x^2 + M_y^2 + M_z^2)^{1/2}, \tag{2.85}$$

which should be a constant. This is indeed the case, although depending on the beam diameter there may be some "missing magnetization" at a domain wall. This is just an artifact which arises when the beam diameter is greater than the wall width and the oppositely directed polarization measured on each side of the wall adds to zero.

When domain walls are examined at high resolution, we obtain [2.191] results like those in Fig. 2.40 for a 180° wall in Fe. In order to understand the measurements, we first show the results of a magnetic microstructure calculation. Figure 2.40a shows a schematic representation of the calculated magnetization distribution in the upper 0.2 μm of an Fe sample. The sample has two domains with magnetization in the $+y$ and $-y$ directions. The cross section in the x–z plane shows how the Bloch wall separating the domains is perpendicular to the surface inside the sample but turns over into a Néel wall at the surface. Line profiles of the relative magnetization M_x/M_s and M_y/M_s along the x direction are shown in Fig. 2.40b. Note the asymmetry of the surface Néel wall. The calculation, broadened to account for the electron beam diameter, shown by the solid line is seen to be in good agreement with the experimental results. When required, SEMPA has very high resolution capability even to the point of investigating domain wall structure.

2.3.1.4 Summary of SEMPA Features and Comparison to Other Imaging Techniques

Several features of SEMPA make it particularly suited to the investigation of exchange coupled layers. The features of SEMPA will be summarized and compared to other techniques used to investigate magnetic microstructure. SEMPA directly measures the direction and relative magnitude of the magnetization *vector*. Most methods used for observing magnetic microstructure are sensitive to the magnetic fields associated with ferromagnetic materials. For example, decoration of domain walls with fine magnetic particles as in the *Bitter* method rely on fringing fields at domain walls [2.192]. In Lorentz microscopy, either in transmission or reflection, magnetic contrast comes from the deflection of the electron beam by the magnetic induction inside the material or emanating from it [2.172, 193]. Only the magneto-optic Kerr effect also directly measures the relative magnetization of the sample by determining the change in the polarization of light upon reflection [2.194, 195].

SEMPA has very high spatial resolution, 50 nm in the instrument used in this work and higher in instruments employing SEMs with field emission cathodes [2.185, 186]. Only transmission Lorentz microscopy has higher spatial resolution, ≤ 10 nm, but the sample must be thinned to less than about 300 nm thick which can change the magnetization distribution. Differential phase contrast microscopy [2.196] and electron holography [2.197] are variations on transmission Lorentz microscopy which have the similarity that their response is to the magnetic flux density integrated over the thickness of the sample. As is demonstrated in Sect. 2.32, the ability of SEMPA to observe the magnetization in thin film structures on an extremely high quality substrate, like the Fe single crystal whisker, is crucial to the investigation.

The surface sensitivity of SEMPA, which is approximately 1 nm and can be an obstacle to overcome in some investigations, is a very useful feature for investigating exchange coupled layers. As will be seen, it is possible to observe the magnetization of the top layer of an Fe/Cr/Fe(0 0 1) sandwich and determine its coupling to the substrate without any interfering signal from the substrate. Furthermore, the sample preparation and measurement can be done *in situ*. In contrast, Kerr microscopy has a probing depth of the order of 10 nm and a contribution to the signal from the substrate cannot be avoided. Also, samples must be coated with a protective layer and removed from the preparation chamber to the optical microscope for optimum resolution (200 nm) Kerr microscopy. A disadvantage of SEMPA is its sensitivity to magnetic fields which makes it impossible to apply a field to obtain a measure of the strength of the coupling as can be done in a magneto-optic Kerr measurement (*Bader* and *Erskine*, Chap. 4).

Another significant advantage of SEMPA is that it is possible to separate magnetic from topographic contrast which can be a source of confusion in other imaging methods. In fact, one can look for correlations that would indicate an influence of the topography on the magnetization. Because the same incident electron beam can also be used for scanning Auger microscopy, it is possible to investigate the relationship between particular chemical features on the surface and the magnetization.

2.3.2 SEMPA Measurements of Exchange Coupled Multilayers

As mentioned in the introduction, multilayer structures of magnetic layers separated by nonmagnetic interlayers exhibit many interesting properties including a "giant" magnetoresistance and long-range oscillatory coupling. Many questions about the mechanism of the exchange coupling, its range, and its strength remain unanswered. In this work we focus primarily on the mechanism of the interaction. Is the pronounced periodicity of the *magnetic* coupling as a function of the interlayer thickness related to the *electronic* structure, for example, to the Fermi surface of the interlayer material? To answer this question we have prepared very high quality "sandwich" structures, in the first instance of Fe/Cr/Fe, and determined the periods of oscillation of the magnetic exchange coupling. A bilayer of magnetic and nonmagnetic material is the basic building block for a multilayered structure that could include hundreds of layers. The three layer sandwich is the limiting case of a multilayer. It has been shown that the exchange coupling strength is independent of the number of bilayers and is the same for superlattice and sandwich structures [2.198].

2.3.2.1 The Fe/Cr/Fe (0 0 1) Sample

Theoretical investigations of magnetic coupling deal with perfect crystals of magnetic and nonmagnetic materials with a sharp interface between them. In making a comparison to theory, it is important that experimental artifacts like imperfect crystallinity, defects, interface roughness, and interdiffusion, be minimized. We describe here in some detail our attempts to grow Fe/Cr/Fe sandwiches which approach the idealized structures of theory. The single crystal Fe whisker substrate, the geometry and growth of the Cr interlayer, and the characterization of the growth and determination of the thickness are all important aspects of the sample preparation for investigation of the magnetic exchange coupling using SEMPA.

The Fe single crystal whiskers have (1 0 0) faces and approximately square cross sections. They are typically several hundred μm in width and a centimeter or two in length. Iron whiskers are extremely high-quality crystals with a very low dislocation density [2.199]. RHEED patterns show an arc of spots expected from a perfect crystal and seen only in measurements of other high-quality crystal faces such as cleaved GaAs and high temperature annealed Si(1 1 1). The whisker surfaces are naturally flat from growth. Scanning tunneling microscopy measurements of Fe(1 0 0) whisker surfaces show that there is a distance of about 1 μm between each single-atom-high step [2.200]. This corresponds to an alignment of the surface to the (1 0 0) plane to better than 0.01° which cannot be achieved on a metal surface by mechanical polishing. Furthermore, the surface is strain free. The absence of strain is also apparent in the magnetic microstructure. Strains in the whisker, for example, near an end of the whisker that is clamped, show up as very irregular domain patterns. The SEMPA measurements of the exchange coupling are facilitated by having a region of the surface consisting of two oppositely directed domains running along the length of the whisker; a SEMPA measurement is used to select suitable whiskers.

The lattice constant of Cr, $a_{Cr} = 0.2885$ nm, is well-matched to that of Fe, $a_{Fe} = 0.2866$ nm, and would be expected to grow in registry with the Fe substrate with a small contraction of the Cr lattice in the plane of the film of somewhat less than 0.7%. Both are body-centered cubic crystals at room temperature. The desired mode of film growth is the Frank–van der Merwe mode in which each layer is completed before the next starts, that is, layer-by-layer growth. In a quasi-equilibrium situation where the temperature is high enough that the deposited atoms can diffuse to low energy sites, the following relation must hold:

$$\gamma_{Cr} + \gamma_i + \gamma_e - \gamma_{Fe} \leq 0, \tag{2.86}$$

where γ_{Cr} and γ_{Fe} are the surface free energies of the film and the substrate, γ_i is the interface free energy, and γ_e is the strain energy [2.201]. For first layer growth, γ_i is the interface free energy for the Fe–Cr interface and is small and negative [2.202], and for subsequent layers it goes to zero for the homoepitaxial growth of Cr on Cr [2.201]. Estimates of the surface free energies vary considerably [2.202, 203] but that of Cr is consistently less than that of Fe, indicating the possibility of layer-by-layer growth of Cr on Fe. The lattice mismatch is small and the strain energy, γ_e, is also small. The growth of Fe on Cr is not expected to be such a good example of layer-by-layer growth. However, some roughness in the final Fe film does not affect the investigation of the coupling as we show that the coupling period is not sensitive to the thickness of the final Fe overlayer.

A careful investigation of the exchange coupling as a function of interlayer thickness requires measuring many identical films differing only by small increments of thickness. The preparation of many films individually is tedious, and reproducibility in relative thickness and film quality is difficult to achieve. An alternative method which has proven highly successful [2.204–207] is to grow a film linearly increasing in thicknesses, that is in a wedge, such that all thicknesses in that range are accessible. A schematic of such a Cr wedge interlayer is shown in Fig. 2.41. We obtained the wedge-shaped Cr interlayer by moving a precision piezo-controlled shutter during the Cr evaporation. The wedge area was typically a few hundred µm wide by a few hundred µm long. Over this small region, it was easier to ensure sample homogeneity and quality than for a large sample. The ability of the SEM to measure such a small sample is one of its great advantages.

The first step in preparing the Fe/Cr/Fe(0 0 1) sample shown in Fig. 2.41 was to clean the Fe whisker [2.208] by 2 keV Ar ion bombardment at 750 °C. Following this initial cleaning, the surface could be recleaned by a brief ion bombardment at room temperature followed by a 800 °C anneal. Sputter damage is removed and a smooth, flat surface is recovered by annealing. After annealing, the principle contaminant, which was oxygen, was below about 0.05 monolayer (ML). The Cr was evaporated from a bead of Cr that had been

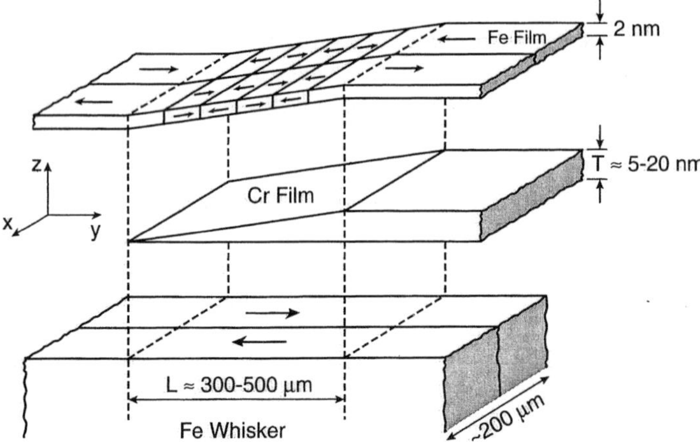

Fig. 2.41. A schematic expanded view of the sample structure showing the Fe(0 0 1) single-crystal whisker substrate, the evaporated Cr wedge, and the Fe overlayer. The arrows in the Fe show the magnetization direction in each domain. The z-scale is expanded approximately 5000 times. (From [2.206])

electroplated on a 0.25 mm diameter W wire [2.209]. Different evaporation rates were used, ranging from 1 to 12 ML per minute. These were achieved with filament currents of 7–9 A and produced minimal outgassing. An Auger spectrum of the Cr wedge just after evaporation revealed 0.01 ML of oxygen. There was no apparent correlation between the time the Cr was exposed to residual gas contaminants before being covered by the Fe overlayer and the behavior of the magnetic coupling. The Fe overlayer was electron beam evaporated from a pure Fe rod at rates of approximately 10 ML per minute to a thickness ranging from about 0.5 to 2 nm. The magnetization of the Fe overlayer could be monitored by SEMPA during evaporation allowing one to observe the appearance of the domain pattern due to the coupling through the Cr to the substrate. This domain pattern was independent of Fe overlayer thickness over the range tested (approximately 0.7–3 nm). Usually the Fe overlayer was evaporated onto the Cr layer at a substrate temperature of 50–100 °C.

The temperature of the Fe whisker substrate during evaporation of the Cr wedge is crucial to the quality of growth of the Cr film which in turn has a profound effect on the interlayer exchange coupling. The first indication of differences in the structural quality of films grown at two different temperatures was from observations of the RHEED patterns. The RHEED pattern of the clean Fe substrate exhibits an arc of spots as expected for an ideal crystal, and Kikuchi lines are also visible. If the Cr is evaporated on the Fe substrate at temperatures of 300–350 °C, the RHEED pattern remains as an arc of spots with an additional very slight streaking. On the other hand, if the substrate is in the neighborhood of 100 °C or below, during the Cr evaporation the sharp RHEED pattern changes to broad steaks with some indication of 3d growth. In a separate experiment, Scanning Tunneling Microscopy (STM) observations have confirmed [2.200] that the high temperature growth proceeds layer-by-layer, but the low temperature growth produces a growth front containing five to six layers.

Measurements of the RHEED intensity as a function of thickness also provide a means to determine the thickness very precisely. The dashed curve in Fig. 2.42 shows the intensity of the specular RHEED beam measured during deposition of the first 15 layers of Cr evaporated in the thick part of a Cr wedge with the Fe substrate at 350 °C. The electron beam was incident 3–4° from the surface and 2° off the [1 0 0] azimuth. The diffraction was near to the out-of-phase condition such that diffraction from one layer high Cr islands interfered destructively with the Fe substrate. When the Cr reaches a half layer coverage, the RHEED intensity is at a minimum and increases to a maximum as the layer fills in to completion. This process is repeated with each layer to give the cusp-like oscillations observed which are indicative of layer by layer growth [2.210]. In the SEM, it is also possible to measure the RHEED intensity after deposition by scanning the SEM beam along the Cr wedge. The solid curve in Fig. 2.42 shows the oscillations in RHEED intensity as the beam is scanned along the first part of the same Cr wedge sample which when measured during deposition gave the dashed curve. Note the similarity in the shape of the oscillations indicating that the growth at each instant in time is frozen at a point in space and revealed in the solid curve. The electron beam has to be scanned beyond the nominal zero thickness value to reach the bare substrate. This is due to the penumbra, the extent of which is known from the extended evaporation source and the distance of the shutter from the whisker. We use these spatial RHEED intensity oscillations to provide a precise measure of the thickness of the Cr film.

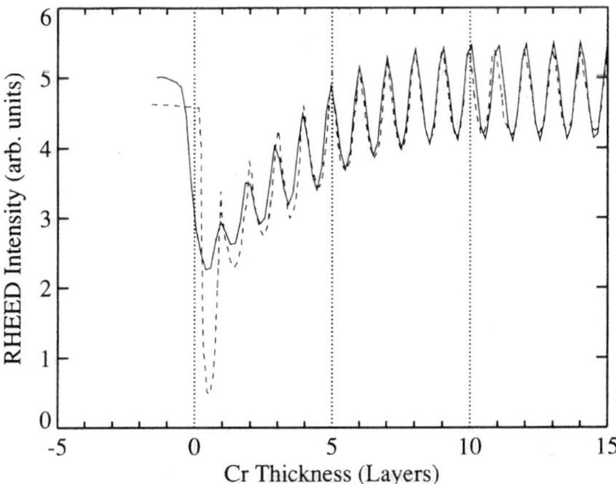

Fig. 2.42. RHEED intensity oscillations measured as a function of time during the growth of a Cr film on Fe(0 0 1) are shown by the dashed line. Spatial RHEED intensity oscillations measured as the electron beam is scanned along the wedge after deposition are shown by the solid line

2.3.2.2 SEMPA Observations of Interlayer Exchange Coupling

SEMPA measures very directly whether the Fe overlayer is ferromagnetically or antiferromagnetically coupled to the Fe whisker through a given thickness. First, the magnetization in the substrate is measured, and then, after the sandwich structure is grown, the magnetization in the Fe overlayer, separated by the Cr interlayer of varying thickness [2.206], is measured. As a first example, Fig. 2.43a shows the SEMPA magnetization image of the clean Fe whisker substrate. In this section we use the coordinates of Fig. 2.41. The region of interest on the whisker has two domains along the length of the whisker. The magnetization, in the upper domain is in the $+y$ direction (white) and that of the lower domain is in the $-y$ direction (black). The domains in the top Fe layer of a Fe/Cr/Fe(0 0 1) sandwich for a Cr wedge evaporated at a substrate temperature of 30 °C are displayed in the magnetization image shown in Fig. 2.43b. The coupling starts off ferromagnetic, that is the magnetization in the Fe overlayer is in the same direction as the substrate below. At a Cr thickness of about three layers, the coupling between the Fe layers changes from ferromagnetic to antiferromagnetic. This continues to reverse through several oscillations as the Cr interlayer thickness increases. The period or wavelength of the oscillations varies from 1.6 to 1.9 nm of Cr thickness, equivalent to a thickness of 11–13 Cr layers. Note that the scale on Fig. 2.43 is the thickness of the Cr interlayer which increases from zero at the left of Fig. 2.43b to 11 nm at the right over a distance of approximately 0.5 mm on the Fe whisker. In the region where the Cr interlayer is thicker the exchange coupling is less well defined as indicated by the irregular domains.

The sensitivity of the exchange coupling to the quality of the Cr interlayer is strikingly demonstrated in Fig. 2.43c which shows the domains in the Fe overlayer of an Fe/Cr/Fe(0 0 1) sandwich in the case where the Cr wedge was grown with the substrate at 350 °C. The coupling is initially ferromagnetic and switches to antiferromagnetic at five layers. However, in contrast to the coupling through a Cr wedge grown at lower temperature, the coupling through Cr grown at higher temperature is seen to change, after the initial ferromagnetic coupling region, with each layer of Cr giving a period of oscillation of nearly two layers. We say "nearly" because between 24 and 25, 44 and 45, and 64 and 65 layers, indicated by arrows at the top of the figure, no reversal takes place. This corresponds to a phase slip resulting from the accumulation of a phase difference owing to the incommensurability of the exchange coupling period and the lattice constant. This oscillatory exchange coupling continues through 75 layers (over 10 nm) of Cr.

The persistence of the short-period oscillations with increasing interlayer thickness is closely correlated with the interlayer roughness as indicated by RHEED intensity oscillations. Growth at lower temperatures, 250 °C for example, allows roughness to build up more in the Cr wedge such that in one instance the short-period oscillations were not observed beyond a thickness of about 30 ML. A particularly vivid example of the correlation between the

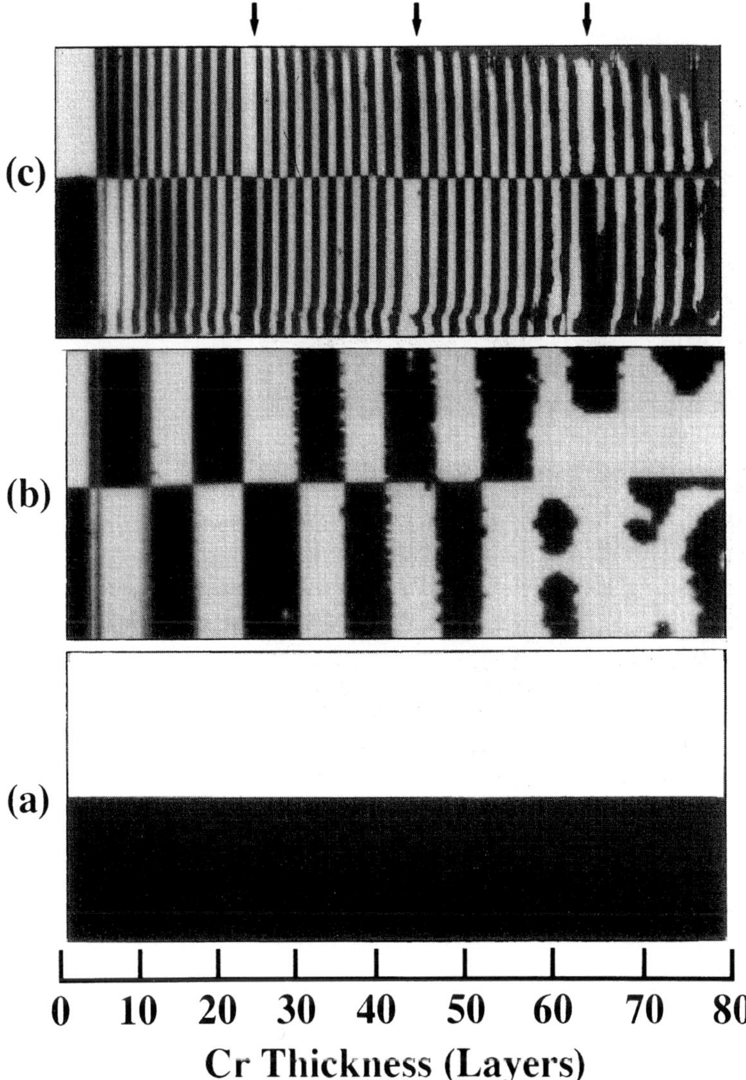

Fig. 2.43. SEMPA image of the magnetization M_y (axes as in Fig. 2.41) showing domains in (**a**) the clean Fe whisker, (**b**) the Fe layer covering the Cr spacer layer evaporated at 30 °C, and (**c**) the Fe layer covering a Cr spacer evaporated on the Fe whisker held at 350 °C. The scale at the bottom shows the increase in the thickness of the Cr wedge in (**b**) and (**c**). The arrows at the top of (**c**) indicate the Cr thicknesses where there are phase slips. The region of the whisker imaged is about 0.5 mm long

RHEED intensity oscillations and the short-period oscillations in the exchange coupling is seen in Fig. 2.44. In Fig. 2.44a an image of the RHEED intensity is shown as the SEM beam is rastered over a Cr wedge grown at 250 °C. We attribute the absence of RHEED oscillations in the lower right part of the image

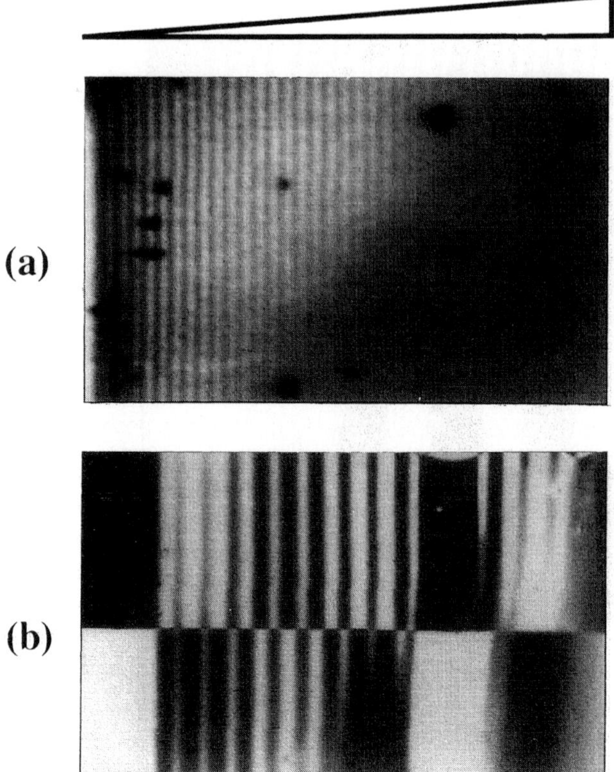

Fig. 2.44. The effect of roughness on the inertlayer exchange coupling is shown by a comparison of (**a**) the oscillations of the RHEED intensity along the bare Cr wedge with (**b**) the SEMPA magnetization image over the same part of the wedge

to roughness of the Cr interlayer, probably induced by damage in the Fe substrate which had been sputtered and annealed many times. However, the important point to note is that in the magnetization image of the Fe overlayer in Fig. 2.44b the coupling reverts to oscillatory coupling with a long-period exactly where the RHEED intensity oscillations are absent.

It is noteworthy that a calculation [2.156] of the exchange coupling in Fe/Cr/Fe predicted short-period oscillations, in addition to the long-period oscillations, before short-period oscillations had been observed experimentally. *Wang* et al. [2.156] pointed out that the apparent discrepancy with experiment could be accounted for by interface roughness corresponding to the displacement of one quarter of the atoms in an interface by one layer. For a position on the wedge n layers thick, this roughness corresponds to 25% of the surface being at $n-1$ layers, 50% at n layers, and 25% at $n+1$ layers. This is equivalent to a three layer growth front of the Cr with 0.1 nm rms roughness. We see the need, in order to observe the short-period oscillations, to approach more closely ideal layer-by-layer growth where one layer is completed before the next begins.

We now take a closer look at the exchange coupling observed in Fig. 2.43c. Measurements of the bare Cr wedge [2.173], before the Fe overlayer was deposited to obtain the magnetization image of Fig. 2.43c, are shown in Fig. 2.45a–c. The RHEED intensity oscillations used to determine the wedge thickness are shown in Fig. 2.45a. Such a line scan is taken from an image of the RHEED intensity as in Fig. 2.44a. Also apparent in this image are features in the topography of the Fe whisker surface which are replicated in the wedge. These features are compared to their counterparts in the SEMPA intensity image to bring the RHEED and SEMPA images into registry. In this way, the RHEED provides an atomic layer scale to determine the wedge thickness. The build up of disorder and roughness with increasing wedge thickness is indicated by the corresponding decrease in the amplitude of the RHEED intensity oscillations in Fig. 2.45a.

The measured spin polarization of secondary electrons from the bare Cr is shown in Fig. 2.45b. The high polarization of electrons from the Fe at the start of the wedge decreases exponentially as the Fe electrons are attenuated by the Cr film of increasing thickness. A fit to the exponential gives a $1/e$ sampling depth for SEMPA in Cr of 0.55 ± 0.04 nm. Subtracting an exponential leaves the polarization of the Cr alone which is shown magnified by a factor of 4 in Fig. 2.45c. Because of the attenuation of electrons coming from layers below the

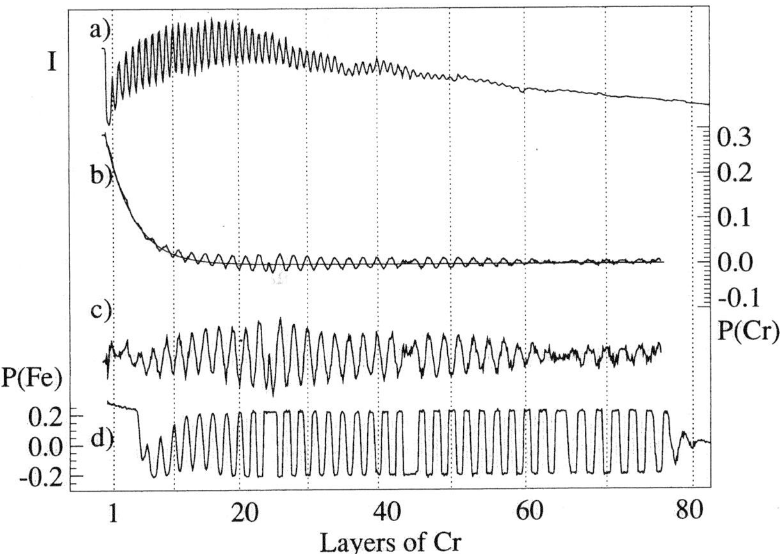

Fig. 2.45. (a) RHEED intensity oscillations determine the thickness of the Cr wedge deposited on the Fe whisker. (b) The spin polarization P(Cr) of secondary electrons emitted from the Cr wedge. (c) Data of (b) after subtracting the exponential shown and multiplying by 4. (d) The spin polarization, P(Fe), from the Fe overlayer deposited on the Cr wedge of (a–c). (From [2.173])

surface, the measured polarization, $P(\text{Cr})$, is dominated by the polarization of the surface layer which is seen to reverse approximately every layer.

When the Fe overlayer was added to this Cr wedge we obtained the image of Fig. 2.43c from which we can also get a profile of the polarization, $P(\text{Fe})$, shown in Fig. 2.45d. The initial coupling between the Fe layers is ferromagnetic and reverses at five layers. The polarization of the Fe overlayer is seen to be opposite to that of the top Cr layer before deposition. This observation is consistent with spin polarized photoemission [2.211] and electron energy loss measurements [2.212] which have found that the Cr interface layer couples antiferromagnetically to Fe. With this coupling at each interface and if the Cr orders antiferromagnetically with alternating planes of aligned spins for layer stacking in the $[0\,0\,1]$ direction, one expects Fe layers separated by an even (odd) number of layers of Cr to be coupled antiferromagnetically (ferromagnetically). However, from Fig. 2.45d we see that Fe separated by seven layers of Cr is coupled antiferromagnetically, opposite to expectations. A close examination of Fig. 2.45c reveals that there is a "defect" in the antiferromagnetic layer stacking of Cr between one and four layers. That is, at a thickness of less than four layers, two adjacent layers of Cr must have parallel moments.

The short-period oscillations in the interlayer exchange coupling in Fe/Cr/Fe have been attributed [2.156] to an RKKY-type interaction through paramagnetic Cr. The asymptotic form of the RKKY interaction at a distance z from a plane of ferromagnetic moments [2.213] is $\sin(k_s z)/(k_s z)^2$ where for a free-electron gas the Fermi surface spanned by the wave vector k_s is just the Fermi sphere so $k_s = 2k_F$. In the case of Fe/Cr/Fe, we are interested in the exchange coupling, $J(nd)$, between two planes of moments separated by the Cr interlayer of thickness nd

$$J(nd) \propto \sin(k_s nd)/(k_s nd)^2. \tag{2.87}$$

Here the distinction is made that the thickness does not vary continuously but in monolayer steps, nd, where d is the layer spacing. The phase slips seen in Fig. 2.43c and Fig. 2.45d occur because the wave vector k_s governing the oscillations is incommensurate with the lattice wave vector, $2\pi/a = \pi/d$. The measure of the incommensurability, δ, gives the fraction of a lattice wave vector by which the spanning wave vector differs from the lattice wave vector, i.e. $k_s = (1 - \delta)\pi/d$. The oscillatory part of the interaction can then be written

$$J(nd) \propto \sin[(\pi/d)(1 - \delta)nd] = -(-1)^N \sin(N\delta\pi). \tag{2.88}$$

The interaction is seen to change sign with each layer and to be modulated by an envelope function with period $N = 2/\delta$. There is a node in the function $\sin(N\delta\pi)$ every δ^{-1} layers. There is an accumulation of phase $\delta\pi$ with each additional layer of Cr with a phase slip of one layer after 20 layers corresponding to $\delta = 0.05$, and $k_s = 0.95\pi/d$.

Chromium is very special in that there is strong "nesting" of the Cr Fermi surface [2.214]. Here an extended region of one part of the Fermi surface is

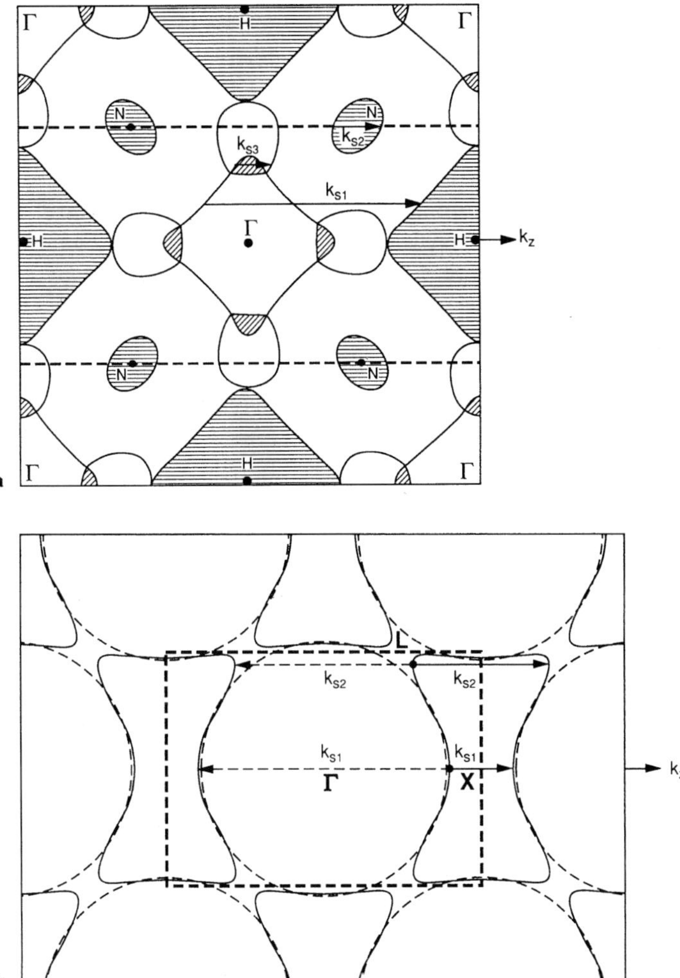

Fig. 2.46. Cuts through the Brillouin zone which contain the z direction of the layer stacking are shown for (**a**) Cr (from [2.214]) and (**b**) Ag (from [2.217]). The heavy dashed lines show the interface-adapted first Brillouin zone. The k_s which give rise to oscillations in the magnetic coupling are shown. The light dashed circles in (**b**) indicate a free electron Fermi surface

parallel to another part separated by the spanning wave vector k_{s1} as shown in Fig. 2.46a. The susceptibility is strongly enhanced at this k_s which leads to strong short-period oscillatory interlayer exchange coupling in the RKKY picture. The same enhanced susceptibility at k_{s1} leads to spin density wave (SDW) antiferromagnetism in bulk Cr below the Néel temperature, T_N [2.215]. This makes it hard to distinguish between an RKKY-like coupling and an explanation where the magnetization in the Fe overlayer is locked to the antiferromagnetism of the Cr. However, our measurements of the short-period oscillations in the interlayer exchange coupling of Fe/Cr/Fe(0 0 1) spanned the temperature range from T_N to $1.8T_N$, over which bulk Cr is *paramagnetic*. This would suggest that either the coupling takes place through paramagnetic Cr or that the presence of the Fe substrates stabilizes antiferromagnetism in Cr even above the bulk Néel temperature.

To further investigate the coupling mechanism, we have analyzed the magnetization images of the bare Cr wedge to obtain $P(Cr)$ of Fig. 2.45c. The same phase slips in the oscillations of the coupling of the Fe overlayer are also observed in $P(Cr)$ at Cr thicknesses of 24–25, 44–45, and 64–65 layers. Thus Cr/Fe(0 0 1) exhibits incommensurate SDW behavior within the Cr film; this behavior has also been observed over the temperature range from T_N to $1.8T_N$. Although thermal fluctuations destroy SDW antiferromagnetism in bulk Cr above T_N, it appears that the Fe substrate establishes a SDW in the Cr film some distance from the interface. There are two closely related ways to view this response in the Cr film. In one view, since even above T_N the magnetic susceptibility is enhanced at the nesting wave vector, an antiferromagnetic response can be induced in the Cr by the presence of the Fe. Alternatively, if the Cr is viewed as paramagnetic, RKKY-like oscillations would be established which would be quite similar to the antiferromagnetic order because both derive from the same strong Fermi surface nesting.

The origin of the mechanism giving rise to coupling with long-period oscillations is less clear. There are two spanning vectors of the Cr Fermi surface, k_{s2} and k_{s3} in Fig. 2.46a, which give periods of 1.35 and 1.62 nm [2.216] which are comparable to the 1.6–1.9 nm observed. However, Cr with its unfilled-bands has at least 11 spanning vectors where there is significant Fermi surface nesting [2.216] and one has to argue why only two periods of oscillation are observed. We have seen how the roughness associated with low temperature growth can destroy short-period oscillations in the magnetic coupling. Furthermore, it must be remembered that the nesting is only part of the story; the matrix elements, which have not as yet been calculated, undoubtedly play an important role.

Although Fe/Cr/Fe has been an important system in which the antiferromagnetic coupling and subsequently oscillations were first observed, deriving conclusions about the mechanism of the exchange coupling is more complicated because of the SDW-antiferromagnetism and the complex Fermi surface of Cr. It is useful then to consider another interlayer material, like Ag, which is not an antiferromagnet and which has a much simpler Fermi surface as shown in Fig. 2.46b. The Fermi surface is nearly spherical with necks at the L points in [1 1 1] directions [2.217]. There are just two nesting vectors which would be expected to lead to coupling in the [0 0 1] direction which can be seen in the (1 1 0) cut through the Brillouin zone shown in the figure. These are shown by the dashed lines, k_{s1} across the diameter of the "sphere" and k_{s2} connecting portions of the Fermi surface at the necks. Because the structure is periodic in the z direction with layer spacing, d, a reciprocal lattice vector $2\pi/d$ can be added or subtracted to k_s, to give for example the k_{s1} and k_{s2} shown by the solid lines in Fig. 2.46b which connect the same points on the Fermi surface as the dashed k_s. The period corresponding to k_{s1} is not a short-period less than $2d$ which would be possible in a free-electron gas, but rather the period is

$$\lambda = 2\pi/[(2\pi/d) - k_s]. \tag{2.89}$$

This difference between the free electron gas and the multilayer structure has been discussed and variously referred to as the vernier effect or aliasing by the periodic lattice planes [2.157, 159, 163]. The periods predicted [2.157] from the bulk Fermi surface of Ag are $\lambda_1 = 5.58d$ and $\lambda_2 = 2.38d$, where for bulk Ag the layer spacing is $d = 0.204$ nm.

We have investigated [2.174] the interlayer exchange coupling in Fe/Ag/Fe(0 0 1) using the procedures described above for Fe/Cr/Fe(0 0 1) differing only in Ag evaporation rate and temperature of the Fe whisker during evaporation. However, the growth is not as good and there is significant mismatch in the growth direction. Nevertheless, the oscillations of the exchange coupling persists to thicknesses of over 50 ML (> 10 nm). We observed 23 reversals in the magnetization. There is variation in the spacing between reversals which immediately shows that more than one period of oscillation was present. To extract these periods from the data, a Fourier transform of the data was carried out which revealed two periods that were used as the initial values for the periods in a fitting program. The data was modeled by adding two sine waves with these periods and with adjustable phases and amplitudes. This continuous function was discretized with the Ag lattice. Then all positive coupling values were set to the same magnetization value and negative coupling values were set to an equal but opposite magnetization, thereby simulating the effect of the Fe overlayer. The coupling period determined from varying the parameters to achieve the best fit are $\lambda_1 = 5.73 \pm 0.05t_0$ and $\lambda_2 = 2.37 \pm 0.07t_0$. The experimentally determined coupling periods are in excellent agreement with the theoretical values of $5.58t_0$ and $2.38t_0$ considering possible uncertainties in the theory and the possibility of slight tetragonal distortions in the Ag film. The SEMPA measurements show that the oscillations of the interlayer exchange coupling in Fe/Ag/Fe(0 0 1) are consistent with theories in which oscillation periods are derived from Fermi surface spanning vectors.

It is interesting to compare our results with interlayers of Cr and Ag in epitaxial structures with results from sputtered multilayers where the interface is not coherent and generally rougher. *Parkin* [2.218] and Sect. 2.4, has reported results of study of multilayer structures of 18 different transition or noble metal spacer materials between Co layers. The sputtered multilayers were polycrystalline, textured (1 1 1), (1 1 0), and (0 0 0 1) for fcc, bcc, and hcp, respectively. For the eight interlayer materials which showed oscillation in the coupling, each material except Cr has a single oscillation period of 1.0 ± 0.1 nm. The similarity of the periods for quite different materials has led many to the conclusion that this long-period coupling does not derive from Fermi surface properties.

In contrast to the studies of sputtered samples, studies of epitaxially grown structures show a variety of periods. Besides our SEMPA studies Fe/Cr/Fe, which we have discussed at length, others have found evidence for two periods of oscillation in Fe/Cr/Fe(0 0 1) [2.204, 207]. Two periods of oscillation have also been found in Co/Cu/Co(0 0 1) [2.219], Fe/Au/Fe(0 0 1) [2.220], and of course in our SEMPA measurements of Fe/Ag/Fe(0 0 1) [2.174]. Additionally, short-period oscillations in the magnetic coupling have been observed for Fe/Cu/Fe(0 0 1) [2.219], Fe/Mo/Fe(0 0 1) [2.221], Fe/Mn/Fe(0 0 1) [2.222], and Fe/Pd/Fe(0 0 1) [2.223]. In each case, the observed periods of oscillation of the magnetic coupling have been related to Fermi surface spanning vectors. This is particularly striking in the case of the (0 0 1) films of the noble metals for which two periods are expected from Fermi surface nesting, as we discussed for Ag. In each case, two periods are observed which agree quite well with Fermi surface predictions.

How can these results from epitaxially grown structures be reconciled with those from sputter deposited multilayers? Recently, *Stiles* [2.216] has calculated the degree of the Fermi surface nesting of the fcc, bcc and hcp transition metals for which oscillatory coupling was observed by *Parkin* in the sputtered multilayers. Because of the complicated nature of the Fermi surfaces and the large number of spanning vectors, it is possible to identify in each case a spanning vector that could give rise to the observed oscillation. If the observed periods can be attributed to Fermi surface spanning vectors, the alternative question is why periods are not observed which correspond to the several other k_s for which there is Fermi surface nesting? First, as we discussed for Cr, the matrix elements which would give the strength of the coupling at each k_s are not known and could be quite small. Second, as illustrated so vividly in the case of Fe/Cr/Fe(0 0 1) by Figs. 2.43, 44, roughness at an interface can completely wash out short-period oscillations.

2.3.2.3 Biquadratic Coupling

The coupling of two Fe layers through Cr has been discussed thus far in terms of the component of magnetization along the whisker, M_y, in the coordinates of Fig. 2.41. The coupling is seen to be ferromagnetic or antiferromagnetic in nature, depending on the Cr layer thickness, leading to images of an oscillatory coupling as seen in Fig. 2.43. As discussed in Sect. 2.3.1, the other in-plane component of magnetization, M_x, is measured simultaneously. Both components are shown in Fig. 2.47 in a region of wedge thickness from 20 to 30 ML which includes a phase slip. At thicknesses where M_y reverses, the other component of magnetization, M_x, is observed. that is, an additional coupling which tries to hold the magnetization of the two Fe layers perpendicular to each other is manifested. One of the many ways of writing the total coupling energy is [2.205, 224]

$$E = A_{12}[1 - m_1 \cdot m_2] + 2B_{12}[1 - (m_1 \cdot m_2)^2], \tag{2.90}$$

where m_1 and m_2 are unit vectors in the direction of magnetization of the two Fe layers. Within a constant, the first term is proportional to $\cos\theta_{12}$ and the second to $\cos^2\theta_{12}$, where θ_{12} is the angle between $\boldsymbol{m}_1$ and $\boldsymbol{m}_2$.

The first term in (2.90) is the bilinear coupling which we have emphasized thus far, where J has been replaced by A_{12}; for positive (negative) A_{12} it is a

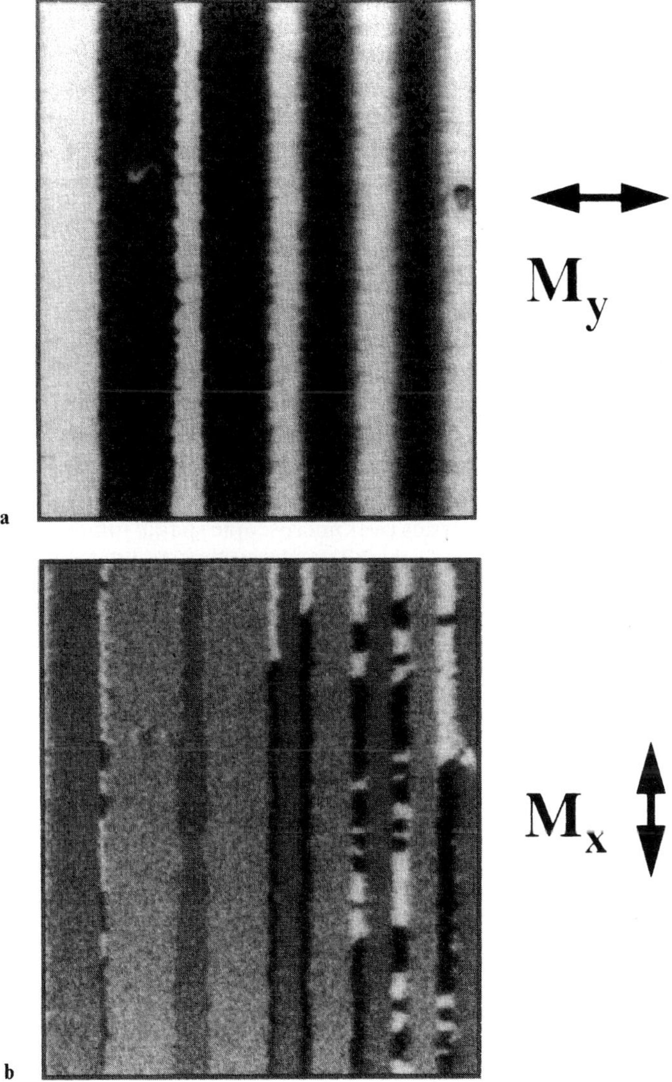

Fig. 2.47. The SEMPA magnetization images (**a**) M_y and (**b**) M_x are shown for that region of a Fe/Cr/Fe(0 0 1) trilayer over which the Cr thickness varies from 20 to 30 layers. A phase slip lies in this thickness range. The biquadratic coupling is evident from the strong contrast in the M_x image. (From [2.206])

minimum for m_1 and m_2 parallel (antiparallel). The second term is the biquadratic term; for negative B_{12} it is a minimum when m_1 and m_2 are perpendicular. The coupling is usually dominated by the bilinear term except at transitions from ferromagnetic to antiferromagnetic coupling where A_{12} goes through zero. At these thicknesses, the biquadratic term dominates and magnetic domains oriented along the x axis which are orthogonal to the domains of the Fe whisker substrate, are observed as in Fig. 2.47. The transition regions are not simply domain walls in the Fe film, but are much wider and scale in width with the slope of the Cr wedge. This perpendicular coupling was first observed for Fe/Cr/Fe and attributed to biquadratic coupling in a magneto-optic Kerr microscopy investigation of Fe/Cr/Fe sandwich structures [2.205].

Slonczewski [2.225] has proposed a theory which attributes the biquadratic coupling to fluctuations in the bilinear coupling caused by fluctuations in the thickness in the Cr interlayer. In general, there are fluctuations in the bilinear coupling near the transition thickness at which the coupling changes sign. There will be both the regions of ferromagnetic coupling, and at slightly different thickness on the other side of the transition, antiferromagnetic coupling. In the case of Cr films grown at elevated temperatures, the fluctuations are just the short period part of the interlayer exchange coupling. Thus, at a Cr thickness of $n + \frac{1}{2}$ layers, where the bilinear coupling makes a transition from ferromagnetic to antiferromagnetic coupling, there are many microscopic regions with n or $n + 1$ layers giving rise to fluctuations owing to the different bilinear coupling at the two thicknesses. The exchange of coupling within the Fe overlayer resists there being many magnetization reversals over microscopic spatial dimension as would be dictated by the fluctuations in the bilinear coupling, and the energy is

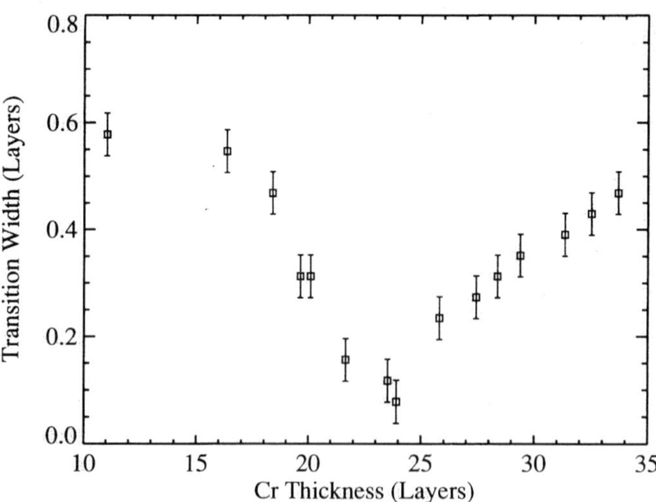

Fig. 2.48. Measurement of the width of the transition region (10–90% of M_y) for the switching of the bilinear coupling as a function of thickness of the Cr film

lowered by the magnetic moments turning in a direction perpendicular to the competing bilinear coupling directions.

In this theory, B_{12} is proportional to $(\Delta J)^2$ where $\pm \Delta J$ represents the fluctuation in the bilinear exchange coupling. One consequence of this fluctuation model of the biquadratic coupling is that this coupling will be small when the strength ΔJ of the short-period interlayer exchange coupling interaction is small. This can be expected to occur at a phase slip where, as seen from (2.88), there is a node in the envelop function $\sin(N\delta\pi)$. The width of the biquadratic coupling regions, in fractions of a Cr layer thickness, is a measure of the relative strength of the biquadratic and bilinear coupling. This width is plotted for several transitions near a phase slip in Fig. 2.48. The diminishing width of the biquadratic coupling region measured near a phase slip is in agreement with the predictions of the *Slozcewski* model [2.225] of the biquadratic coupling.

2.3.2.4 Summary

Since the first observations [2.153] of oscillations in the exchange coupling and magnetoresistance in multilayers of Fe or Co separated by non-ferromagnetic spacer layers, it was clear that the oscillations, and in particular the periods of the oscillations, stand as beacons to guide theories of the coupling mechanism. SEMPA is especially well-suited to make a quantitative comparison of the periods of the long range oscillatory coupling with the extremal features of the spacer layer Fermi surface. The high spatial resolution of SEMPA permits the use of small, high-quality specimens. Obtaining high-quality trilayer samples on regions of nearly perfect Fe single crystal whiskers has allowed the observation of oscillation of the coupling in Fe and Ag over many periods. The surface sensitivity of SEMPA is exploited in these studies to interrogate only the top layer of a tri-layer structure. The possibility to prepare specimens *in situ* and bring other techniques to bear in the SEM, such as RHEED and Auger spectroscopy, are all important capabilities. SEMPA measurements have given the most precise determination of the periods of the oscillation of the interlayer magnetic coupling in the Fe/Cr/Fe(0 0 1) and Fe/Ag/Fe(0 0 1) systems, the two systems that have been studied by SEMPA to date. Our experimental results on Cr and Ag interlayers support theories of interlayer exchange coupling based on Fermi surface properties. The long-range coupling between the magnetic layers is determined by the electronic response of the spacer layer. The indirect exchange coupling takes place through the electrons at the Fermi surface, the same electrons which are involved in the magneto-transport and which define the Fermi surface. SEMPA measurements have led to an increased understanding of exchange coupling of magnetic layers.

Acknowledgements. This work was supported by the Technology Administration of the U.S. Department of Commerce and the Office of Naval Research. The Fe whiskers were grown at Simon Fraser University under an operating grant from the National Science and Engineering Research Council of Canada.

2.4 Giant Magnetoresistance and Oscillatory Interlayer Coupling in Polycrystalline Transition Metal Multilayers

S.S.P. PARKIN

Metallic multilayers have attracted much attention over the past several decades, in a large part because of the possibility of creating artificial metals with potentially new properties or new combinations of properties [2.226–229]. Magnetic multilayers are of particular interest because of the importance of magnetic materials for many technological applications. The simplest such multilayered structure is comprised of alternating thin layers of magnetic and non-magnetic metals, as shown schematically in Fig. 2.49. It would not be surprising, given the delocalized nature of metallic electrons, that the magnetic layers would be magnetically coupled via the conduction electrons of the non-magnetic layer. (We will term this layer the *spacer* layer.) The nature of the magnetic coupling via the spacer layer material has been a subject of intense interest for more than 30 years. In recent years this interest has focussed on multilayers comprised of transition metals (TM) and noble (NM) metals. This chapter is devoted to the magnetic and transport properties of metallic magnetic multilayers composed of transition and noble metals.

Prior to studies of magnetic coupling in magnetic multilayers, indirect magnetic exchange coupling was extensively investigated in dilute alloys comprised of low concentrations of transition metal atoms with localized magnetic moments randomly distributed in metallic hosts, for example, Mn or Fe atoms dissolved in Cu or Au. It was found that the localized magnetic moments are exchange coupled via a spin polarization of the conduction electrons of the host metal [2.230–232]. The spin polarization was inferred from, for example, Cu^{63} nuclear magnetic resonance (NMR) measurements in which satellites were observed surrounding the main NMR line [2.232]. The satellites, corresponding

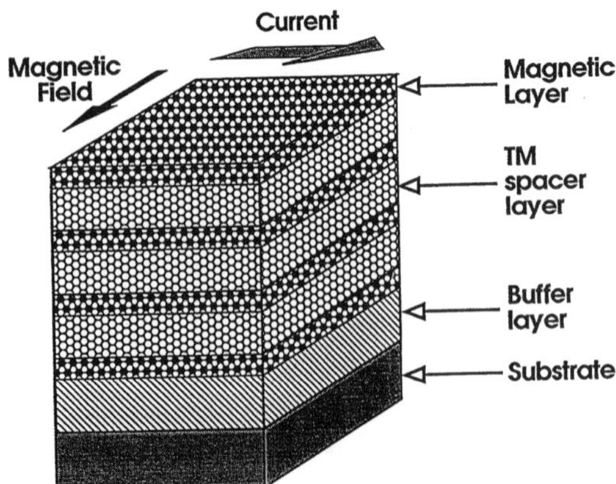

Fig. 2.49. Schematic diagram of a multilayer comprised of alternating magnetic and non-magnetic layers, grown on a buffer layer. The magnetoresistance measurements described here were made with the current and magnetic field in the plane of the layers with the magnetic field either parallel or orthogonal to the current

to successive spherical shells of Cu atoms surrounding the magnetic impurities, are shifted alternately to higher and lower magnetic resonance fields resulting from oscillations in the spin polarization of the Cu conduction electrons. (A schematic figure of the spin polarization is shown in Fig. 2.50.) For higher concentrations of magnetic impurities the oscillating spin polarization is manifested as an oscillating exchange interaction, alternating between ferromagnetic and antiferromagnetic coupling depending on the separation of the magnetic impurities. This coupling is of the well known Ruderman–Kittel–Kasuya–Yosida (RKKY) form [2.233].

Metal multilayers constructed of thin magnetic layers separated by thin non-magnetic layers [2.226] would at first sight appear to be much simpler systems in which to study magnetic coupling mediated by non-magnetic metals. However, in experiments from the 1960s until about 1988, the coupling of such magnetic layers through most metals, including Cu and Au, was found to be ferromagnetic in sign with a strength that apparently decayed exponentially with increasing separation of the magnetic layers [2.234, 235]. These experiments were in contradiction with most theoretical models which predicted an oscillating exchange interaction analogous to that found in the dilute magnetic alloys [2.236]. Whilst an oscillatory magnetic coupling of the RKKY form was observed in multilayers composed of the rare earth metals Gd and Y [2.237],

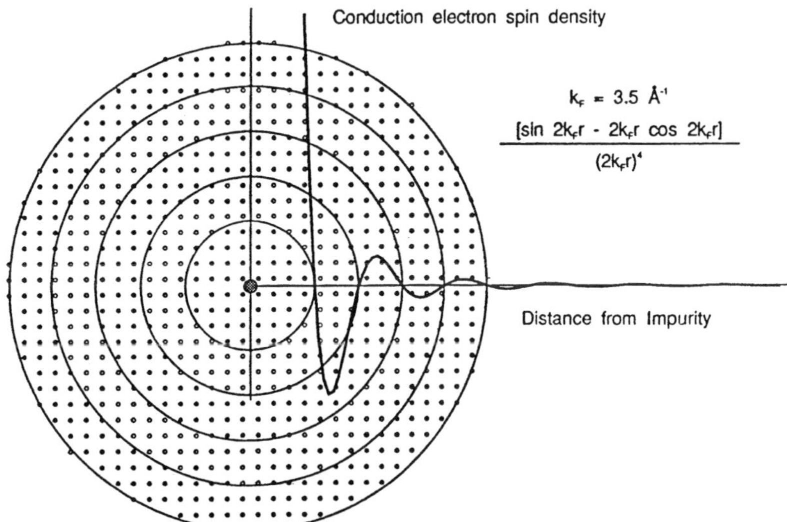

Fig. 2.50. Schematic representation of the spin polarization induced in a paramagnetic metal by a localized magnetic impurity. The open and closed circles represent spin of opposite polarization. The functional dependence of the induced spin polarization in a nearly free electron metal within the RKKY model is shown. The period of oscillation of the spin polarization is related to the inverse Fermi wavelength of the metal. The period is given by $\lambda_F/2$, where λ_F is the Fermi wavelength. ($k_F = 2\pi/\lambda_F$)

only recently has an oscillating magnetic exchange coupling been found in transition metal multilayers. The first observations were made in Fe/Cr and Co/Ru multilayers [2.238], and subsequently in Co/Cu [2.239] and later in the majority of transition metal multilayered systems [2.240]. Oscillatory coupling was first found in transition metal (TM) multilayers grown by conventional sputter deposition techniques and only later in single crystalline multilayers prepared in ultra high vacuum (UHV) deposition systems using electron beam or thermal evaporation cells. This was surprising since there was a belief that multilayered films prepared by sputtering would have interfaces so disordered as to make the observation of coupling unlikely.

In this last subsection of this chapter the magnetic and related properties of *polycrystalline* magnetic multilayered structures will be described. This paper will concentrate on introducing the basic features of such multilayered systems as well as simple phenomenological descriptions of their behavior. Please note that this subsection is not intended to be a comprehensive review nor a detailed historical treatment of the subject. In the following subsections the properties of single crystalline multilayered structures will be briefly discussed and finally detailed theoretical models developed to account for giant magnetoresistance and oscillatory interlayer coupling will be presented, directly complementing the treatments in the preceeding sections of this chapter.

2.4.1 Preparation of Multilayers

A wide variety of deposition methods have been used to prepare magnetic thin films and multilayers. These include electrochemical deposition techniques [2.241, 242], as well as a wide variety of vacuum deposition techniques [2.243]. The latter fall into two main categories. Sputter depositon involves the use of highly energetic but otherwise inert particles to knock off by bombardment atoms of the material of interest from a target comprised of this material. The energetic particles are created by ionizing typically argon or some other rare gas atoms and accelerating the ions into the target. The atoms knocked free from the target have energies typically in the range from 2 to 30 eV. However the energy of these atoms will be reduced to a greater or lesser extent, prior to deposition on the substrate, by collisions with the sputtering gas. This process depends upon a variety of factors including the sputtering gas pressure and the detailed construction of the sputtering apparatus, for example, the target-substrate distance. The presence of the sputtering gas in the chamber (the pressure typically ranges from $\simeq 1 \times 10^{-4}$ Torr to 10 to 100 mTorr) precludes the use of most *in situ* characterization techniques to study the growth and structure of the film. However, sputter deposition is a relatively simple and inexpensive technique suitable for the growth of most metals. There are many different types of sputter sources, specially designed for different applications. Magnetron sputter guns are designed with strong permanent magnets to give rise to a magnetic field to confine the plasma close to the target material and away from the substrate. This reduces damage to the substrate and film from energetic ion bombardment. An

important advantage is that this also allows the use of lower sputtering gas
pressures while maintaining relatively high growth rates. The concentration of
impurities in the deposited films from residual gases in the system, such as
oxygen or nitrogen, depend on the film growth rate versus the residual gas
pressures in the deposition system. Thus magnetron sputter deposition leads to
films with comparatively low residual gas impurity levels.

A principle advantage of sputter deposition is the ease with which many
different materials can be deposited at relatively high deposition rates. In almost
all cases the deposited structures are polycrystalline with crystallites oriented in
many different directions (Fig. 2.51a). However, typically one crystallographic
orientation is preferred, leading to crystallographically textured films. For
example, for fcc metals, the films are usually textured in the (1 1 1) orientation
whereas for bcc and hcp metals, sputtered films are typically textured in the
(1 1 0) and (0 0 0 1) directions respectively.

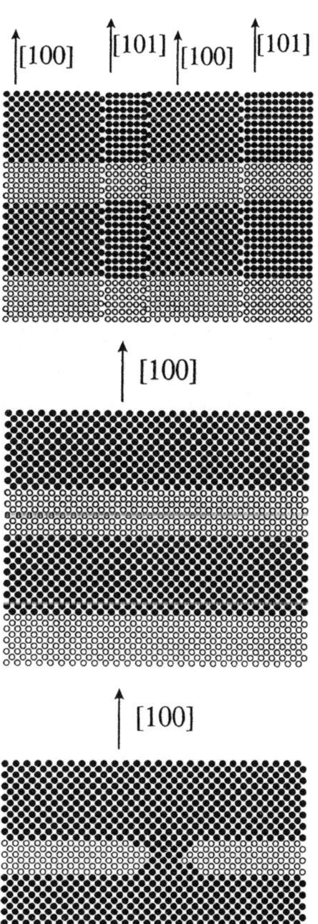

Fig. 2.51. Schematic representation of the structure of multilayers prepared by sputter deposition and MBE. In the former
the multilayers consist of crystallites which may be oriented
along different crystallographic directions. In the latter single
crystalline films can be prepared for appropriate combinations
of magnetic and non-magnetic materials. Also shown is a
schematic representation of a single crystalline multilayer in
which there is a "pinhole" of magnetic material leaking through
the spacer layer. Such a pinhole or equivalent defects can give
rise to strong direct ferromagnetic bridging of adjacent magnetic
layers

Alternative deposition techniques include vapor deposition from thermal (Knudsen) cells or electron beam evaporators. These techniques are usually carried out in ultra high vacuum systems with base pressures as low as 10^{-11} Torr. Such systems are often referred to as molecular beam epitaxy (MBE) systems. In general film depositon rates in such systems are usually much lower than for sputtering systems. The evaporated material also typically has much lower energies, $\simeq 0.1$ eV per atom, than from sputtering processes. However, as noted above there will be considerable thermalization of the sputtered material by collisions with the sputtering gas species. The lower deposition rates in MBE systems, as well as their increased complexity, means that film production is perhaps 20 times lower as compared with sputter deposition tools. However, MBE systems are important for the growth of highly oriented single crystalline films. This is accomplished by matching the lattice of the film layers of interest with an appropriate substrate material. In many cases no direct lattice match is possible but a variety of techniques have been developed for the growth of particular materials using an additional seed layer between the substrate and film, as discussed in Chap. 1 Volume II, and in [2.244, 245].

2.4.2 Antiferromagnetic Coupling and Giant Magnetoresistance in Fe/Cr Multilayers

2.4.2.1 Antiferromagnetic Coupling

The first evidence for antiferromagnetic (AF) coupling of magnetic layers via a transition metal was made in crystalline bcc (1 0 0) Fe/9 Å Cr/Fe sandwiches [2.246] using Brillouin light scattering (BLS) and magneto-optical Kerr hysteresis loops. Interest in the Fe/Cr system was heightened by the subsequent observation that the resistance of antiferromagnetically coupled (1 0 0) Fe/Cr/Fe sandwiches [2.247] and (1 0 0) Fe/Cr multilayers [2.248] decreases enormously with the application of a magnetic field. Indeed the changes are so large that the phenomenon has been termed *giant magnetoresistance* (GMR). The same phenomena were subsequently observed in sputtered Fe/Cr multilayers [2.238]. Typical magnetization and resistance versus field loops for a sputtered Fe/Cr multilayer are shown in Fig. 2.52, which also includes a schematic diagram of the magnetic structure of the Fe layers in zero field and large positive and negative fields.

The antiferromagnetic coupling of the Fe layers results in a net zero magnetic moment in small magnetic fields, as shown in the magnetic hysteresis loop in Fig. 2.52. However, the application of a magnetic field sufficiently large to overcome the AF coupling causes the magnetic moments of the Fe layers to become aligned with the field. Consider an Fe/Cr/Fe sandwich in which the magnetic moments, M_1 and M_2 of the Fe layers, are coupled with an interlayer exchange constant, A_{12}. The exchange coupling energy per unit area can then be

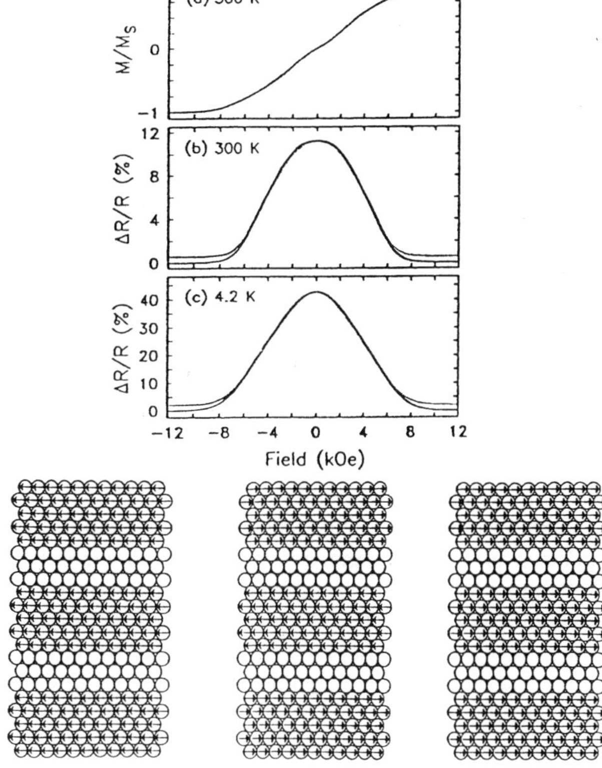

Fig. 2.52. Room temperature in plane magnetic hysteresis loop (**a**), and magnetoresistance versus field curves at (**b**) room temperature and (**c**) 4.2 K for an Fe/Cr multilayer of the form Si(1 0 0)/Cr(30 Å)/[Fe(20 Å)/Cr(9 Å)]$_{30}$/Cr(35 Å). Full and dashed lines in (**b**) and (**c**) correspond to current orthogonal and parallel to the applied field, respectively. The magnetic arrangement of the Fe layers is shown schematically for large negative, zero and large positive applied fields.

written, within a Heisenberg model, as $E = A_{12} \cos \theta_{12}$, where θ_{12} is the angle between M_1 and M_2. In a magnetic field, ignoring any magnetic anisotropy, the total energy of the sandwich per unit area $E(H)$ will be $E(H) = A_{12} \cos \theta_{12} - M_s t_{\mathrm{Fe}} H (\cos \phi_1 + \cos \phi_2)$ where M_s is the magnetization of the Fe layers, t_{Fe} is the thickness of each Fe layer, H is the magnetic field and ϕ_1 and ϕ_2 are the angles between H and the magnetization of each layer. Minimizing the energy for a given field leads to a saturation field $H_s = 2A_{12}/M_s t_{\mathrm{Fe}}$ where the Fe layers become parallel to one another ($\theta_{12} = 0$). Thus the strength of the antiferromagnetic interlayer coupling can easily be obtained from simple magnetic hysteresis loops such as shown in Fig. 2.52.

The saturation field is expected to increase with the inverse thickness of the magnetic layer. This is observed experimentally, to a good approximation, for Fe layers varying from $\simeq 10$ Å to several hundred angstroms thick. Note that by comparison it is usually impossible to reorient the magnetic moments in typical bulk antiferromagnets in fields readily available from usual electro- or superconducting magnets, although there are some exceptions. In Fe/Cr multilayers and others discussed in later sections, the interlayer coupling is weak compared to the intralayer exchange coupling within the Fe layers by a factor of more than 100. This together with the very large magnetic moments on each Fe layer leads to much lower saturation fields compared to bulk antiferromagnets.

Note that the magnetization of the Fe/Cr/Fe sandwich increases linearly with field as $M_s^2 t_{Fe} H / A_{12}$ in the simple model described above. This is in reasonable agreement with magnetization curves on polycrystalline multilayer samples (Fig. 2.52). Inclusion of magnetic anisotropy will lead to more complicated field dependences of the magnetization [2.249] such as is often found for single crystalline multilayers (*Fert* and *Bruno*, Sect. 2.2). It has been implicitly assumed that the magnetization of each Fe layer behaves as a single magnetic entity. However the magnetic stiffness of the Fe layers must also be included and will also influence the magnetic hysteresis loop of the multilayer.

The dependence of saturation field on the number of Fe layers, N, has been studied in detail for Fe/Cr structures [2.250]. The saturation field of a multilayer is expected to be twice as large as that of a sandwich since each magnetic layer has twice as many neighboring Fe layers. This is indeed observed: the saturation field varies as $(1 - 1/N)$ where N is the number of Fe layers. More importantly this result shows that there is no significant dependence of A_{12} on the length of the multilayer in contrast to speculations based on studies of Fe/Cr sandwiches and multilayers by different groups.

Confirmation of the antiferromagnetic alignment of the Fe layers in Fe/Cr suggested by magnetization loops has been carried out by polarized neutron reflectivity measurements. A magnetic Bragg peak corresponding to the AF magnetic unit cell is observed in small magnetic fields at twice the chemical superlattice period. As the magnetic field is increased the intensity of the AF magnetic peak decreases, disappearing altogether above the saturation field [2.250].

2.4.2.2 Giant Magnetoresistance

The variation of resistance of Fe/Cr multilayers and sandwiches with magnetic field is correlated with the change in the magnetic arrangement of the Fe layers, as demonstrated in Fig. 2.52. The MR is defined with respect to the resistance at high field. This definition of the magnetoresistance [2.238] is now widely accepted but differs from that used by some other groups [2.248]. The resistance of the structure is increased when neighboring Fe layers are arranged antiparallel to one another in small magnetic fields compared to parallel alignment of the Fe layers in large fields. In a first approximation the resistance of the structure varies with the angle between the magnetization of adjacent magnetic layers as $\cos\theta_{12}$. Since the net moment of the structure M, varies as $\cos(\theta_{12}/2)$ the

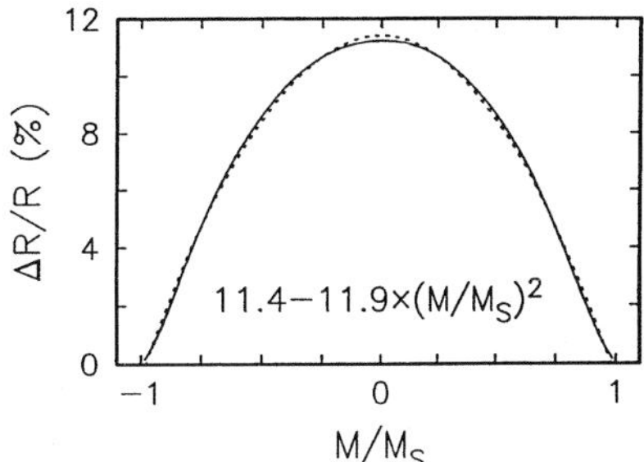

Fig. 2.53. Dependence of saturation magnetoresistance (MR) on normalized magnetization (M/M_s) at room temperature of the same Fe/Cr multilayer shown in Fig. 2.52. The data has been fitted with a curve of the form, $MR = 11.4 - 11.9 \times (M/M_s)^2$

resistance of the structure will consequently vary with M as $1 - (M/M_s)^2$. Figure 2.53 shows the dependence of magnetoresistance on magnetization for an Fe/Cr multilayer at room temperature. The MR does indeed vary as the square of the magnetization, demonstrating that the resistance does indeed vary as $\cos\theta_{12}$.

2.4.3 Magnetoresistance of Ferromagnetic Metals

2.4.3.1 Anisotropic Magnetoresistance

The magnitude of the giant magnetoresistance effect is significantly larger than the magnetoresistance of typical magnetic metals at room temperature. The dependence of resistance on magnetic field is shown in Fig. 2.54 for $\simeq 1000\,\text{Å}$ thick magnetron sputtered films of Fe, Co and Ni and various Ni alloys deposited at room temperature on Si(1 0 0) substrates. The variation with field depends on the orientation of the field with respect to the measuring current. For a field parallel to the current (longitudinal magnetoresistance), the resistance increases at low fields, whereas for a field perpendicular to the current (perpendicular magnetoresistance), the resistance decreases with increasing field. The difference in resistance at high fields represents the anisotropic magnetoresistance effect (AMR) common to all ferromagnetic metals [2.251, 252]. The variation in resistance is related to the variation of the magnetization of the ferromagnetic film. At low fields the magnetization is broken up into randomly oriented magnetic domains which are swept away by application of relatively small fields. In this state of technical saturation, the resistance is anisotropic and depends on the orientation of the magnetization with respect to the current. An anisotropic scattering mechanism such as that provided by spin–orbit coupling must be invoked to account for the AMR [2.251, 253]. Further discussion of AMR is out of the scope of this article but an excellent recent review can be found in [2.252].

In contrast to the AMR the GMR is isotropic, as demonstrated in Fig. 2.52 for an Fe/Cr multilayer for which resistance versus field curves for fields parallel and orthogonal to the current are included. The small anisotropy in resistance at

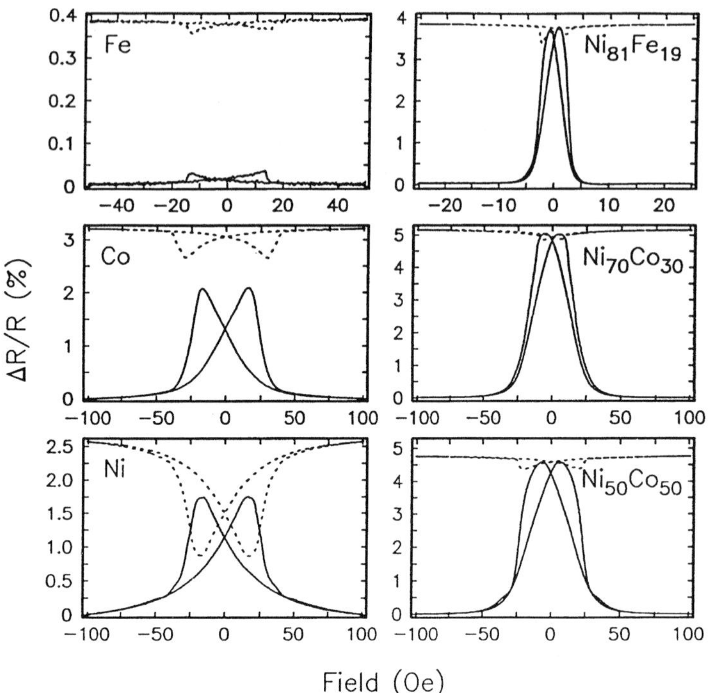

Fig. 2.54. Examples of the anisotropic magnetoresistance effect in sputtered polycrystalline films of Fe, Co, Ni and Ni$_{81}$Fe$_{19}$, Ni$_{70}$Co$_{30}$ and Ni$_{50}$Co$_{50}$. The full and dotted lines correspond to magnetic field applied orthogonal and parallel to the current respectively in the plane of the films. The films in each case are $\simeq 1000$ Å thick

high fields is due to the AMR of the individual Fe layers. The AMR effect is much smaller than the GMR. For the sample of Fig. 2.52, the AMR has values of 0.53% at 300 K and 2.1% at 4.2 K as compared to GMR values (for fields orthogonal to the current) of $\simeq 11.3\%$ at room temperature and 42.7% at 4.2 K. Thus the ratio of the GMR to AMR is about 21 at 300 K and 20 at 4.2 K. It is interesting to note that the temperature dependence of the AMR is very similar to that of the GMR, increasing by about a factor of four as the temperature is decreased from 300 to 4.2 K. In contrast, the magnitude of the antiferromagnetic coupling between the Fe layers is much less dependent on temperature. Substantial AF coupling persists to temperatures as high as 350 °C. For higher temperatures, above approximately 375 °C, the Fe and Cr layers dissolve into one another destroying the multilayer structure.

2.4.3.2 Resistance of Magnetic Metals – Mott Mechanism

A schematic diagram of the density of states of the sp- and d-bands of ferromagnetic Fe, Co and Ni is shown in Fig. 2.55 [2.254]. Usually it is assumed

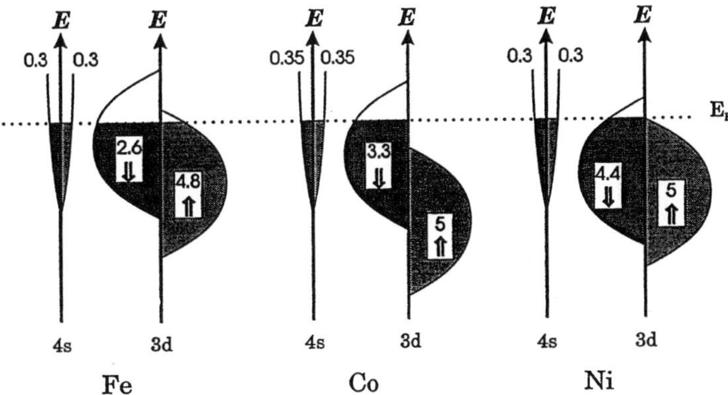

Fig. 2.55. Schematic diagram of the densities of states in the sp- and d-bands of ferromagnetic Fe, Co and Ni. The total numbers of electrons in the down-spin (left) and up-spin (right) bands are also shown (after [2.254])

that in ferromagnetic metals the conductivity is primarily carried by electrons from the sp-bands which are broad and, as a consequence, have low effective masses. In contrast, the d-bands are narrow and have high effective masses. Dating back to *Mott* [2.255], it is commonly assumed that there are two largely independent conduction channels, corresponding to the up-spin and down-spin sp electrons. Only at temperatures high compared to the ferromagnetic ordering temperature will spin–flip scattering processes cause mixing of the electrons within these two spin channels. The d-bands play a very important role in providing final states into which the sp electrons can be scattered. The scattering mechanisms include all of the usual scattering mechanisms in metals, including scattering from impurities, structural defects, phonons and magnons, etc. The density of states at the Fermi level for the up-spin and down-spin d electron bands can be very different (Fig. 2.55), particularly for the strong ferromagnetic metals, Co and Ni. This means that the scattering rates into these states will be significantly different for the two conduction channels. Consequently this leads to the possibility of substantially different mean free paths $\lambda^{\pm}$ and conductivities $\sigma^{\pm}$ in the two channels. In cobalt, for example, the density of states at the Fermi level is ten times higher for down-spin electrons as compared to up-spin electrons [2.256]. For detailed reviews see, for example, [2.255–257].

Various theories of the origin of the giant magnetoresistance effect in magnetic multilayers have been proposed as discussed in Sect. 2.2 [2.258–265]. Many of these models are based on ideas developed to account for the resistivity of ferromagnetic metals. The simplest model is an equivalent resistor network model [2.254, 263], shown schematically in Fig. 2.56 for a multilayer (after [2.254]). Each of the ferromagnetic and non-magnetic spacer layers consists of two resistors corresponding to the two conductivity channels associated with the up- and down-spin electrons. In the ferromagnetic layers, the resistivity is

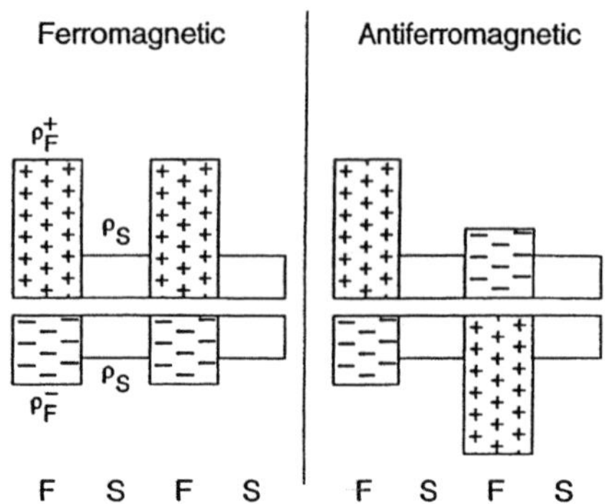

Fig. 2.56. Simple resistor network model of the giant MR in a magnetic multilayer comprised of ferromagnetic layers (F) in which the resistivity, $\rho^\pm$, has different values in the up- and down-spin channels, and spacer layers (S) with a spin independent resistivity, ρ_s. (after [2.254])

spin-dependent, $\rho_F^\pm$, whereas in the spacer layers the resistivity in the two channels is identical, ρ_s. The resistance of the multilayer is then equivalent to that of a total of eight resistors, with four resistors in each channel. The net resistivities of the two channels can be treated as resistors in parallel.

Adding up the resistors within a given channel is more complicated, but there are two simple cases [2.254]. For short mean free paths compared to the thickness of the layers the resistors are independent and should themselves be added in parallel. Under these circumstances it is obvious that the resistance in the ferromagnetic and antiferromagnetic configurations is the same and consequently there is no magnetoresistance defined as, $\Delta R/R = (R_{AF} - R_F)/R_F$, where R_{AF} and R_F are the resistances corresponding to the AF and F configurations. Another straightforward case is when the mean free paths are long compared to the layer thicknesses in the multilayer. Then the resistivity is an average of the resistivity of the various layers in the multilayer in proportion to the thicknesses of the corresponding layers. Note that for the F configuration only two resistivities must be averaged but in the AF configuration there are four. Taking these averages and subsequently adding these resistivities of the two spin channels in parallel leads to the result that $\Delta R/R = [(\alpha^+ - \alpha^-)^2]/[4(\alpha^+ + d/t)(\alpha^- + d/t)]$, where d and t are the thicknesses of the spacer and ferromagnetic layers and $\alpha^+ = \rho_F^+/\rho_s$ and $\alpha^- = \rho_F^-/\rho_s$. The magnetoresistance in this model depends on two parameters, α^+/α^- and $t\alpha^-/d$. This model shows, not surprisingly, that the magnitude of $\Delta R/R$ is strongly dependent on the scattering asymmetry between the spin conduction channels in the ferromagnetic layers. Of course, it is irrelevant in which spin channel the scattering is stronger. This highly simplified model also predicts that for a constant ratio, α^+/α^-, the MR decreases monotonically with increasing spacer layer thickness, falling off as $1/d^2$ for large d. As discussed later in Sect. 2.4.5c the MR is actually found to decrease exponentially with d for large d. The reason for this discrepancy is that the resistor network model is no longer applicable for d large compared to the mean free path in the spacer layer.

A simple resistor network model can easily give values of MR exceeding 100% for ratios of α^+/α^- of $\simeq 8$ to 10 [2.254]. A basic assumption of this model is that the spin dependent scattering giving rise to the MR originates purely within the interior of the magnetic layers, i.e. bulk scattering, although the model can be readily generalized to allow for spin dependent scattering at the interfaces, by adding additional resistors in the network. The relative contributions of spin dependent scattering from bulk scattering and from spin dependent scattering at the interfaces between the magnetic and spacer layers is a subject of great current interest. As discussed in Sect. 2.4.7, experiments strongly suggest that interfacial scattering is of overwhelming importance. Nevertheless, we note that, as shown by the resistor network model, the magnitude of the MR is expected to be related to the ratio of the scattering rates within the two conduction channels no matter where the spin dependent scattering takes place. The scattering asymmetries have been indirectly determined from measurements of the resistivity of magnetic ternary alloys [2.256, 266]. However, no correlation between the magnitude of the scattering asymmetries from studies of bulk magnetic alloys and the magnitude of the MR in magnetic multilayers has yet been found. If scattering at the interfaces between the magnetic and non-magnetic components is giving rise to the giant MR effect this would not be a surprising result. More detailed models of the giant MR effect are given in Sects. 2.1 and 2.2.

2.4.4 Oscillatory Interlayer Coupling

2.4.4.1 Oscillatory Interlayer Coupling – An Example

The first studies of the dependence of interlayer coupling on spacer layer thickness in Fe/Cr were carried out with single crystalline multilayers [2.247, 248, 267]. The conclusion of this early work was that the coupling remained antiferromagnetic for all Cr layer thicknesses except for very thin Cr layers. Studies on sputtered polycrystalline Fe/Cr films contrasted this with evidence for oscillations in the strength of the antiferromagnetic interlayer coupling with increasing thickness of Cr [2.238]. This was the first report of oscillatory interlayer coupling via a transition metal. Later work on single crystal Fe/Cr/Fe wedges has not only confirmed the existence of long-period oscillations but has shown the presence of additional short-period oscillations with oscillation periods of just two Cr monolayers as discussed in detail in Sect. 2.3 [2.268–270]. Whilst early speculations on the origin of the antiferromagnetic coupling in Fe/Cr multilayers and sandwiches were based on the unique magnetic character of Cr, the discovery of similar and stronger antiferromagnetic coupling in Co/Ru ruled out such models [2.238].

Magnetization versus in-plane field loops are shown in Fig. 2.57 for a series of sputtered $Ni_{80}Co_{20}/Ru$ multilayers as a function of Ru spacer layer thickness. The loops clearly show an oscillatory variation of saturation field with Ru thickness. For the Ru spacer layer thicknesses of 4, 12, 24 and 37 Å shown in Fig. 2.57 the magnetization of the multilayer is saturated in very low fields of

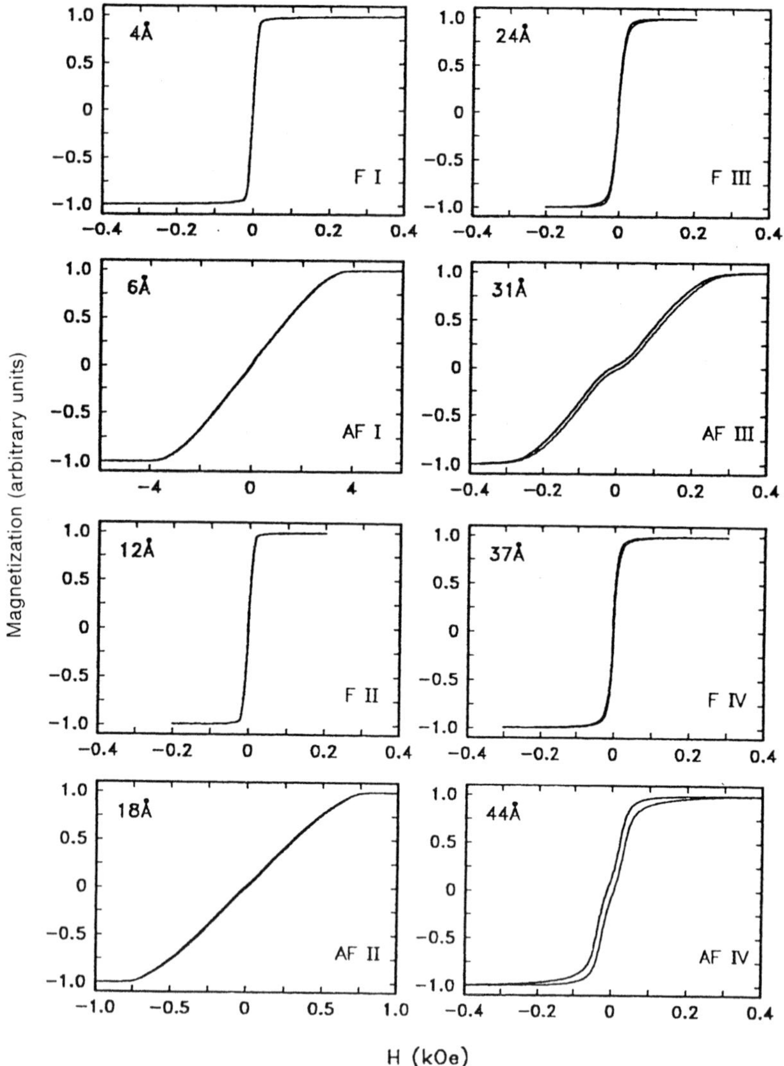

Fig. 2.57. Typical examples of magnetization versus in-plane magnetic field at room temperature for several $Ni_{80}Co_{20}/Ru$ multilayers as a function of increasing Ru spacer layer thickness. The structures are of the form, $100\,\text{Å}$ $Ru/[30\,\text{Å}$ $Ni_{80}Co_{20}/Ru(t_{Ru})]_{20}/50\,\text{Å}$ Ru   with t_{Ru} = 4, 6, 12, 18, 24, 31, 37 and 44 Å

$\simeq 10$ Oe. For intermediate Ru thicknesses the saturation fields are larger, although decaying with increasing Ru thickness. A detailed dependence of saturation field on Ru thickness is shown in Fig. 2.58 for $Ni_{81}Fe_{19}/Ru$ (permalloy) multilayers. Five oscillations in the saturation field are shown in Fig. 2.58 with an oscillation period of $\simeq 11$ Å. In the limit of very thin Ru the coupling is

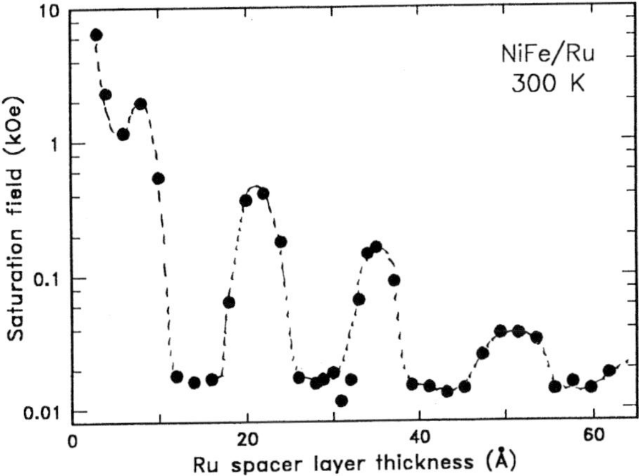

Fig. 2.58. Dependence of saturation field on Ru spacer layer thickness for several series of $Ni_{81}Fe_{19}/Ru$ multilayers with structure, 100 Å Ru/[30 Å $Ni_{81}Fe_{19}/Ru(t_{Ru})$]$_{20}$, where the topmost Ru layer thickness is adjusted to be $\simeq 25$ Å for all samples

antiferromagnetic. Even for Ru layers just $\simeq 3$ Å thick, strong AF coupling is observed. In contrast, in this limit the coupling is ferromagnetic for multilayers containing $Ni_{80}Co_{20}$, as shown in Fig. 2.58. Whereas the sign and magnitude of the coupling for very thin Ru layers is extremely sensitive to the composition of the magnetic layer, the period of the oscillation is independent of the magnetic material.

2.4.4.2 Oscillatory Coupling – A General Phenomenon

Antiferromagnetic coupling and oscillations in the coupling have been found in numerous transition metals [2.240] as well as a number of noble metals including Cu [2.239, 271–274]. Properties of the interlayer coupling in a series of sputter deposited Co-based multilayers for a variety of transition metal spacer layers multilayers are summarized in Fig. 2.59 [2.240]. These experiments have demonstrated that antiferromagnetic coupling and oscillations in the magnetic coupling is not limited to a small subset of multilayers but is a general property of most transition metal and noble metals. It is interesting to speculate why these oscillations were not observed in earlier studies since magnetic multilayers have been under extensive investigation for many years. Indeed, up until very recently, many of the same metals shown in Fig. 2.59 were considered to give rise to ferromagnetic coupling [2.235]. The most likely explanation is that the early work concentrated on structures prepared by MBE techniques. It seems that such structures often contain structural defects that give rise to direct ferromagnetic coupling of the magnetic layers. For example, there may be pinholes of the

The data of Fig. 2.59 is presented below in the periodic-table layout of the original. Each element cell gives, where coupling data exist, A_1 (Å) and ΔA_1 (Å) in the upper row and J_1 (erg/cm^2) and P (Å) in the lower row, with the most stable crystal structure indicated by the symbol beside the element.

Element	A_1 (Å)	ΔA_1 (Å)	J_1 (erg/cm^2)	P (Å)	Structure	Note
Ti					hcp	No Coupling
V	9	3	0.1	9	bcc	
Cr	7	7	.24	18	bcc	
Mn					complex cubic	Antiferro-Magnet
Fe					bcc	Ferro-Magnet
Co					hcp	Ferro-Magnet
Ni					fcc	Ferro-Magnet
Cu	8	3	0.3	10	fcc	
Zr					hcp	No Coupling
Nb	9.5	2.5	.02	*	bcc	
Mo	5.2	3	.12	11	bcc	
Tc					hcp	
Ru	3	3	5	11	hcp	
Rh	7.9	3	1.6	9	fcc	
Pd					fcc	Ferromagnetic Coupling
Ag					fcc	+
Hf					hcp	No Coupling
Ta	7	2	.01	*	bcc	
W	5.5	3	.03	*	bcc	
Re	4.2	3.5	.41	10	hcp	
Os					hcp	
Ir	4	3	1.85	9	fcc	
Pt					fcc	Ferromagnetic Coupling
Au					fcc	+

Oscillatory exchange coupling period is P (Å),

Coupling strength at first antiferromagnetic peak is J_1 (erg/cm^2).

Position of first antiferromagnetic peak is A_1 (Å).

Width of first antiferromagnetic peak is ΔA_1 (Å).

+No coupling is observed with Co

Fig. 2.59. Compilation of data on various polycrystalline Co/TM multilayers with magnetic layers comprised of Co and spacer layers of the transition and noble metals (from [2.240]). Periodic Table of A_1 (Å), the spacer layer thickness corresponding to the position of the first peak in antiferromagnetic exchange coupling strength as the spacer layer thickness is increased; J_1(erg/cm^2), the magnitude of the antiferromagnetic exchange coupling strength at this first peak; ΔA_1(Å), the approximate range of spacer layer thickness of the first antiferromagnetic region; and P(Å), the oscillation period. The most stable crystal structure of the various elements are included for reference. Note that no dependence of the coupling strength on crystal structure nor any correlation with electron density (pror$_{\rm W-s}^{-3}$) is found.
* For the elements Nb, Ta and W, only one AF coupled spacer layer thickness region was observed, so it was not possible to directly determine P. For Ag and Au no oscillatory coupling was observed in Co-based multilayers. Pd and Pt show strong ferromagnetic coupling with no evidence for oscillatory coupling from spin-engineered structures (Sect. 2.5.3c)

magnetic material in the spacer layer bridging the spacer layer (an example is shown schematically in Fig. 2.51).

As can be seen from Fig. 2.59 the period of the oscillatory coupling is similar for most metals, with the exception of Cr for which the period is significantly longer. In no case has any evidence been found for a significant dependence of the oscillation period on the magnetic material, although, as mentioned above, the phase of the oscillation is sensitive to the magnetic material [2.238, 240, 275]. An example is shown in Figs. 2.57, 58 in which the sign of the coupling for very thin Ru layers is opposite for magnetic layers of $Ni_{80}Co_{20}$ (ferromagnetic coupling) and $Ni_{81}Fe_{19}$ (antiferromagnetic coupling). The phase of the oscillation also varies with the spacer layer material for the same magnetic material (Fig. 2.59).

The coupling strength falls off rapidly with increasing spacer layer thickness d [2.238–240, 270, 275], therefore coupling strengths must be compared for the same equivalent spacer layer thickness. By assuming that the coupling strength falls off as $1/d^2$, where d is the spacer layer thickness, as, for example, exhibited by the data in Fig. 2.48, values of the interlayer coupling strengths for the same d can be calculated. These values (for $d = 3$ Å) are plotted versus the number of d electrons in Fig. 2.60. Figure 2.60 shows that the coupling strength systematically varies throughout the periodic table from small values for small d-band filling in the 5d metals to larger values for large d-band filling in the 3d metals. No evidence for significant magnetic interlayer coupling is found for Ti, Zr and Hf, but the trends in coupling strength displayed in Fig. 2.60 would in any case suggest weak coupling. In other cases, in particular, Ag and Au, it appears that the presence of antiferromagnetic coupling is often obscured by structural defects. Moreover the coupling is very weak for these metals. For Pd and Pt the coupling is strongly ferromagnetic with a coupling strength for thin layers considerably larger than that found for any other transition or noble metal. This is probably because both Pd and Pt are readily magnetically polarized by the magnetic layers with the development of very large moments on the Pd and Pt layers at the interfaces with the magnetic layers. Indeed, no compelling evidence for oscillations in the interlayer coupling for Pd or Pt has yet been reported for multilayers containing Co. Spin-engineered structures similar to those described in Sect. 2.5.4c show only strong ferromagnetic

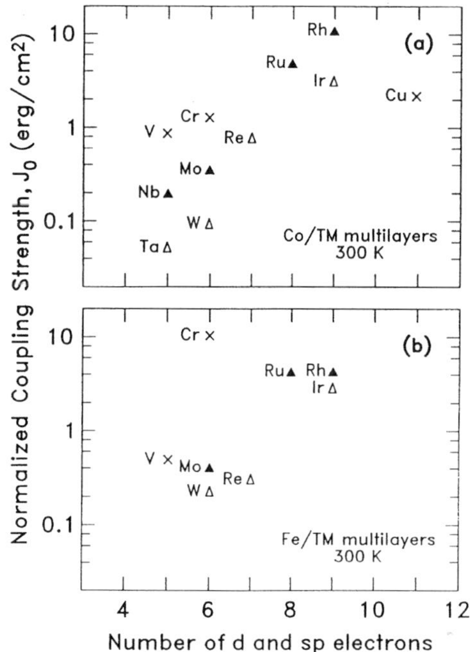

Fig. 2.60. Dependence of normalized exchange coupling constant on the 3d, 4d and 5d transition metals (TM) in (a) Co/TM and (b) Fe/TM multilayers

coupling which decays exponentially with increasing Pd or Pt thickness [2.276]. Evidence for the possibility of weak periodic variations in the strength of the ferromagnetic coupling has been reported in single crystal Fe/Pd/Fe trilayers, as discussed in Chap. 3.1 and [2.277].

Within RKKY and related models the oscillation period is related to the inverse length of wave vectors which span or nest the Fermi surface according to appropriate rules [2.261, 278]. Since the topology of the Fermi surfaces of the elements shown in Fig. 2.59 which display oscillatory coupling are very different, one would conclude that the period of the oscillations should vary widely from element to element. Indeed these have different crystal structures and moreover these films are polycrystalline with different orientations of the structures along the film growth axis. In a particular film structure the crystallites are oriented in a variety of directions, although usually there is some preferential orientation. Thus, the common oscillation period of $\simeq 10$ Å exhibited by these metals, with the exception of Cr, is quite surprising. Similarly the strength of the interlayer coupling should depend, within RKKY-like models, on the details of the Fermi surface topology, so it is surprising that the strength varies so systematically throughout the periodic table. Various models of the magnetic coupling have been developed [2.254, 261, 278–286], which are discussed in detail in earlier subsections of this chapter.

2.4.4.3 Spin Engineering – Direct Measurement of Ferromagnetic Coupling

A number of techniques have been used to study interlayer exchange coupling. These include Brillouin light scattering (BLS) (as discussed by *Cochran* in Chap. 3.2, and [2.287–289], ferromagnetic resonance (FMR) (*Heinrich*, Sect. 3.1, and [2.274]) and spin-polarized low energy electron diffraction (SPLEED) [2.290, 291]. In BLS and FMR the coupling strength is deduced from its effect on the measured frequency of excited spin wave modes. Although coupling strengths of both signs can be found, such data is more complicated to interpret than, for example, magnetic hysteresis loops. Moreover the sensitivity of such techniques is often limited to relatively large interlayer exchange coupling strengths. As discussed in Sect. 2.3 SPLEED has been used in the following manner. By taking advantage of its extreme surface sensitivity, the direction of magnetization in remanence of the topmost layer of a previously magnetized asymmetric sandwich structure is determined relative to that of the lower layer. However, since the measurement is restricted to zero field the magnitude of the interlayer coupling cannot be determined and the existence of ferromagnetic coupling can only be inferred.

The ferromagnetic exchange coupling strength can be directly measured from simple magnetic hysteresis loops by *spin-engineering* appropriate structures [2.275]. A magnetic sandwich is used and it is comprised of soft ferromagnetic layers in which the magnetization of one of the magnetic layers is pinned anti-parallel to the applied magnetic field. The pinning is accomplished by an additional magnetic layer strongly antiferromagnetically coupled to the back of one of the soft layers through a second thin metallic layer. Paradoxically the magnetic moments of the two soft layers become anti-parallel on application of a field.

A schematic structure of a spin-engineered structure designed to measure the ferromagnetic coupling strength in a $Ni_{80}Co_{20}(t_F)/Ru(d)/Ni_{80}Co_{20}(t_F)$ sandwich is shown in Fig. 2.61. One of the $Ni_{80}C_{20}$ layers, F_I, is antiferromagnetically coupled via a second thin Ru layer of thickness, t_P, to a third magnetic layer, in this case cobalt. The coupling between Co and $Ni_{80}C_{20}$ via Ru is several times larger than the coupling between two $Ni_{80}C_{20}$ layers via Ru for equivalent Ru thicknesses. Moreover, Co is AF coupled to $Ni_{80}C_{20}$ in the limit of ultrathin Ru layers with a coupling strength that rapidly increases as the Ru layer thickness is decreased to the point ($\simeq 3$ Å) at which direct coupling through pinholes in the Ru layer overwhelms the AF coupling. Consequently, the $Ni_{80}Co_{20}$ layer, F_I, is extremely strongly antiferromagnetically coupled to the Co layer. In contrast, the AF coupling between the $Ni_{80}C_{20}$ layers in the same limit is very small. Finally, the thickness of the Co layer is chosen such that the magnetic moment of the Co layer is approximately equal to the sum of the magnetic moments of the two $Ni_{80}C_{20}$ layers. Under these circumstances, neglecting anisotropy, the net moment of the structure will be approximately zero in zero field for ferromagnetic A_{12} (Fig. 2.61).

Figure 2.62 depicts magnetic hysteresis loops for four structures of the form $Si/Ru(85$ Å$)/[Co(15$ Å$)/Ru(6$ Å$)/Ni_{80}Co_{20}(15$ Å$)/Ru(d)/Ni_{80}Co_{20}(15$ Å$)]_5$. The $Ni_{80}Co_{20}$ layers and Co layers are each $\simeq 15$ Å thick. The structures contain five identical repeats of the five layer unit shown in Fig. 2.61 separated from each other by a thick Ru layer, $\simeq 85$ Å thick, through which there is negligible exchange coupling. The magnetic hysteresis loops are consistent with the expected spin arrangement shown in Fig. 2.61 and directly give evidence for ferromagnetic A_{12} for Ru layer thicknesses near 3, 13 and 26 Å. In particular, as shown in Fig. 2.62, for these Ru layer thicknesses the magnetic hysteresis loops at low fields exhibit a characteristic shape requiring the application of a field of

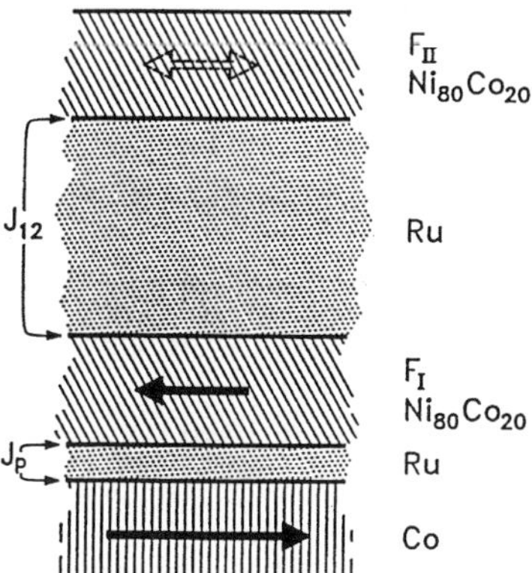

Fig. 2.61. Schematic diagram of a spin-engineered structure. The exchange coupling, J_{12}, between two $Ni_{80}Co_{20}$ layers is measured by pinning the moment of one of the $Ni_{80}Co_{20}$ layers (F_I) antiparallel to a Co layer. The moment of the Co layer is set equal to the sum of the moments of the two $Ni_{80}Co_{20}$ layers

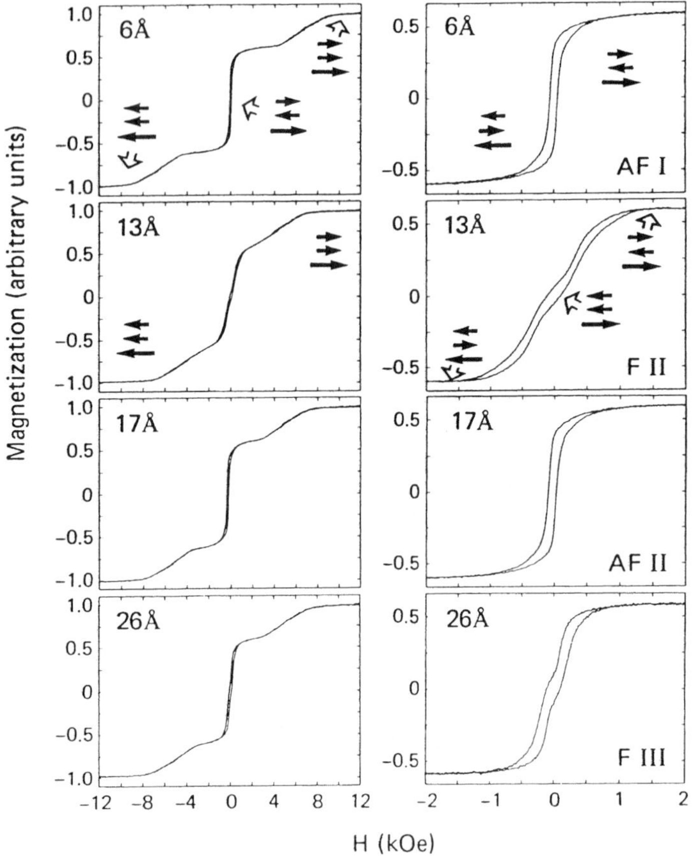

Fig. 2.62. Magnetization versus field curves for four samples of the form Si/Ru(85 Å)/[Co(15 Å)/Ru(6 Å)/$Ni_{80}Co_{20}$(15 Å)/Ru(d)/$Ni_{80}Co_{20}$(15 Å)]$_5$ for $d = 6, 13, 17$ and 26 Å. The low field data is shown in more detail on the right hand side of the figure. Data is shown for representative samples for the first and second antiferromagnetic regions and second and third ferromagnetic regions. The arrangement of the moments of the Co and $Ni_{80}Co_{20}$ F_I and F_{II} layers as the field is varied is shown schematically for F and AF coupled samples

up to 1.3 kOe to reach the intermediate plateau in magnetization found in all of the samples. This plateau at approximately half of the total moment of the structure is consistent with parallel alignment of F_{II} and the Co layer. For intermediate Ru thicknesses the plateau is attained in much smaller fields determined by the magnetic coercivity of the magnetic layers, consistent with antiferromagnetic A_{12} (Fig. 2.61).

The magnitude of the AF coupling was directly measured from the saturation field of a second series of simple bilayer multilayers of the form [$Ni_{80}Co_{20}$(30 Å)/Ru(t_{Ru})]$_{20}$. The strength of the ferro- and antiferro magnetic interlayer exchange coupling is thus given, respectively, by $2n_i|A_{12}| = H_s M t_F$, where H_s is

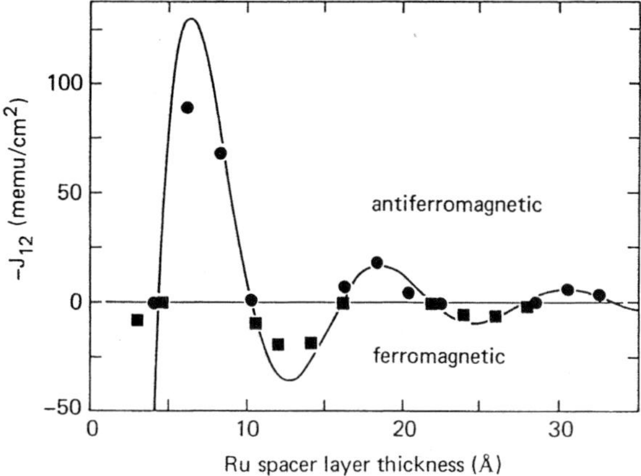

Fig. 2.63. Interlayer exchange coupling strength, J_{12}, for coupling of $Ni_{80}Co_{20}$ layers through a Ru spacer layer. J_{12} is defined per unit area of the interface and is determined from magnetization curves of structures of the form **(a)** $Si/Ru(85 Å)/[Co(15 Å)/Ru(6 Å)/Ni_{80}Co_{20}(15 Å)/Ru(t_s)/Ni_{80}Co_{20}(15 Å)]_5$ for *ferromagnetic* coupling, and **(b)** $Si/Ru(105 Å)[Ni_{80}Co_{20}(30 Å)/Ru(t_{Ru})]_{20}/Ru(105 Å)$ for *antiferromagnetic coupling*. The data points are shown as **(a)** squares and **(b)** circles. For each structural type only **(a)** *ferromagnetic* or **(b)** *antiferromagnetic* coupling can be measured. Data points are not shown for structures for which no coupling could be determined. The solid line corresponds to a fit to the data of a RKKY form

the field required to attain the plateau in the spin-engineered structures and complete saturation in the bilayer multilayers. The coefficient, n_i, is 1 and 2, respectively, for these different structures since each $Ni_{80}Co_{20}$ layer is coupled to just one $Ni_{80}Co_{20}$ layer in the spin-engineered multilayers but to two in the bilayer multilayers (neglecting end effects in the latter [2.250]).

Values of A_{12} determined from the saturation field as described above (corrected for coercivity) are plotted versus Ru layer thickness for both series of structures in Fig. 2.63. The exchange coupling is clearly demonstrated to oscillate through zero. Moreover, as shown in Fig. 2.63, the dependence of A_{12} is well described by a RKKY-like exchange coupling of the form A_{12} prop $\sin(\phi + 2\pi t_{Ru}/\lambda_F)/t_{Ru}^p$, where $p \simeq 1.8$ and $\lambda_F \simeq 11.5$ Å. The value of p is in good agreement with theoretical predictions of 2 for the planar geometry [2.236]. The value of λ_F is much longer than the Fermi wavelength for Ru. However, λ_F will be determined by the detailed shape of the Fermi surface [2.292] which will inevitably give rise to longer length scales (*Fert* and *Bruno*, Sect. 2.2, and *Hathaway*, Sect. 2.1).

2.4.5 Giant Magnetoresistance of Cu-Based Multilayers

The ability to prepare large numbers of structures by sputtering enabled a rapid survey of a multitude of metal multilayers which culminated in the discovery of the enormous giant magnetoresistance effects in Co/Cu multilayers exceeding 65% at room temperature [2.271]. Such values are 10 to 30 times larger than typical anisotropic magnetoresistance values in ferromagnetic alloy films.

2.4.5.1 Influence of Structure on Giant Magnetoresistance

The largest GMR effects have been found in antiferromagnetically coupled polycrystalline Co/Cu multilayers [2.271]. The properties of such magnetic multilayers are often sensitive to growth conditions including, for example, the deposition method, the temperature of growth, the substrate material and the buffer layer, if any, between the substrate and multilayer. The structure and physical properties of multilayers containing Cu layers have been found to be particularly sensitive to deposition conditions, both for sputtered and MBE prepared multilayers. This is demonstrated in Fig. 2.64 which shows room temperature magnetoresistance data for several nominally identical magnetron sputtered Co/Cu multilayers deposited on 50 Å thick Fe or Cu buffer layers. (The growth of the structures is described elsewhere [2.239].) The change in resistance, ΔR, is normalized to the resistance of the multilayer at high field, R. The magnitude of the saturation magnetoresistance, $\Delta R/R$, is very large and is almost 50% in the structure grown on an Fe buffer layer with an Fe capping layer. As can be seen by comparing structures with the same underlayer, changing the capping layer from Fe to Cu considerably reduces the MR. However, the capping layer is not expected to significantly alter the structure or properties of the multilayer. The reduction in MR can be simply accounted for by the higher electrical conductivity of Cu compared to Fe which results in a significant shunting of the sensing current through the capping layer. This reduces the proportion of current passing through the multilayer itself and so reduces the MR.

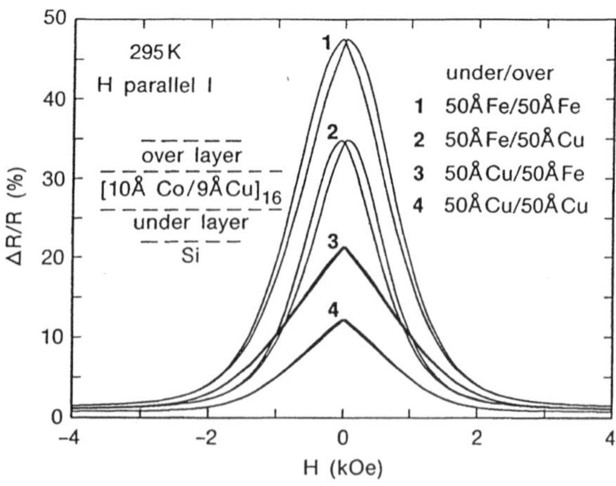

Fig. 2.64. Room temperature resistance versus field curves for four samples of the form Si(1 0 0)/buffer layer/[10 Å Co/9 Å Cu]₁₆/capping layer with 50 Å thick buffer and capping layers of respectively (1) Fe and Fe, (2) Fe and Cu, (3) Cu and Fe and (4) Cu and Cu

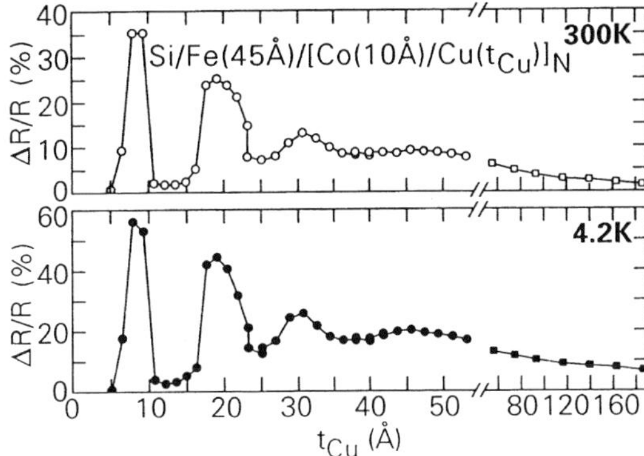

Fig. 2.65. Dependence of saturation transverse magnetoresistance on Cu spacer layer thickness for a family of related superlattice structures of the form $Si/Fe(40\,\text{Å})/[Co(10\,\text{Å})/Cu(t_{Cu})]_N$. An additional Cu layer was deposited on each film structure such that the uppermost Cu layer was $\simeq 55\,\text{Å}$ thick. The number of bilayers in the superlattice, N, is 16 for t_{Cu} below 55 Å (fcir, bcirc) and eight for t_{Cu} above 55 Å ($\square$, fsqu)

In contrast, changing the underlayer material significantly alters the MR, for the same degree of shunting of the sensing current. The latter is realized, to a first approximation, by preparing structures with the same net thicknesses of the Fe, Co and Cu layers in the structure. For example, by comparing the MR data for structures 2 and 3 in Fig. 2.64, which have underlayers/capping layers of 50 Å Fe/50 Å Cu and 50 Å Cu/50 Å Fe, respectively, it is clear that the structure grown using an Fe buffer layer displays a significantly higher MR. The reason for this is evident from both magnetic and structural characterization of these samples. Magnetic studies show incomplete antiferromagnetic coupling of the Co layers for the multilayer grown on copper buffer layers. The magnitude of the GMR is directly related to the degree of AF coupling, therefore this accounts for the reduced MR. Both cross section transmission electron microscopy (XTEM) images as well as Auger sputter depth profiling show the presence of Cu more than 100 Å beneath the silicon surface. These studies suggest that the Cu underlayer reacts with the silicon substrate. The XTEM studies clearly show that the reaction of the Cu with the silicon results in rumpled Co and Cu layers as compared with growth on Fe buffer layers. Indeed varying the buffer layer is an important method to control the structural morphology of the multilayer and thus influence the magnitude of the GMR.

2.4.5.2 Oscillatory Dependence of GMR on Cu Layer Thickness

The dependence of saturation magnetoresistance on Cu layer thickness is shown in Fig. 2.65 for Co/Cu and in Fig. 2.66 for $Ni_{81}Fe_{19}/Cu$ multilayers. The former are prepared on Fe and the latter on $Ni_{18}Fe_{19}$ buffer layers. In both cases, at

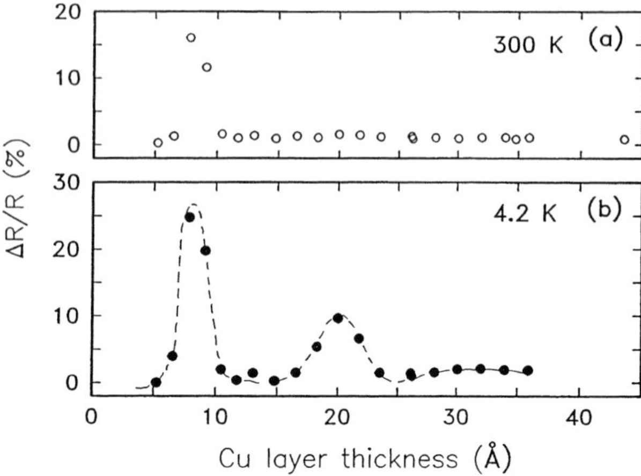

Fig. 2.66. Dependence of saturation magnetoresistance versus Cu spacer layer thickness for multilayers of the form, $\text{Si}(1\,0\,0)/50\,\text{Å }\text{Ni}_{81}\text{Fe}_{19}/[15\,\text{Å }\text{Ni}_{81}\text{Fe}_{19}/\text{Cu}(t_{\text{Cu}})]_{14}/25\,\text{Å Ru}$, at (**a**) 300 K and (**b**) 4.2 K. Ranges of Cu spacer layer thicknesses for which the $\text{Ni}_{81}\text{Fe}_{19}$ layers are coupled antiferromagnetically are shown as AF1, AF2 and AF3

4.2 K well defined oscillations in the MR are found as a function of Cu spacer layer thickness. It was first shown for Fe/Cr and Co/Ru multilayers that the oscillations in MR are clearly connected to oscillations in the exchange coupling mediated by the non-magnetic layer. In these systems, enhanced MR, larger than the anisotropic MR of the magnetic material itself, is found only in the antiferromagnetically coupled multilayers whereas no significant enhancement of the MR is observed in multilayers with substantial ferromagnetic interlayer coupling. In the latter case the relative orientations of the local magnetizations in adjacent magnetic layers are unaffected by the magnetic field.

For the Co/Cu system similar oscillations are found at all temperatures from below 4.2 K to above 400 K. At still higher temperatures the structures are unstable and the Co and Cu layers interdiffuse into one another. For the permalloy/Cu multilayers only a single oscillation is observed at room temperature for magnetron sputtered multilayers. This can be readily explained as follows. For the permalloy/Cu system the exchange coupling is significantly weaker than in Co/Cu. At 4.2 K the antiferromagnetic coupling, $A_{12}(\text{AF})$, for $t_{\text{Cu}} = \simeq 8\,\text{Å}$, is about three to six times smaller than that in Co/Cu. Note that $A_{12}(\text{AF})$ is related to the saturation field, H_s by $A_{12}(\text{AF}) \simeq -H_s M_s t_F/4$, where M_s and t_F are the saturation magnetization and thickness, respectively, of the magnetic layers. If we assume that the structures are not perfect and that there are defects in the system, for example, pinholes of magnetic material extending across the Cu layers from one magnetic layer to the next, or equivalently, necks in the Cu spacer layers where the layer is locally thin, such defects may give rise to ferromagnetic coupling of the magnetic layers. Not only is the magnetic coupling weak in the permalloy/Cu system but it has a much stronger temperature dependence than in Co/Cu. In the latter system the AF coupling at the first AF peak, $A_{12}^1(AF)$, weakens by only $\simeq 25\%$ between 4.2 and 300 K, whereas in the former $A_{12}^1(AF)$ changes by a factor of 2.5. Thus, it seems reasonable to argue that in NiFe/Cu multilayers at low temperatures where the AF coupling is considerably stronger it is likely that more oscillations in coupling will be observed than at higher temperatures where the coupling may be weak compared to direct ferromagnetic coupling via defects.

2.4.5.3 Dilution of GMR with Cu Layer Thickness

Figure 2.65 shows the variation of magnetoresistance in Co/Cu multilayers for Cu layer thicknesses extending out to 180 Å. Actually, substantial magnetoresistance is observed for Cu layers with thicknesses even several times thicker. Typical resistance versus field curves for Co/Cu multilayers are shown in Fig. 2.67 for Cu layers ranging up to 425 Å thick. The shape of these curves is distinct from that of the bell-shaped curves shown in Fig. 2.64 (Note that the permalloy/Cu structures display characteristic triangularly-shaped resistance versus field curves often indicative of incomplete antiferromagnetic coupling of the magnetic layers.) The structures shown in Fig. 2.67 exhibit a double-peaked MR curve with maximum resistance at fields corresponding to the $\pm H_c$, where H_c is the coercive or switching field at which the magnetization passes through zero. The interlayer exchange coupling decreases with increasing Cu thickness much faster than the GMR effect, such that the exchange coupling fields are much weaker than H_c for the structures shown in Fig. 2.67. Thus, the MR in Fig. 2.67 results from the random arrangement of magnetic domains in successive magnetic layers.

A schematic diagram of the likely magnetic domain structure is shown in Fig. 2.68. Lorentz electron microscopy of the remanent magnetic state of

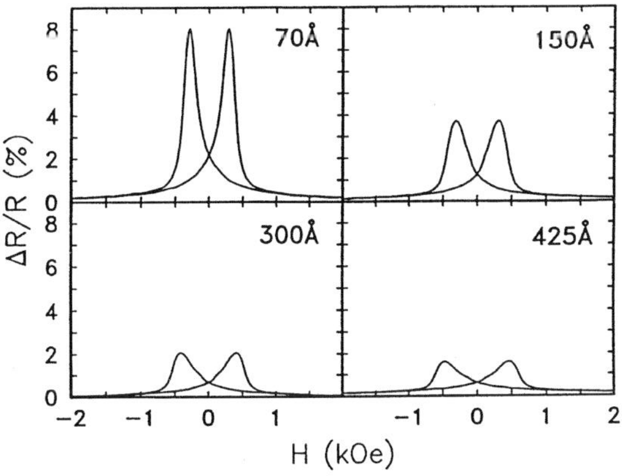

Fig. 2.67. Resistance versus field curves for four Co/Cu multilayers of the form, Si(1 1 1)/Ru(50 Å)/[Co(11 Å)/Cu(t_{Cu})]₆/Ru(15 Å) with Cu spacer layer thicknesses, t_{Cu}, of 70, 150, 300 and 435 Å

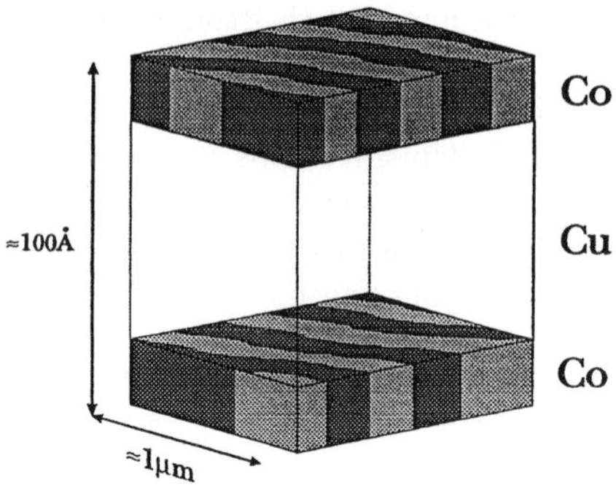

Fig. 2.68. A schematic diagram of the arrangement of the magnetic domains in the remanent magnetic state of a Co/Cu multilayer. The darker and lighter shaded regions correspond to longitudinal magnetic domains aligned parallel and antiparallel to the magnetic field direction

polycrystalline Co films a few hundred angstroms thick shows longitudinal magnetic domains aligned along the magnetizing field direction [2.293]. For Co/Cu multilayers in which the interlayer coupling is weak it is likely that there will be domains in neighboring magnetic layers with their magnetization axes aligned non-parallel to one another. The magnetic configuration with the highest degree of anti-parallelism of magnetic domains will arise when the net magnetic moment is zero, i.e. at $\pm H_c$. The degree of antiparallelism will depend on the nature, (i.e. the symmetry and the strength) of the magnetic anisotropy within the magnetic layers themselves, as well as any magnetic coupling (exchange or magnetostatic) between the magnetic layers.

Perhaps it seems surprising that the giant MR effect was not observed in magnetic multilayer systems long before the observation of antiferromagnetic coupling since the data in Fig. 2.67 suggest that strong antiferromagnetic coupling is not a necessary requirement for the observation of giant MR. Indeed enhanced MR, although small (1–3%), was reported in uncoupled single crystalline Co/Au/Co sandwiches [2.294] prior to the observation of giant MR in antiferromagnetically coupled Fe/Cr multilayers and sandwiches. It was only later recognized that this was most probably a manifestation of the same GMR phenomenon [2.295]. In any case it appears that the magnitude of the giant MR effect is tied to that of the interlayer coupling and the systems which exhibit the largest interlayer coupling also exhibit the largest giant MR effect.

The dependences of saturation magnetoresistance on Cu thickness, t_{Cu}, for thick Cu spacer layers is shown in detail in Fig. 2.69 for Co/Cu multilayers grown on Ru buffer layers. Well defined oscillations in MR are observed for thinner Cu layers. For Cu layers thicker than $\simeq 60$ Å the MR decays. The dependence of MR on t_{Cu} is straightforward. At 4.2 K (Fig. 2.69d), the MR decays approximately as $1/t_{Cu}$. The GMR phenomenon is usually discussed in terms of spin dependent scattering within the interior of the magnetic layers (*bulk* scattering) or at the interfaces between the magnetic and non-magnetic layers (*interfacial* scattering), thus this can be readily understood [2.296] as *dilution* of the spin dependent scattering regions as the measuring current, which is parallel to the layers, is shunted away from these regions through the Cu

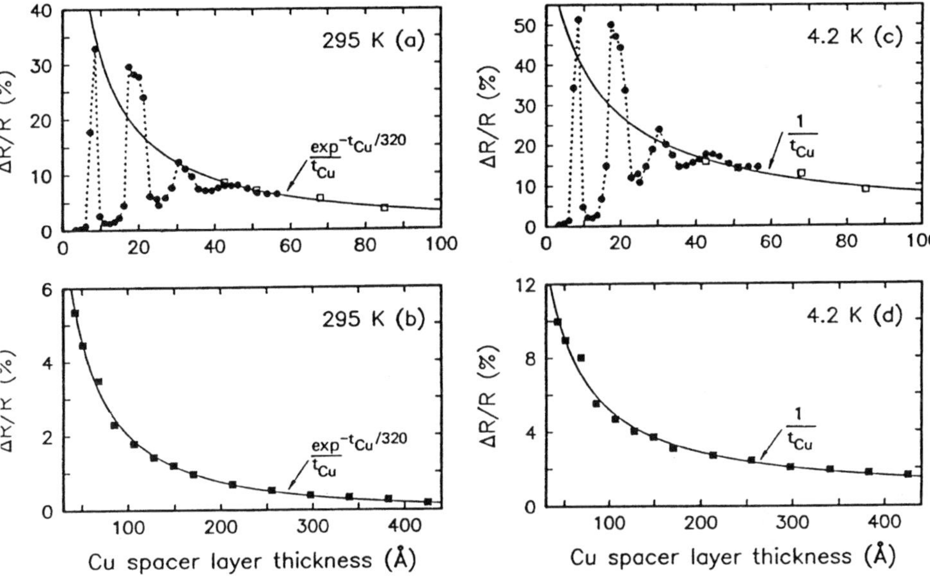

Fig. 2.69. Saturation magnetoresistance versus Cu spacer layer thickness for several series of structures of the form, Si(1 1 1)/Ru(50 Å)/[Co(11 Å)/Cu(t_{Cu})]$_N$/Ru(15 Å). The number of Co/Cu periods, N, is 20 (solid circles) and six (open and closed squares). Data are shown for temperatures of (a) and (b) 295 K, and (c) and (d) 4.2 K. Since the MR increases with N, in figures (a) and (c) the data for the structures with $N = 6$ has been scaled by a factor of 1.6 to make comparison with the $N = 20$ data easier. The curves through the data are of the form $1/t_{Cu} \exp - (t_{Cu}/\lambda_{Cu})$ at 295 K and $1/t_{Cu}$ at 4.2 K. Note the actual curves shown in the figure are of the exact form, $\Delta R/R = 289/(4.3 + t_{Cu}) \exp - (t_{Cu}/318)$ and $\Delta R/R = 0.28 + 554/(13 + t_{Cu})$ at 295 and 4.2 K respectively, and are also scaled by a factor of 1.6 in (a) and (c)

layers. Furthermore, since GMR is found only in systems comprised of at least two magnetic layers separated by a non-magnetic layer, the effect requires the flow of electrons from one magnetic layer to neighboring layers. Scattering within the spacer layers will diminish the flow of electrons and so reduce the magnitude of the GMR. Such scattering should be related to *volume* scattering within the interior of the spacer layers. Therefore it can be described by a scattering length, λ, which should be simply related to the mean free path of thick Cu layers where surface scattering is small compared to volume scattering. Taking into account both volume scattering and *dilution* one expects [2.296] the GMR to decay as $\simeq 1/t_{Cu} \exp - (t_{Cu}/\lambda_{Cu})$. Figure 2.69b shows that such a functional form well describes the dependence of MR on Cu layer thickness at 295 K in Co/Cu for Cu layer thicknesses ranging from 50 to more than 500 Å. The value of λ of $\simeq 320$ Å compares with $\simeq 390$ Å in single crystalline Cu. From measurements of the dependence of MR on Cu thickness at various temperatures λ_{Cu} is found to have a strong temperature dependence increasing as the temperature is reduced.

2.4.6 Low Field Giant Magnetoresistance Structures

Although very large GMR values are obtained in Co/Cu and other Cu-based multilayers at room temperature the magnetic fields required are nevertheless large. For Co/Cu the saturation fields at the first antiferromagnetic peak are $\simeq 10$ kOe for Co layers, 10 Å thick. Lower saturation fields are possible by increasing the thickness of the Co layers but the MR decreases at approximately the same rate as the saturation field giving approximately constant MR per unit field values. By taking advantage of the rapid decrease of interlayer coupling with increasing Cu spacer layer thickness it is possible to obtain MR values exceeding $\simeq 35\%$ at room temperature for saturation fields as low as 100–200 Oe [2.297]. The change in resistance per unit field is still low compared, for example, to thin films of permalloy (Fig. 2.54). For technological applications of GMR such as for magnetoresistive read heads in magnetic storage applications, e.g., computer disk drives, GMR structures must outperform permalloy and related alloys.

The very first observation of GMR was made not in magnetically coupled multilayers but in sandwiches of Co/Au/Co [2.294, 295]. The two Co layers are of different thicknesses but sufficiently thin that they exhibit considerable perpendicular magnetic anisotropy. The anisotropy is large enough that the magnetic moments of both Co layers are aligned normal to the layers in small fields. Since the anisotropy has a strong dependence on film thickness, by carefully choosing the Co layer thicknesses, the magnetic moments of the two Co layers can be arranged to switch their moments in substantially different magnetic fields. This leads to an antiparallel alignment of the Co moments for fields intermediate between the respective switching fields of the two Co layers. The resistance of the structure is higher in this field regime compared to fields for which the moments of the Co layers are parallel. Similar structures using two different magnetic layers chosen to have different in-plane magnetic anisotropies such as Permalloy/Cu/Co/Cu [2.298] and Fe/Cu/Co/Cu multilayers [2.299] have also been considered. None of these structures, however, gives MR/H results better than Co/Cu multilayers.

A structure which does give improved performance and which is conceptually very similar to such two magnetic component multilayers is shown in Fig. 2.70. The structure, and exchange-biased sandwich (EBS), is of the form $F_I/S/F_{II}/FeMn$. The structure contains two ferromagnetic layers, F_I and F_{II} and a single non-magnetic spacer layer, S, in which one of the magnetic layers, F_{II}, is exchange coupled to an antiferromagnetic layer of FeMn [2.300]. The structure takes advantage of a phenomenon first discovered more than 30 years ago in oxidized Co particles [2.301] and subsequently extensively studied in a number of thin film systems [2.302–304]. This phenomenon, often referred to as exchange anisotropy, arises from an interfacial magnetic exchange coupling between an antiferromagnetic layer and a ferromagnetic layer. Under appropriate conditions the exchange anisotropy results in a unidirectional anisotropy in the F layer such that its magnetic hysteresis loop is centered about a non-zero

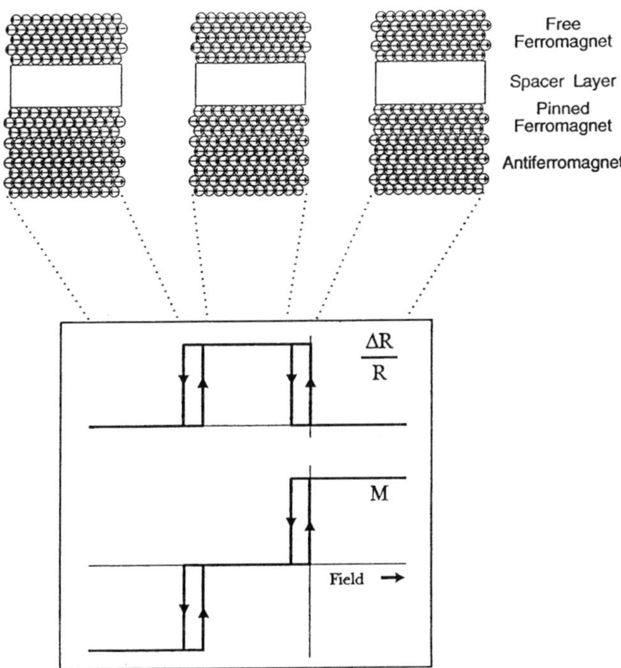

Fig. 2.70. Schematic diagram of an exchange-biased sandwich structure

field, a bias field, H_B. The latter imposes a unidirectional magnetic anisotropy on F_{II} such that its magnetic hysteresis loop is centered about a non-zero field, H_B. In contrast, providing the magnetic coupling of F_I and F_{II} via the spacer layer is weak enough, the magnetic hysteresis loop of F_I is centered close to zero field. The moments of F_I and F_{II} are thus aligned anti-parallel for some field range intermediate between zero and H_B.

A resistance versus field curve is shown in Fig. 2.71a for a typical EBS structure where F_I and F_{II} are $Ni_{81}Fe_{19}$ and S is Cu. The current and field are aligned along the unidirectional anisotropy direction. The structure displays a giant MR effect exactly analogous to that in multilayers with a higher resistance for fields where F_I and F_{II} are antiparallel. As is found for multilayered structures [2.239, 305] replacing the $Ni_{81}Fe_{19}$ layers with Co layers of the same thickness increases the magnetoresistance of the structure by approximately a factor of two, as presented in Fig. 2.71b.

2.4.7 Interfacial Origin of Giant Magnetoresistance

The detailed origin of the *giant magnetoresistance* (GMR) effect has been a subject of some controversy. As discussed in Sect. 2.4.3 the GMR effect is believed to be related to the *contrast* between the scattering lengths in the up-

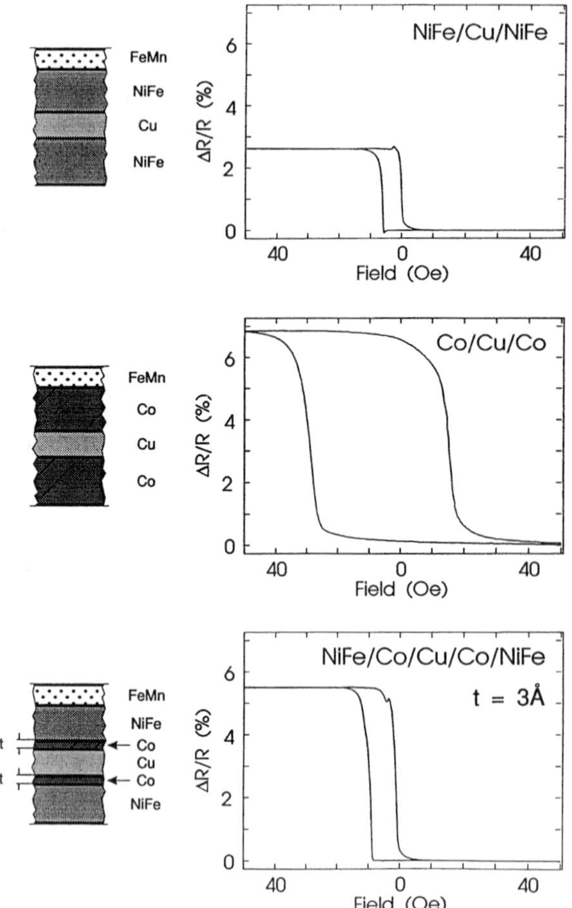

Fig. 2.71. Room temperature resistance versus field curves for (**a**) $Si/Ni_{81}Fe_{19}(53)/Cu(32\ \text{Å})/Ni_{81}$-$Fe_{19}(22\ \text{Å})/FeMn(90\ \text{Å})/Cu(10\ \text{Å})$ and (**b**) the same structure with the permalloy layers replaced by Co and (**c**) the same structure as in (**a**) but with 3.0 Å thick Co layers added at each $Ni_{81}Fe_{19}/Cu$ interface. (Note the thicknesses of the $Ni_{81}Fe_{19}$ layers have correspondingly been reduced by 3.0 Å)

spin and down-spin conduction channels. A question of particular importance is the whereabouts, of the conduction electron scattering, whether within the bulk of the magnetic layers or at the magnetic/non-magnetic interfaces, which gives rise to the effect. In this section we discuss several experiments to probe this question.

2.4.7.1 Dependence of GMR on Magnetic Layer Thickness

For predominant "interface scattering" of the electrons, one expects the GMR to be reduced by *dilution* of the interfacial regions, for example, by increasing the thickness of the magnetic or non-magnetic layers for thicknesses above the width of the interfacial scattering regions. For Fe/Cr, Co/Ru and Co/Cu and $Ni_{91}Fe_{19}$/Cu multilayers, the GMR effect takes its maximum value for magnetic (Fe, Co or $Ni_{81}Fe_{19}$) layer thicknesses of just 8–10 Å. For structures designed to have negligible current shunting through the non-magnetic layers and any buffer or capping layers, the GMR falls off simply as the inverse magnetic layer thickness for thicker layers. These results are consistent with predominant "interface scattering" within less than 10 Å of the interfaces. However, such a length scale is so short that the properties of the magnetic layers themselves are likely to change when they become so thin.

2.4.7.2 Interfacial "Dusting"

A simple method [2.306, 307] to examine the importance of interfacial scattering is to modify the magnetic/non-magnetic interfaces by adding a third material. It is important to consider cases where the MR might be expected to increase since it is very easy to reduce the MR of a structure by inserting material at the interfaces which gives rise to increased spin independent scattering or equivalently resistivity. This is the case for almost any non-magnetic material.

A particularly interesting case is that of $Ni_{81}Fe_{19}$/Cu multilayers in which thin layers of Co are inserted at each $Ni_{81}Fe_{19}$–Cu interface. Since the magnetoresistance of $Ni_{81}Fe_{19}$/Cu multilayers is about half that of similar Co/Cu multilayers with layers of comparable thickness we might expect a significant increase in the MR of the permalloy/Cu multilayers if interfacial scattering is significant. Figure 2.72 demonstrates that indeed the properties of $Ni_{81}Fe_{19}$/Cu multilayers are dramatically modified by inserting thin Co layers at the $Ni_{81}Fe_{19}$/Cu interfaces. Also included in Fig. 2.72 are the resistance versus field curves at temperatures of 300 and 4.2 K for three $Ni_{81}Fe_{19}$/Cu structures, $Si/Ru(34\ Å)/Ni_{81}Fe_{19}(10\ Å)/Co(t_i)/[Cu(19\ Å)/Co(t_i)/Ni_{81}Fe_{19}(10\ Å)/Co(t_i)]_{19}/Ru(14\ Å)$, with Co interface layer thicknesses, t_i, of 2.6, 3.5 and 4.4 Å. For comparison, resistance versus field data are also shown in Fig. 2.72 for two $Ni_{81}Fe_{19}$/Cu multilayers without Co interface layers of the form, $Si/Ni_{81}Fe_{19}(50\ Å)/[Ni_{81}Fe_{19}(15\ Å)/Cu(20\ Å)]_{14}/Ru(25\ Å)$. All of these structures contain Cu spacer layers which correspond to the second antiferromagnetic oscillation in Figs. 2.65, 66.

The addition of the thin Co layers restores the antiferromagnetic coupling, absent at room temperature in the pure $Ni_{81}Fe_{19}$/Cu structures at this Cu layer thickness, and, perhaps more remarkably, dramatically increases the MR of the multilayers. The room temperature data in Fig. 2.72 clearly show that the interlayer exchange coupling is determined by the character of the ferromagnetic/spacer layer interfaces. However, since the MR is only present for antiferromagnetically coupled structures, perhaps it is not surprising that the reestablishment of antiferromagnetic coupling in these structures is accompanied by a giant MR effect. In contrast, the low temperature data, for which there is

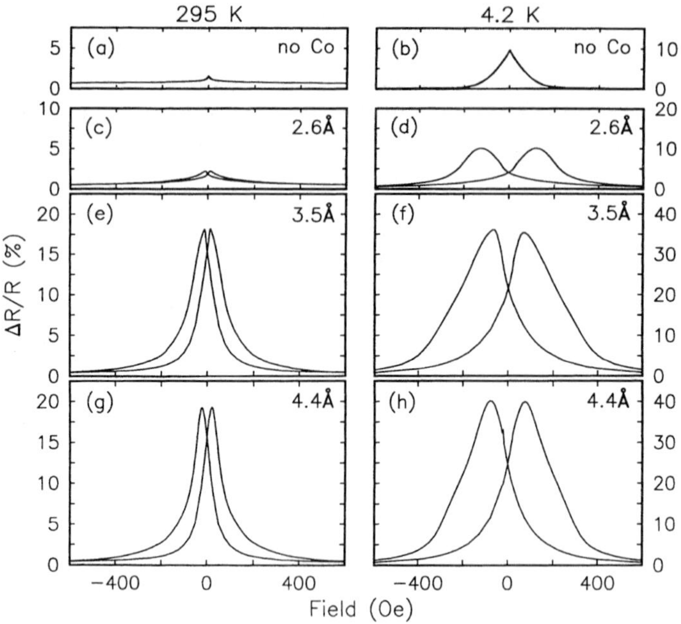

Fig. 2.72. Resistance versus in-plane field for structures of the form **(a)** and **(b)**, $Si/Ni_{81}Fe_{19}(50\ \text{Å})/[15\ \text{Å}\,Ni_{81}Fe_{19}(15\ \text{Å})/Cu(20\ \text{Å})]_{14}/Ru(25\ \text{Å})$ and **(c)**, **(d)**, **(e)**, **(f)**, **(g)** and **(h)**, $Si/Ru(34\ \text{Å})/Ni_{81}Fe_{19}(10\ \text{Å})/Co(t_i)/[Cu(19\ \text{Å})/Co(t_i)/Ni_{81}Fe_{19}(10\ \text{Å})/Co(t_i)]_{19}/Ru(14\ \text{Å})$ with t_i = **(c)** and **(d)** 2.6 Å, **(e)** and **(f)** 3.5 Å, and **(g)** and **(h)** 4.4 Å. Data in the left hand column of the figure are taken at 295 K and data in the right hand column, are measured at 4.2 K

antiferromagnetic coupling even in the pure $Ni_{81}Fe_{19}/Cu$ structures, unambiguously demonstrates that the addition of Co layers just 3–4 Å thick dramatically increases the magnetoresistance, almost quadrupling it.

2.4.7.3 Length Scale of Interfacial Scattering

By using exchange-biased sandwich structures (described in Sect. 2.4.6) instead of multilayers, the dependence of the MR on the thickness of the inserted interfacial layer for interface layers as thin as $\simeq 0.5$ Å can be examined in detail [2.306]. From such data a characteristic length can be ascribed to the thickness of the interfacial layer required to establish the character of the interface and the magnitude of the MR. This length is extremely short at just 2 to 3 Å at room temperature for a wide range of magnetic/non-magnetic material combinations. Such experiments seem to unambiguously demonstrate the interfacial origin of the giant MR effect.

Similar to the studies described in the previous section, thin magnetic layers of a different character are inserted at the magnetic/non-magnetic interfaces in the exchange biased sandwich (EBS) structures described in Sect. 2.4.6. Again, useful EBS structures are those comprised of Co/Cu/Co and Permalloy/Cu/Permalloy. As discussed in earlier sections for the multilayered structures described above [2.271, 305], replacing the NiFe layers in a NiFe/Cu/NiFe sandwich with Co layers of the same thickness increases the magnetoresistance of the structure by approximately a factor of two. The importance of interface scattering can be evaluated by introducing thin layers of, for example, Co at the NiFe/Cu interfaces in NiFe/Cu/NiFe sandwiches. If spin dependent interface scattering is the dominant mechanism giving rise to giant MR, thin layers of Co will result in a large increase in MR. In contrast, if bulk scattering is at the origin of giant MR, much thicker layers of Co will be required to substantially alter the MR effect.

Figure 2.71c shows that "dusting" of the NiFe/Cu interfaces with thin Co layers just 3.0 Å thick almost doubles the MR of the NiFe/Cu/NiFe EBS making it comparable to that of the EBS in which the NiFe layers are completely replaced by Co. Figure 2.73a describes in detail, the dependence of the saturation magnetoresistance on the thickness of the Co interface layer, t_i.

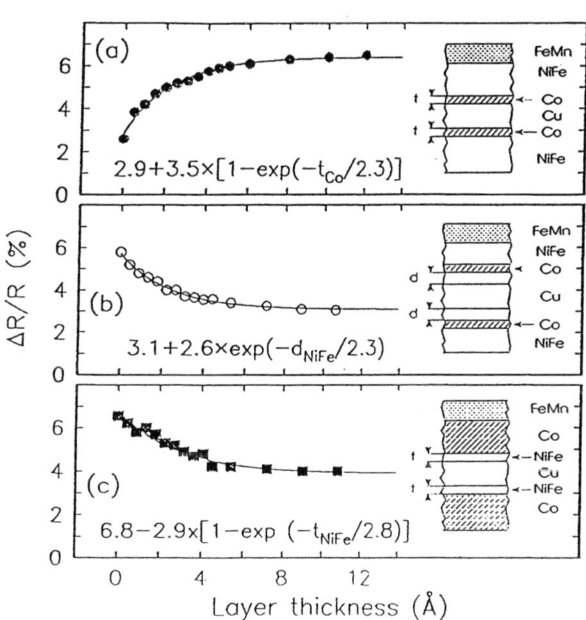

Fig. 2.73. Dependence of room temperature saturation magnetoresistance on (**a**) Co interface layer thickness, t_{Co}, in sandwiches of the form $Si/Ni_{81}Fe_{19}(53-t_i)/Co(t_i)/Cu(32)/Co(t_i)/Ni_{81}Fe_{19}(22-t_i)/FeMn(90)/Cu(10)$, (**b**) distance of a 5 Å thick Co layer from the $Ni_{81}Fe_{19}/Cu$ interfaces in sandwiches of the form $Si/Ni_{81}Fe_{19}(49-d)/Co(5)/Ni_{81}Fe_{19}(d)/Cu(30)/Ni_{81}Fe_{19}(d)/Co(5)/Ni_{81}Fe_{19}(18-d)/FeMn(90)/Cu(10)$, and (**c**) $Ni_{81}Fe_{19}$ interface layer thickness, t_i, in sandwiches of the form $Si/Co(57-t_i)/Ni_{81}Fe_{19}(tNi_{81}Fe_{19})/Cu(24)/Ni_{81}Fe_{19}(t_i)/Co(29-t_i)/FeMn(100)/Cu(10)$. Note layer thicknesses are in angstroms

The dependence is well described by a function of the form, $\Delta R/R = a + b \times [1 - \exp(-t_i/\xi)]$, where, ξ, is extremely small and is $\simeq 2.3$ Å. Note that the thickness of the NiFe layers has been reduced by approximately the thickness of the Co layers inserted at the interfaces and that the sheet resistance of the structures shown in Fig. 2.73a varies by less than 5% from the most resistive to the least resistive.

Since it is possible, if bulk scattering were important, that the MR of such a NiFe/Cu/NiFe EBS may be increased no matter where the additional Co layers are introduced, a companion set of structures to those shown in Fig. 2.73a were prepared in which 5 Å thick Co layers, initially positioned at the NiFe/Cu interfaces, are gradually moved into the interior of the NiFe layers. As can be seen from Fig. 2.73b the MR rapidly decreases with increasing separation d of the thin Co layers from the NiFe/Cu interfaces. The dependence of MR on d is well described by $\Delta R/R = a + b \times \exp(-d/\xi)$, where ξ is $\simeq 2.3$ Å. The MR rapidly saturates at a value corresponding to that of the pure NiFe/Cu/NiFe EBS structure. Finally in Fig. 2.73c data is shown for a series of Co/Cu/Co/ FeMn exchange-biased sandwiches in which thin NiFe layers are introduced at the Co/Cu interfaces. In this case the MR which is initially high is rapidly decreased by introduction of the NiFe layers, attaining a value comparable to that of a NiFe/Cu/NiFe sandwich. Again the length scale associated with the decay in MR is very short and in this case was determined to be $\simeq 2.8$ Å.

A wide variety of structures comprising many different combinations of magnetic layers and magnetic interface layers were studied. In each case the saturation magnetoresistance found was determined by the character of the magnetic/non-magnetic interface which was established within a characteristic length, $\xi \simeq 2$ to 3 Å. The possibility of alloy formation between the interface layers inserted in the sandwiches and the host magnetic layers was examined by introducing interface layers comprised of Co–Fe and Co–Ni alloys of various compositions. In these cases ξ was similarly short but the increased or decreased magnetoresistance values obtained were those corresponding to the respective alloy material. A variety of spacer layers distinct from Cu were also studied. The results are very similar to those for Cu spacer layers.

These experiments have shown that very thin interface layers, just one or two atomic layers thick, determine the magnitude of the enhanced MR effect, clearly demonstrating the predominant role of interface scattering. This is further emphasized by the fact that as the temperature is reduced the length scale, ξ, *decreases*. Moreover these studies highlight the degree of control required in designing and engineering structures for studies of the giant MR effect.

2.4.8 Giant Magnetoresistance in Systems Other than Multilayers

From the discussion in Sects. 2.4.5c and 2.4.6 it is clear that the basic requirement for the observation of giant MR is a structure in which there are magnetic entities whose magnetic moments can be varied as a function of applied field.

There are several systems other than magnetic multilayers which exhibit such properties. One example is discussed in this section.

2.4.8.1 Granular Alloys

A similar GMR effect to that found in magnetic multilayers has recently been reported in films comprised of small magnetic particles in metallic hosts. Examples include Co–Cu [2.308–311] and Co–Ag [2.310, 312]. Results have been reported for films prepared by sputter deposition [2.308, 309, 312] and MBE [2.310, 311]. A schematic diagram of a film structure prepared by MBE is shown in Fig. 2.74. The film is crystallographically ordered along the [1 1 1] direction by use of a thin Pt seed layer on a sapphire (0 0 0 1) substrate [2.313]. There are several important differences between films prepared by sputtering and MBE. In contrast to films prepared by sputtering it is expected that slow co-evaporation under uhv conditions and at moderate growth temperatures will lead to spontaneous phase separation of, for example, Co and Cu or Co and Ag. This follows since for temperatures below $\simeq 400\,^{\circ}$C both Co–Cu and Co–Ag are mutually insoluble, and at such growth temperatures there should be adequate surface diffusion for phase separation. An important consequence is that whereas for sputter-deposition substantial GMR is usually only obtained after post-growth annealing of the film, for MBE preparation no post-anneal step is required. Another important difference is that whereas sputtered films exhibit no substantial magnetic anisotropy for fields aligned parallel and perpendicular to the surface of the film [2.312], MBE prepared films exhibit substantial anisotropy, particularly at low temperatures [2.310, 311]. Typical magnetization and resistance versus field curves at 4.2 K for a 940 Å thick single crystalline $Co_{26}Ag_{74}$ granular alloy prepared at 250 °C are shown in Fig. 2.75. Substantial anisotropy in the dependence of MR and magnetization on the field are found. Possible sources of the magnetic anisotropy are discussed elsewhere [2.310].

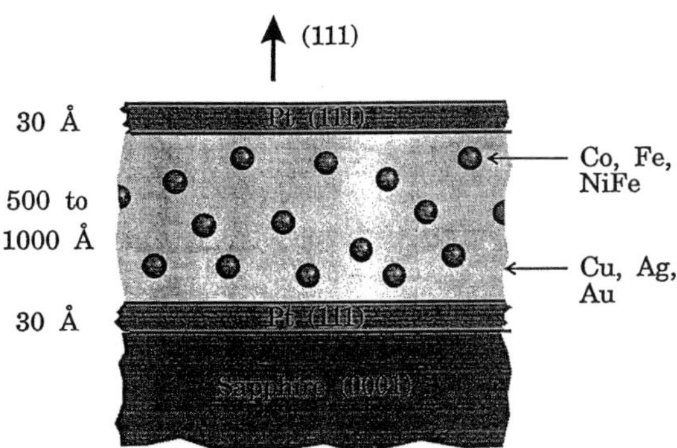

Fig. 2.74. Schematic diagram of a crystalline granular alloy containing small magnetic particles of, for example, Fe, Co or a Ni–Fe alloy, embedded in a metallic host of, for example, Cu, Ag or Au

The film in Fig. 2.75 exhibits a saturation magnetoresistance of more than 70% at 4.2 K, while at room temperature the GMR is almost 25%. Note that the resistance of the Co–Ag film takes its largest values for fields equal to $\pm H_c$ where H_c is the coercive field where the magnetization of the film passes through zero. This is true for both orientations of the field shown in the figure for which very different coercive fields are observed.

The variation of resistance with field shown in Fig. 2.75 is very similar to that described in Sect. 2.4.5c for Co/Cu multilayers with Cu layers sufficiently thick that the Co layers are magnetically decoupled [2.239, 296]. By analogy with the earlier work on multilayers it seems clear that the resistance of the alloy film is related to the relative orientation of the magnetizations on adjacent Co particles. The resistance is highest when the magnetic moments of the particles are arranged with the highest degree of anti-parallel alignment with respect to one another. The resistance is minimized when the particles or layers are magnetized parallel to one another. For both alloy films and decoupled multilayers, it is expected that the magnetizations of adjacent Co particles or layers will have the highest degree of anti-parallel alignment when the net magnetization is zero; this accounts for the peaks in resistance for these magnetic configurations (i.e. at $H = \pm H_c$). Figure 2.75 also demonstrates that the resistance of the film at low fields depends on the magnetic history of the sample. Note that in Fig. 2.75a the resistance in zero field is lower than the peak resistance, but that in Fig. 2.75b the

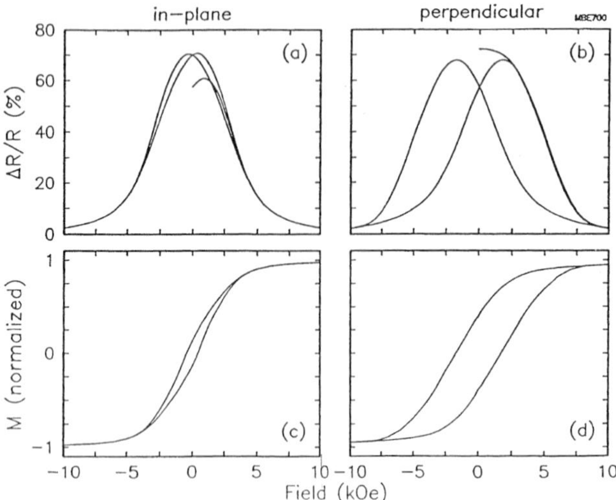

Fig. 2.75. Field dependence, at 4.2 K, of saturation magneto-resistance, $\Delta R/R$, and normalized magnetization for a 940 Å thick (1 1 1) oriented $Co_{0.26}Ag_{0.74}$ film. (a) $\Delta R/R$, for a field orthogonal to the measuring current, in the plane of the film. (b) $\Delta R/R$, for a field perpendicular to the film plane. (c) Magnetization, normalized to the saturation value, for an in-plane field as in (a). (d) Magnetization, normalized to the saturation value, for a perpendicular field as in (b). These data are from [2.309]

zero field resistance is higher than the peak resistance. The data in Fig. 2.75b were obtained immediately after field cooling, whilst the data in Fig. 2.75a were subsequently obtained by rotating the sample at 4.2 K in zero field. For crystalline Co–Cu films the resistance in zero field can easily be varied by more than a factor of two simply by varying the magnetic and temperature history of the sample.

It is interesting to examine the relationship of magnetoresistance to magnetization in the granular films. The data in Fig. 2.75 is replotted in Fig. 2.76 as MR versus magnetization. The data now are very similar to that for the Fe/Cr multilayer shown in Fig. 2.53. The simple dependence of MR on $(M/M_s)^2$ discussed earlier for antiferromagnetically coupled multilayers is not surprisingly not exactly followed in granular alloys where the magnetic structure is more complicated. However from symmetry the MR clearly has to be an even function of M/M_s and the curves of MR vs M/M_s are well described by a function of the form $a - b(M/M_s)^2 - c(M/M_s)^4$. Note the very similar variation of MR on M/M_s for a field aligned parallel and perpendicular to the film as compared to the very different variations of M and MR on H.

At present there is very little detailed work on the structure of granular alloy films since it is much more difficult to determine the structure of such films compared to multilayers. However, the determination of the structure is important in order to understand the magnetoresistance of these alloys. The structure up until recently has largely been inferred from electron microscopy studies. Such studies are most sensitive to large particles and since only a small portion of the film is examined it is difficult to obtain useful information on the average

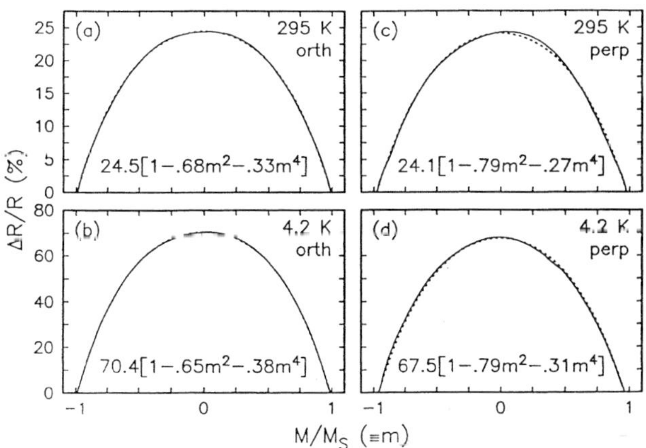

Fig. 2.76. Magnetoresistance versus magnetization for the same film shown in Fig. 2.75 for a field aligned in-plane and perpendicular to the film plane at room temperature and 4.2 K. The in-plane MR data were measured with the field orthogonal to the current. The measured data and the second order polynomial fits to the data are shown as solid and dotted curves respectively. The form of the fitted curves is included in the figure

particle size. Single crystalline films are particularly important for the structural determination of granular alloys using X-ray scattering techniques since the interpretation is greatly simplified.

The Co–Cu alloy system is particularly important since Co–Cu multilayers exhibit the largest GMR of any system. However, the weak contrast between Co and Cu which have similar lattice parameters and similar atomic numbers means the scattering contrast in standard X-ray scattering experiments and electron imaging techniques is small. Moreover, following the discussion in Sect. 2.4.7, the MR is expected to be dominated by the smallest magnetic clusters with the largest surface to volume ratio. Thus a structural characterization technique sensitive to very small particles is needed. Such a technique is grazing incidence, anomalous small angle X-ray scattering (SAXS) which has been used to examine the structure of MBE Co–Cu alloy films [2.311] as well as Co–Ag films [2.310]. SAXS is sensitive to clusters with sizes ranging from $\simeq > 10\,\text{Å}$ to several hundred angstrom in diameter. A grazing incidence geometry is used in which both the incident and scattered X-ray beams make small angles with respect to the film plane to enhance scattering from the film as compared to the substrate [2.314]. The intensity of the scattered X-ray beam is measured with respect to the magnitude of the in-plane scattering wave vector close to the nearly specularly reflected beam. Surface diffuse scattering from roughness at the film-air interface gives rise to an intense background signal. The experiments are carried out at a synchrotron such that by tuning the X-ray energy to eliminate the Co–Cu contrast the background signal can be independently determined. A second measurement is made with the energy tuned for maximum Co–Cu scattering contrast. After background subtraction the data can be analyzed to give a particle size. More details are given in [2.310, 311]. For the Co–Ag film shown in Fig. 2.75 SAXS measurements gave a characteristic Co cluster diameter and separation of $\simeq 25\,\text{Å}$ and $\simeq 76\,\text{Å}$ respectively.

The size of the magnetic clusters in granular alloys can be varied by annealing the films after growth. Detailed SAXS studies have been carried out on a series of identical $\simeq 840\,\text{Å}$ thick $Co_{16}Cu_{84}$ alloy films prepared simultaneously at $200\,°C$ and subsequently annealed at temperatures ranging up to $550\,°C$. The SAXS data show that the as-deposited films contain Co clusters with a characteristic diameter of $21\,\text{Å}$. The MR of these films at 4.2 K is $\simeq 35\%$. With thermal annealing the particle diameter is increased to $\simeq 250\,\text{Å}$ and the MR drops to $\simeq 1\%$. The detailed dependence of magnetoresistance on Co cluster size is shown in Fig. 2.76. Assuming the dominant role of interfacial spin dependent scattering as discussed above for magnetic multilayered structures, simple phenomenological arguments would suggest that the MR should scale approximately as the cluster surface to volume ratio. In contrast, if there were significant bulk spin dependent scattering, the GMR would depend only weakly on cluster size. The MR data in Fig. 2.76 scale approximately as the inverse cluster size consistent with the predominant contribution from interfacial scattering. The data can be compared with detailed theoretical models of GMR developed for the special geometry of granular alloys [2.315]. The best fit to the

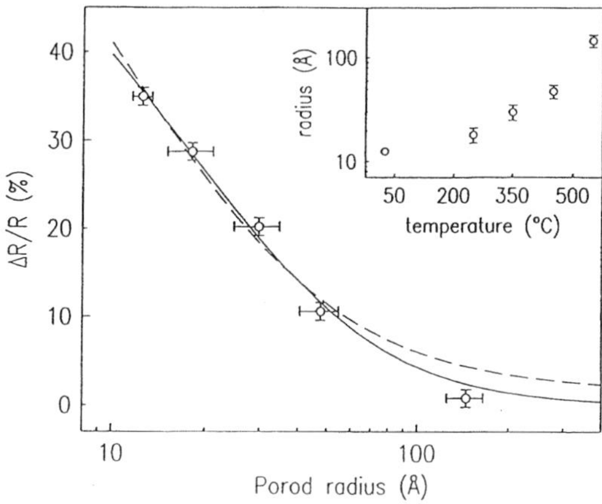

Fig. 2.77. Saturation MR versus the radii of Co particles measured in identical films of crystalline (1 1 1) oriented $Co_{16}Cu_{84}$ films deposited at 200 °C and subsequently annealed at temperatures ranging up to 550 °C. The inset shows the variation of the measured radii with the anneal temperature. The solid line corresponds to a fit to the data with only interfacial spin dependent electron scattering, while the broken line includes both interfacial and bulk spin dependent scattering as discussed in the text. The Porod radius is the radius inferred from grazing incidence small angle X-ray scattering measurements. These data are from [2.311]

data in Fig. 2.76 is obtained in such a theory with no bulk spin dependent scattering. This fit is shown in the figure as a solid line for which the ratio of the interfacial spin dependent mean free paths corresponding to the spin-up and spin-down electron channels, $\lambda_s^- / \lambda_s^+ \simeq 10$. Also shown (dashed line) is a curve corresponding to the introduction of a very small bulk spin-dependent scattering, which degrades the quality of the fit.

2.4.9 Conclusions

Two important properties exhibited by magnetic multilayers have been described in this section. These properties, closely connected with one another, are oscillations between antiferromagnetic and ferromagnetic interlayer exchange coupling and giant magnetoresistance. The discussion has concentrated on polycrystalline multilayers. The initial discovery of these properties in polycrystalline multilayers was surprising. It was unexpected that simple sputter deposition techniques could produce multilayers with thin layers sufficiently well defined to display such properties. These simple growth techniques have proved to be very useful for the exploration of different material systems and for the examination of the dependences of interlayer coupling and giant MR on details of the structures. The ability to survey many different materials has led to the discovery of extraordinarily large room temperature magnetoresistance in Co/Cu multilayers. The systematics of the oscillatory coupling phenomenon, namely an oscillation period common to most simple metals, and a coupling strength that increases exponentially with d-band filling for any of the 3d, 4d and 5d transition metal series, are important for determining which theoretical models are the correct models.

The magnitude of the giant MR effect has been shown to have simple dependences on magnetic and non-magnetic layer thicknesses in magnetic multilayers. It has been demonstrated that materials engineering, in particular, the insertion of ultra thin layers at the magnetic/non-magnetic interfaces, is useful in understanding the role of interface versus bulk spin dependent scattering. In contrast to early experiments and theories that suggested an important role for bulk spin dependent scattering, we have shown that interface scattering is the dominant scattering mechanism underlying giant magnetoresistance. Using this understanding we have engineered structures which exhibit large changes in resistance at very low fields. Such structures have technological potential.

The giant MR effect is not confined to simple magnetic multilayers but is exhibited by a variety of structures. One example discussed consists of two component alloys comprised of small magnetic particles in a non-magnetic matrix. These systems display changes in resistance of a similar size to those shown by magnetic multilayers comprised of the same materials. Other examples include conventional antiferromagnetic metals, some of which exhibit enormous negative magnetoresistance at temperatures close to the Néel temperature, where the magnetic sub-systems can be manipulated in comparatively small fields.

The discussion was mostly phenomenological in this section. In the preceeding sections detailed models of both giant MR and oscillatory interlayer coupling are presented.

Acknowledgements. I am especially grateful to K.P. Roche for technical support. I thank many colleagues, too numerous to mention here, for many useful discussions.

References

Section 2.1

2.1   M.A. Ruderman, C. Kittel: Phys. Rev. **96**, 99 (1954)
2.2   T. Kasuya: Progr. Theoret. Phys. (Japan) **16**, 45 (1956)
2.3   K. Yosida: Phys. Rev. **106**, 893 (1957)
2.4   J. Kondo: "Theory of Dilute Magnetic Alloys", in *Solid State Physics*, ed. by F. seitz, D. Turnbull, H. Ehrnereich, Vol. 23, 183 (Academic Press, New York, 1969)
2.5   A.J. Freeman: "Energy Band Structure, Indirect Exchange Interactions and Magnetic Ordering", in *Magnetic Properties of Rare Earth Metals*, ed. by R.J. Elliot (Plenum Press, New York, 1972) p. 245
2.6   T. Kasuya: "S–d and s–f Interaction and Rare Earth Metals", in *Magnetism*, ed. by G.T. Rado, H. Suhl, Vol. IIB (Academic Press, New York, 1966) p. 215
2.7   J.H. Van Vleck: Rev. Mod. Phys. **34**, 681 (1962)

2.8   Y. Yafet: J. Appl. Phys. **61**, 4058 (1987)
2.9   W.M. Fairbairn, S.Y. Yip: J. Phys. Condens. Matter **2**, 4197 (1990)
2.10 S. Baltensperger, J.S. Helman: Appl. Phys. Lett. **57**, 2954 (1990)
2.11 K.B. Hathaway, J.R. Cullen: J. Magn. Magn. Mat. **104–107**, 1840 (1992)
2.12 M.B. Stearns: J. Magn. Magn. Mat. **5**, 167 (1977)
2.13 J.C. Slonczewski: Phys. Rev. B **39**, 6995 (1989)
2.14 A. Bardasis, D.S. Falk, R.A. Ferrell, M.S. Fullenbaum, R.E. Prange, D.L. Mills: Phys. Rev. Lett. **14**, 289 (1965)
2.15 R.P. Erickson, K.B. Hathaway, J.R. Cullen: Phys. Rev. B **47**, 2626 (1993)
2.16 J. Barnas: J. Magn. Magn. Mat. **111**, L215 (1992)
2.17 M. Huberman (to be published)
2.18 D.M. Edwards, J. Mathon, R.B. Muniz, M.S. Phan: J. Phys. Condens. Matter **3**, 4941 (1991)
2.19 D.M. Deavon, D.S. Rokhsar, M. Johnson: Phys. Rev. B **44**, 5977 (1991)
2.20 C. Chappert, J.P. Renard: Europhys. Lett. **15**, 553 (1991)
2.21 R. Coehoorn: Phys. Rev. B **44**, 9331 (1991)
2.22 P. Bruno, C. Chappert: Phys. Rev. Lett. **67**, 1602 (1991)
2.23 M. Stiles: Bull. Am. Phys. Soc. **37**, 255 (1992)
2.24 Y. Wang, P.M. Levy, J.L. Fry: Phys. Rev. Lett. **65**, 2732 (1990)
2.25 F. Herman, R. Schrieffer: Phys. Rev. B **46**, 5806 (1992)
2.26 P.W. Anderson: Phys. Rev. **124**, 41 (1961)
2.27 P.A. Wolff: Phys. Rev. **124**, 1030 (1961)
2.28 For example, A.J. Heeger: "Localized Moments and Nonmoments in Metals: The Kondo Effect", in *Solid State Physics*, ed. by F. Seitz, D. Turnbull, H. Ehrenreich, Vol. 23 (Academic Press, New York, 1969) p. 284; also [2.29]
2.29 J.R. Schrieffer: J. Appl. Phys. **38**, 1143 (1967)
2.30 C.E.T. Goncalves Da Silva, L.M. Falicov: J. Phys. C **5**, 63 (1972)
2.31 C. Lacroix, J.P. Gavigan: J. Magn. Magn. Mat. **93**, 413 (1991)
2.32 P. Bruno: J. Magn. Magn. Mat. **116**, L13 (1992)
2.33 B. Caroli: J. Phys. Chem. Solids **28**, 1427 (1967)
2.34 N. Garcia, A. Hernando: J. Magn. Magn. Mat. **99**, L12 (1991)
2.35 J.C. Slonczewski: Phys. Rev. Lett. **67**, 3172 (1991)
2.36 D.M. Edwards, J. Mathon, R.B. Muniz, Murielle Villeret, J.M. Ward: Proceedings of the NATO Advanced Research Workshop "Magnetic Properties and Structure in Systems of Reduced Dimension" Cargese, June 1992
2.37 J.C. Slonczewski: Proceedings of the First International Symposium on Metallic Multilayers, Kyoto, March 1993, to be published in J. Magn. Magn. Mat.
2.38 J. Barnas, P. Grunberg: J. Magn. Magn. Mat. **121**, 326 (1993)
2.39 P. Bruno: J. Magn. Magn. Mat. **121**, 248 (1993)
2.40 D. Stoeffler, F. Gautier: Prog. Theor. Phys. Suppl. No. **101**, 139 (1991); also D. Stoeffler, K. Ounadjela, F. Gautier: J. Magn. Magn. Mat. **93**, 386 (1991) and D. Stoeffler, F. Gautier: Phys. Rev. B **44**, 10389 (1991)
2.41 H. Hasegawa: Phys. Rev. B **42**, 2368 (1990); Phys. Rev. B **43**, 10803 (1991)
2.42 P.M. Levy, K. Ounandjela, S. Zhang, Y. Wang, C.B. Sommers, A. Fert: J. Appl. Phys. **67**, 5914 (1991)
2.43 F. Herman, J. Sticht, M. Van Schilfgaarde: Mat. Res. Soc. Symp. Proc. **231**, 195 (1992)
2.44 J.R. Cullen, K.B. Hathaway: Phys. Rev. B **47**, 14998 (1993)
2.45 Z.Q. Qiu, J.E. Mattson, C.H. Sowers, U. Welp, S.D. Bader, H. Tang, J.C. Walker: Phys. Rev. B **45**, 2252 (1992)

Section 2.2

2.46 R.W. Erwin, J. Borchers, M.B. Salamon, S. Sinha, J.J. Rhyne, J.E. Cunningham, C.P. Flynn: Phys. Rev. Lett. **56**, 259 (1986); C.F. Majkrzak, J.W. Cable, J. Kwo, M. Hong, D.B. McWhan, Y. Yafet, J.W. Waszczak, C. Vettier: Phys. Rev. Lett. **56**, 2700 (1986)

2.47 J. Grünberg, R. Schreiber, Y. Pang, M.B. Brodsky, H. Sowers: Phys. Rev. Lett. **57**, 2442 (1986)
2.48 C. Carbone, S.F. Alvarado: Phys. Rev. B **36**, 2433 (1987)
2.49 S.S.P. Parkin, N. More, K.P. Roche: Phys. Rev. Lett. **64**, 2304 (1990)
2.50 H. Sato, P.A. Schroeder, J.M. Slaughter, W.P. Pratt Jr., W. Abdul Razzaq: Superlattices Microstructure **4**, 45 (1987); E. Velu, C. Dupas, D. Renard, J.P. Renard, J. Seiden, Phys. Rev. B **37**, 668 (1988)
2.51 M.N. Baibich, J.M. Broto, A. Fert, F. Nguyen Van Dau, F. Petroff, P. Etienne, G. Creuzet, A. Friederich, J. Chazelas: Phys. Rev. Lett. **61**, 2472 (1988)
2.52 G. Binash, P. Grünberg, F. Saurenbach, W. Zinn: Phys. Rev. B **39**, 4828 (1989)
2.53 S.S.P. Parkin, A. Mansour, G.P. Felcher: Appl. Phys. Lett. **58**, 1473 (1991)
2.54 J. Unguris, R.J. Celotta, D.T. Pierce: Phys. Rev. Lett. **67**, 140 (1991)
2.55 S.T. Purcell, W. Folkerts, M.T. Johnson, N.W.E. McGee, K. Jager, J. Aan de Stegge, W.P. Zeper, P. Grünberg: Phys. Rev. Lett. **67**, 903 (1991)
2.56 M. Rührig, R. Schäfer, A. Hubert, R. Mosler, J.A. Wolf, S. Demokritov, P. Grünberg: Phys. Status Solidi. A **125**, 635 (1991)
2.57 S. Demokritov, J.A. Wolf, P. Grünberg, W. Zinn: Mat. Res. Soc. Symp. Proc. Vol. **231**, 133 (1992)
2.58 S.T. Purcell, M.T. Johnson, N.W.E. McGee, R. Coehoorn, W. Hoving: Phys. Rev. B **45**, 13064 (1992)
2.59 S.S.P. Parkin, D. Mauri: Phys. Rev. B **44**, 7131 (1991)
2.60 M.E. Brubaker, J.E. Mattson, C.H. Sowers, S.D. Bader: Appl. Phys. Lett. **58**, 2306 (1991)
2.61 S.S.P. Parkin: Phys. Rev. Lett. **67**, 3598 (1991)
2.62 Y.Y. Huang, G.B. Felcher, S.S.P. Parkin: J. Magn. Magn. Mat. **99**, L31 (1991)
2.63 Z. Celinski, B. Heinrich: J. Magn. Magn. Mat. **99**, L25 (1991)
2.64 A. Cebollada, J.L. Martinez, J.M. Gallego, J.J. de Miguel, R. Miranda, S. Ferrer, F. Batallan, G. Fillion, J.P. Rebouillat: Phys. Rev. B **39**, 9726 (1989)
2.65 B. Heinrich, Z. Celinski, J.F. Cochran, W.B. Muir, J. Rudd, Q.M. Zhong, A.S. Arrott, K. Myrtle, J. Kirschner: Phys. Rev. Lett. **64**, 673 (1990)
2.66 D. Pescia, D. Kerkmann, F. Schumann, W. Gudat: Z. Phys. B **78**, 475 (1990)
2.67 J.F. Cochran, J. Rudd, W.B. Muir, B. Heinrich, Z. Celinski: Phys. Rev. B **42**, 508 (1990)
2.68 W.R. Bennett, W. Schwarzacher, W.F. Egelhoff, Jr.: Phys. Rev. Lett. **65**, 3169 (1990)
2.69 J.J. de Miguel, A. Cebollada, J.M. Gallego, R. Miranda, C.M. Schneider, P. Schuster, J. Kirschner: J. Magn. Magn. Mat. **93**, 1 (1991)
2.70 S.S.P. Parkin, R. Bhadra, K.P. Roche: Phys. Rev. Lett. **66**, 2152 (1991)
2.71 D.H. Mosca, F. Petroff, A. Fert, P.A. Schroeder, W.P. Pratt, Jr., R. Laloee, S. Lequien: J. Magn. Magn. Mat. **94**, L1 (1991)
2.72 B. Rodmacq, P. Mangin, C. Vettier: Europhys. Lett. **15**, 503 (1991)
2.73 B. Heinrich, J.F. Cochran, M. Kowalewski, J. Kirschner, Z. Celinski, A.S. Arrott, K. Myrtle: Phys. Rev. B **44**, 9348 (1991)
2.74 C.A. dos Santos, B. Rodmacq, M. Vaezzadeh, B. George: Appl. Phys. Lett. **59**, 126 (1991)
2.75 F. Pétroff, A. Barthélémy, D.H. Mosca, D.K. Lottis, A. Fert, P.A. Schroeder, W.P. Pratt, R. Laloee, S. Lequien: Phys. Rev. B **44**, 5355 (1991)
2.76 A. Fuss, S. Demokritov, P. Grünberg, W. Zinn: J. Magn. Magn. Mat. **103**, L221 (1992)
2.77 M.T. Johnson, S.T. Purcell, N.W.E. McGee, R. Coehoorn, J. aan de Stegge, W. Hoving: Phys. Rev. Lett. **68**, 2688 (1992)
2.78 W.F. Egelhoff, Jr., M.T. Kief: Phys. Rev. B **45**, 7795 (1992)
2.79 J.P. Renard, P. Beauvillain, C. Dupas, K. Le Dang, P. Veillet, E. Vélu, C. Marlière, D. Renard, J. Mag. Mat. **115**, L147 (1992); D. Greig, M.J. Hall, C. Hammond, B.J. Hickey, H.P. Ho, M.A. Howson, M.J. Walker, N. Wiser, D.G. Wright, J. Magn. Magn. Mat., **110**, L239 (1992); J. Kohlhepp, S. Cordes, H.J. Elmers, U. Gradmann, J. Magn. Magn. Mat. **11**, L231 (1992); A. Kamijo, H. Igaraschi: Jpn. J. Appl. Phys. **31**, L1050 (1992)
2.80 W. Folkerts: J. Magn. Magn. Mat. **94**, 302 (1991); J. Barnas, P. Grünberg: J. Magn. Magn. Mat. **99**, 57 (1991)

2.81 P. Grünberg: J. Appl. Phys. **57**, 3673 (1985); B. Heinrich, S.T. Purcell, J.R. Dutcher, K.B. Urquhart, J.F. Cochran, A.S. Arrott: Phys. Rev. B **38**, 12879 (1988); M. Vohl, J. Barnas, P. Grünberg: Phys. Rev. B **39**, 12003 (1989); B. Hillebrands: Phys. Rev. B **41**, 530 (1990)

2.82 D. Stoeffler, F. Gautier: Prog. Theor. Phys. Suppl. **101**, 139 (1990); Phys. Rev. **44**, 10389 (1991)

2.83 K. Ounadjela, C.B. Sommers, A. Fert, D. Stoeffler, F. Gautier, V.L. Moruzzi: Europhys. Lett. **15**, 875 (1991)

2.84 F. Herman, J. Sticht, M. van Schilfgaarde: J. Appl. Phys. **69**, 4783 (1991); Mat. Res. Soc. Symp. Proc. Vol. **231**, 195 (1992)

2.85 Y. Yafet: Phys. Rev. B **36**, 3948 (1987)

2.86 C. Chappert, J.P. Renard: Europhys. Lett. **15**, 553 (1991)

2.87 P. Bruno, C. Chappert: Phys. Rev. Lett. **67**, 1602 (1991); Phys. Rev. Lett. **67**, 2592 (Erratum) (1991); Phys. Rev. B **46**, 261 (1992)

2.88 R. Coehoorn: Phys. Rev. B **44**, 9331 (1991)

2.89 M.A. Ruderman, C. Kittel: Phys. Rev. **96**, 99 (1954)

2.90 Y. Wang, P.M. Levy, J.L. Fry: Phys. Rev. Lett. **65**, 2732 (1990)

2.91 C. Lacroix, J.P. Gavigan: J. Magn. Magn. Mat. **93**, 413 (1991)

2.92 D.M. Edwards, J. Mathon, R.B. Muniz, M.S. Phan: Phys. Rev. Lett. **67**, 493 (1991)

2.93 D.M. Deaven, D.S. Rokhsar, M. Johnson: Phys. Rev. B **44**, 5977 (1991)

2.94 N. Garcia, A. Hernando: J. Magn. Magn. Mat. **99**, L12 (1991)

2.95 K.B. Hathaway, J.R. Cullen: J. Magn. Magn. **104–107**, (1992)

2.96 P. Bruno: J. Magn. Magn. Mat. **116**, L13 (1992)

2.97 M.R. Halse: Philos. Trans. R. Soc. London, Ser. A **265**, 507 (1969)

2.98 C. Dupas, P. Beauvillain, C. Chappert, J.P. Renard, F. Trigui, P. Veillet, E. Velu, D. Renard: J. Appl. Phys. **67**, 5680 (1990)

2.99 B. Dieny, V.S. Speriosu, S.S.P. Parkin, B.A. Gurney: Phys. Rev. B **43**, 1297 (1991)

2.100 T. Shinjo, H. Yamamoto: J. Phys. Soc. Jpn. **59**, 3061 (1990)

2.101 A. Fert, I.A. Campbell: J. Phys. F**6**, 849 (1976)

2.102 J.W.F. Dorleijn, A.R. Miedema: J. Phys. F**5**, 487 (1975); J. Phys. F**7**, L23 (1977)

2.103 I.A. Campbell, A. Fert: "Transport Properties in Ferromagnets", in *Ferromagnetic Materials*, ed. by E.P. Wohlfarth (North Holland, Amsterdam, 1982) p. 769

2.104 B. Loegel, F. Gautier: J. Phys. Chem. Solids **32**, 2723 (1971); J. Durand: "Transport Phenomena and Nuclear Magnetic Resonance in Ferromagnetic Metals and Alloys", Thesis (Strasbourg, 1973)

2.105 H. Hayakawa, J. Yamashita: Prog. Theor. Phys. **54**, 952 (1975)

2.106 R.E. Camley, J. Barnas: Phys. Rev. Lett. **63**, 664 (1989)

2.107 J. Barnas, A. Fuss, R.E. Camley, P. Grünberg, W. Zinn: Phys. Rev. B **42**, 8110 (1990)

2.108 F. Trigui, E. Velu, C. Dupas: J. Magn. Magn. Mater. **93**, 421 (1991)

2.109 B. Dieny: Europhysics Lett. **17**, 261 (1992)

2.110 B. Dieny: J. Phys. C **4**, 8009 (1992); and private communication

2.111 A. Barthélémy, A. Fert: Phys. Rev. B **43**, 13124 (1991)

2.112 D.M. Edwards, R.B. Muniz, J. Mathon: IEEE Trans. Magn. **27**, 3548 (1991)

2.113 B.L. Johnson, R.E. Camley: Phys. Rev. B **44**, 9997 (1991)

2.114 M.B. Stearns: J. Magn. Magn. Mat. **104–107**, 1745 (1992)

2.115 R.Q. Hood, L.M. Falicov: Phys. Rev. B **46**, 8283 (1992)

2.116 A.C. Erlich, D.J. Gillespie: J. Appl. Phys. **73**, 5536 (1993)

2.117 Z. Tesanovic, M.V. Jaric, S. Maekawa: Phys. Rev. Lett. **57**, 2760 (1986)

2.118 G. Fishman, D. Calecki: Phys. Rev. Lett. **62**, 1302 (1989)

2.119 A. Fert: "Transport Properties of Thin Metallic Films and Multilayers", in *Science and Technology of Nanostructured Magnetic Materials*, ed. by G.C. Hadjipanayis, G.A. Prinz (Plenum Press, London, 1991) p. 221

2.120 P.M. Levy, S. Zhang, A. Fert: Phys. Rev. Lett. **65**, 1643 (1990); S. Zhang, P.M. Levy, A. Fert: Phys. Rev. B **45** (1992)

2.121 S. Zhang, P.M. Levy: Mat. Res. Soc. Symp. Proc. Vol. **231**, 255 (1992)

2.122 A. Vedyayev, B. Dieny, N. Ryshanova: Europhys. Lett. **19**, 329 (1992)

2.123 A. Fert, A. Barthélémy, P. Etienne, S. Lequien, R. Loloee, D.K. Lottis, D.H. Mosca, F. Petroff, W.P. Pratt, P.A. Schroeder: J. Magn. Magn. Mater. **104–107**, 1712–1716 (1992)

2.124 J. Inoue, S. Maekawa: Prog. Theor. Phys. **106**, 187 (1991)

2.125 B. Rodmacq, B. George, M. Vaezzadeh, Ph. Mangin: Phys. Rev. B **46**, 1206 (1992)

2.126 M.A.M. Gijs, M. Okada: Phys. Rev. B **46**, 2908 (1992)

2.127 D.H. Mosca: "Giant Magnetoresistance in Magnetic Materials", Thesis (Porto Alegre-Orsay, 1991)

2.128 F. Petroff, A. Barthélémy, A. Fert, P. Etienne, S. Lequien: J. Magn. Magn. Mater. **93**, 95 (1991)

2.129 E.E. Fullerton, D.M. Kelly, J. Guimpel, I.K. Schuller, Y. Bruynseraede: Phys. Rev. Lett. **68**, 859 (1992)

2.130 Y. Obi, K. Takanashi, Y. Mitami, N. Tsuda, H. Fujimori: J. Magn. Magn. Mat. **104–107**, 1747 (1992); K. Takanashi, Y. Obi, Y. Mitami, H. Fujimori: J. Phys. Soc. Jpn. **61**, 1169 (1992)

2.131 S.S.P. Parkin, Z.G. Li, D.J. Smith: Appl. Phys. Lett. **58**, 2710 (1991); also D. Greig, M.J. Hall, C. Hammond, B.J. Hickey, H.P. Ho, M.A. Howson, M.J. Walker, N. Wisen, D.G. Wright: J. Magn. Magn. Mat. **110**, 239 (1992); M.E. Tomlinson, R.J. Pollard, D.G. Lord, P.J. Grundy: J. Magn. Magn. Mat. **111**, 79 (1992)

2.132 R.J. Highmore, W.C. Shih, R.E. Sonekh, J.E. Evetts: J. Magn. Magn. Mat. **115** (1992)

2.133 Y. Saito, S. Hashimoto, K. Inomata: Appl. Phys. Lett. **60**, 2436 (1992)

2.134 J. Kohlhepp, S. Cordes, H.J. Elmers, U. Gradmann: J. Magn. Magn. Mat. **111**, 231 (1992)

2.135 B.A. Gurney, D.R. Wilhuit, V.S. Speriosu, I.L. Sanders: IEEE Trans. Magn. **26**, 2747 (1990)

2.136 P. Baumgart, B.A. Gurney, D. Wilhuit, T. Nguyen, B. Dieny, V.S. Speriosu: J. Appl. Phys. **69**, 4792 (1991); B.A. Gurney, P. Baumgart, D.R. Wilhuit, B. Dieny, V.S. Speriosu: J. Appl. Phys. **70**, 5867 (1991)

2.137 R. Nakatami, K. Okuda: InterMag Conf. (St Louis, 1992)

2.138 B. Dieny, V.S. Speriosu, J.P. Nozières, B.A. Gurney, A. Vedyayev, N. Ryzhanova: *Magnetism and Structure in Systems of Reduced Dimension*, ed. by R.F.C. Farrow et al. (Plenum Press, New York, 1993) p. 279

2.139 J.M. George, A. Barthélémy, F. Petroff, T. Valet, A. Fert: Mat. Res. Symp. Proc. Vol. 313 (Materials Research Society 1993) p. 737

2.140 S. Zhang, P.M. Levy: Phys. Rev. B **43**, 11048 (1991)

2.141 J.L. Duvail, D.K. Lottis, A. Fert: Conference on Magnetism and Magnetic Materials 1994, to appear in J. Appl. Phys. 1994

2.142 W.P. Pratt, S.F. Lee, J.M. Slaughter, P.A. Schroeder, J. Bass: Phys. Rev. Lett. **66**, 3060 (1991); S.F. Lee, W.P. Pratt Jr, R. Loloee, P.A. Schroeder, J. Bass: Phys. Rev. B **46**, 548 (1992)

2.143 M. Johnson: Phys. Rev. Lett. **67**, 3594 (1991); M. Johnson, R.H. Silsbee: Phys. Rev. B **35**, 4959 (1987)

2.144 S. Zhang, P.M. Levy: J. Appl. Phys. **69**, 4786 (1991)

2.145 A. Fert, T. Valet: J. Magn. Magn. Mat. **121**, 378 (1993); T. Valet, A Fert: Phys. Rev. B **48**, 7099 (1993)

2.146 P.A. Schroeder: *Magnetism and Structure in Systems of Reduced Dimension*, ed. by R.F.C. Farrow et al. (Pleneum Press, New York, 1993) p. 129; S.F. Lee, W.P. Pratt, Q. Yang, D. Holody, R. Loloee, P.A. Schroeder, J. Bass: J. Magn. Magn. Mat. **118**, L1 (1993)

2.147 F. Nguyen Van Dau, A. Fert, M. Baibich: J. Phys. (Paris) **49**, C8-1663 (1988)

2.148 S. Zhang, P.M. Levy: Mat. Res. Symp. Proc. Vol. 313, (Materials Research Society 1993) p. 53

Section 2.3

2.149 C.F. Majkrzak, J.W. Cable, J. Kwo, M. Hong, D.B. McWhan, Y. Yafet, J.V. Waszczak, C. Vettier: Phys. Rev. Lett. **56**, 2700 (1986)

2.150 P. Grünberg, R. Schreiber, Y. Pang, M.B. Brodsky, H. Sowers: Phys. Rev. Lett. **57**, 2442 (1986)

2.151 M.N. Baibich, J.M. Broto, A. Fert, F. Nguyen Van Dau, F. Petroff, P. Etienne, G. Creuzet, A. Friederich, J. Chazelas: Phys. Rev. Lett. **61**, 2472 (1988)

2.152 G. Binasch, P. Grünberg, F. Saurenbach, W. Zinn: Phys. Rev. B **39**, 4828 (1989)

2.153 S.S.P. Parkin, N. More, K.P. Roche: Phys. Rev. Lett. **64**, 2304 (1990)

2.154 Y. Yafet: J. Appl. Phys. **61**, 4058 (1987)

2.155 C. Kittel: "Indirect Exchange Interactions in Metals", in *Solid State Physics*, ed. by F. Seitz, D. Turnbull, H. Ehrenreich (Academic, New York, 1968), Vol. 22, p. 1

2.156 Y. Wang, P.M. Levy, J.L. Fry: Phys. Rev. Lett. **65**, 2732 (1990)

2.157 P. Bruno, C. Chappert: Phys. Rev. Lett. **67**, 1602 (1991); Phys. Rev. B **46**, 261 (1992)

2.158 F. Herman, J.R. Schrieffer: Phys. Rev. B **46**, 5806 (1992)

2.159 R. Coehoorn: Phys. Rev. B **44**, 9331 (1991)

2.160 W. Baltensperger, J.S. Helman: Appl. Phys. Lett. **57**, 2954 (1990)

2.161 D.M. Edwards, J. Mathon, R.B. Muniz, M.S. Phan: J. Phys. Condens. Matter **3**, 4941 (1991)

2.162 K.B. Hathaway, J.R. Cullen: J. Magn. Magn. Mat. **104–107**, 1840 (1992)

2.163 D.M. Deaven, D.S. Rokhsar, M. Johnson: Phys. Rev. B **44**, 5977 (1991)

2.164 D. Stoeffler, F. Gautier: Prog. Theor. Phys. Suppl. **101**, 139 (1990)

2.165 H. Hasegawa: Phys. Rev. B **42**, 2368 (1990)

2.166 F. Herman, J. Sticht, M. Van Schilfgaarde: Mat. Res. Soc. Symp. Proc. **231**, 195 (1992)

2.167 J. Unguris, D.T. Pierce, A. Galejs, R.J. Celotta: Phys. Rev. Lett. **49**, 72 (1982)

2.168 E. Kisker, W. Gudat, K. Schröder: Solid State Commun. **44**, 591 (1982)

2.169 H. Hopster, R. Raue, E. Kisker, G. Guntherodt, M. Campagna: Phys. Rev. Lett. **50**, 70 (1983)

2.170 D.R. Penn, S.P. Apell, S.M. Girvin: Phys. Rev. Lett. **55**, 518 (1985); Phys. Rev. B **32**, 7753 (1985)

2.171 J. Glazer, E. Tosatti: Solid State Commun. **52**, 905 (1984)

2.172 J.I. Goldstein, D.E. Newbury, P. Echlin, D.C. Joy, C. Fiori, E. Lifshin: *Scanning Electron Microscopy and X-ray Microanalysis* (Plenum, New York, 1984)

2.173 J. Unguris, R.J. Celotta, D.T. Pierce: Phys. Rev. Lett. **69**, 1125 (1992)

2.174 J. Unguris, R.J. Celotta, D.T. Pierce: J. Magn. Magn. Mat., **127**, 205 (1993)

2.175 R.J. Celotta, D.T. Pierce: *Microbeam Analysis-1982*, ed. by K.F.J. Heinrich (San Francisco Press, San Francisco) p. 469

2.176 K. Koike, H. Matsuyama, K. Hayakawa: Scanning Micros. Suppl. **1**, 241 (1987)

2.177 G.G. Hembree, J. Unguris, R.J. Celotta, D.T. Pierce: Scanning Micros. Suppl. **1**, 229 (1987)

2.178 M.R. Scheinfein, J. Unguris, M.H. Kelley, D.T. Pierce, R.J. Celotta: Rev. Sci. Instrum. **61**, 2501 (1990)

2.179 J. Unguris, M.R. Scheinfein, R.J. Celotta, D.T. Pierce: "Scanning Electron Microscopy with Polarization Analysis: Studies of Magnetic Microstructure", in *Chemistry and Physics of Solid Surfaces VIII*, ed. by R. Vanselow, R. Howe (Springer, Berlin, Heidelberg, 1990) p. 239

2.180 J. Unguris, D.T. Pierce, R.J. Celotta: Rev. Sci. Instrum. **57**, 1314 (1986)

2.181 D.T. Pierce, R.J. Celotta, M.H. Kelley, J. Unguris: Nucl. Instrum. Meth. A **266**, 550 (1988)

2.182 J. Kessler: *Polarized Electrons*, 2nd ed. (Springer, Berlin, Heidelberg, 1985)

2.183 K. Koike, K. Hayakawa: Jpn. J. Appl. Phys. **23**, L187 (1984)

2.184 J. Unguris, G.G. Hembree, R.J. Celotta, D.T. Pierce: J. Microscopy **139**, RP1 (1985)

2.185 R. Allenspach, M. Stampanoni, A. Bischof: Phys. Rev. Lett. **65**, 3344 (1990)

2.186 H.P. Oepen, J. Kirschner: Scanning Micr. **5**, 1 (1991)

2.187 R. Jungblut, C. Roth, F.U. Hillebrecht, E. Kisker: Surf. Sci. **269/270**, 615 (1992)

2.188 M.R. Scheinfein, D.T. Pierce, J. Unguris, J.J. McClelland, R.J. Celotta: Rev. Sci. Instrum. **60**, 1 (1989)

2.189 D.T. Pierce, J. Unguris, R.J. Celotta: MRS Bulletin **13**, 19 (1988)

2.190 J. Unguris, M.R. Scheinfein, R.J. Celotta, D.T. Pierce: Appl. Phys. Lett. **55**, 2553 (1989)

2.191 M.R. Scheinfein, J. Unguris, J.L. Blue, K.J. Coakley, D.T. Pierce, R.J. Celotta, P.J. Ryan: Phys. Rev. **43**, 3395 (1991)

2.192 F. Bitter: Phys. Rev. **38**, 1903 (1931)

2.193 J.P. Jacubovics: "Lorentz Microscopy and Applications (TEM and SEM)", in *Electron Microscopy in Materials Science Part IV*, ed. by E. Ruedl, U. Valdre (Commission of European Communities, Brussels, 1973) p. 1303

2.194 W. Rave, R. Schafer, A. Hubert: J. Magn. Magn. Mater. **65**, 7 (1987)

2.195 B.E. Argyle, B. Petek, D.A. Herman, Jr.: J. Appl. Phys. **61**, 4303 (1987)

2.196 J.N. Chapman, S. McVitie, J.R. McFadyen: Scanning Micros. Suppl. **1**, 221 (1987)

2.197 A. Tonomura: J. Appl. Phys. **61**, 4297 (1987)

2.198 S.S.P. Parkin, A. Mansour, G.P. Felcher: Appl. Phys. Lett. **58**, 1473 (1991)

2.199 P.D. Gorsuch: J. Appl. Phys. **30**, 837 (1959)

2.200 D.T. Pierce, J. Stroscio, J. Unguris, R.J. Celotta: Phys. Rev. B **49** (1994)

2.201 E. Bauer, J.H. van der Merwe: Phys. Rev. B **33**, 3657 (1986)

2.202 A.R. Miedema: Z. Metallk. **69**, 287 (1978); **69**, 455 (1978)

2.203 L.Z. Mezey, J. Giber: Jpn. J. Appl. Phys. **21**, 1569 (1982)

2.204 S. Demokritov, J.A. Wolf, P. Grünberg, W. Zinn: in Proc. Mater. Res. Soc. Symp. **231**, 133 (1992)

2.205 M. Rührig, R. Schäfer, A. Hubert, R. Mosler, J.A. Wolf, S. Demokritov, P. Grünberg: Phys. Status Solidi (a) **125**, 635 (1991)

2.206 J. Unguris, R.J. Celotta, D.T. Pierce: Phys. Rev. Lett. **67**, 140 (1991)

2.207 S.T. Purcell, W. Folkerts, M.T. Johnson, N.W.E. McGee, K. Jager, J. aan de Stegge, W.B. Zeper, W. Hoving, P. Grünberg: Phys. Rev. Lett. **67**, 903 (1991)

2.208 S.T. Purcell, A.S. Arrott, B. Heinrich: J. Vac. Sci. Technol. B **6**, 794 (1988)

2.209 J.J. McClelland, J. Unguris, R.E. Scholten, D.T. Pierce: J. Vac. Sci. Technol., A **11**, 2863 (1993)

2.210 M.G. Lagally, D.E. Savage, M.C. Tringides: in *Reflection High-Energy Electron Diffraction and Reflection Electron Imaging of Surfaces*, ed. by P.K. Larsen, P.J. Dobson, NATO ASI Series B188 (Plenum Press, New York, 1988) p. 139

2.211 F.U. Hillebrecht, C. Roth, R. Jungblut, E. Kisker, A. Bringer: Europhys. Lett. **19**, 711 (1992)

2.212 T.G. Walker, A.W. Pang, H. Hopster, S.F. Alvarado: Phys. Rev. Lett. **69**, 1121 (1992)

2.213 Y. Yafet: Phys. Rev. B **36**, 3948 (1987)

2.214 D.G. Laurent, J. Calloway, J.L. Fry, N.E. Brener: Phys. Rev. B **23**, 4977 (1981)

2.215 E. Fawcett: Rev. Mod. Phys. **60**, 209 (1988)

2.216 M. Stiles: Phys. Rev. B **48**, 7238 (1993)

2.217 D. Shoenberg, D.J. Roaf, Philos. Trans. R. Soc. London **255**, 85 (1962)

2.218 S.S.P. Parkin: Phys. Rev. Lett. **67**, 3598 (1991)

2.219 M.T. Johnson, S.T. Purcell, N.W.E. McGee, R. Coehoorn, J. aan de Stegge, W. Hoving: Phys. Rev. Lett. **68**, 2688 (1992)

2.220 A. Fuss, S. Demokritov, P. Grünberg, W. Zinn: J. Magn. Magn. Mater. **103**, L221 (1992)

2.221 Z.Q. Qiu, J. Pearson, A. Berger, S.D. Bader: Phys. Rev. Lett. **68**, 1398 (1992)

2.222 S.T. Purcell, M.T. Johnson, N.W.E. McGee, R. Coehoorn, W. Hoving: Phys. Rev. B **45**, 13064 (1992)

2.223 Z. Celinski, B. Heinrich: J. Magn. Magn. Mater. **99**, L25 (1991)

2.224 P. Grünberg, S. Demokritov. A. Fuss, R. Schreiber, J.A. Wolf, S.T. Purcell: J. Magn. Magn. Mater. **104–107**, 1734 (1992)

2.225 J.C. Slonczewski: Phys. Rev. Lett. **67**, 3172 (1991)

Section 2.4

2.226 T. Shinjo, T. Takada: "Metallic Superlattices", in *Ferromagnetic Materials*, Vol. 3, ed. by E.P. Wohlfarth (Elsevier, Amsterdam, 1987)

2.227 I.K. Schuller: "The Physics of Metallic Superlattices: An Experimental Point of View", in *Physics, Fabrication, and Applications of Multilayered Structures*, ed. by P. Dhez, C. Weisbuch (Plenum, New York, 1988) p. 139

2.228 *Magnetic Properties of Low-Dimensional Systems II*, ed. by L.M. Falicov, F. Meija-Lira, J.L. Moran-Lopez (Springer, Berlin, Heidelberg, 1990)

2.229 L.M. Falicov, D.T. Pierce, S.D. Bader, R. Gronsky, K.B. Hathaway, H.J. Hopster, D.N. Lambeth, S.S.P. Parkin, G. Prinz, M. Salamon, I.K. Schuller, R.H. Victoria: J. Mat. Res. **5**, 1299 (1990)

2.230 R.E. Walstedt, J.H. Wernick: Phys. Rev. Lett. **20**, 856 (1968)

2.231 J.B. Boyce, C.P. Slichter: Phys. Rev. B **13**, 379 (1976)

2.232 L.R. Walker, R.E. Walstedt: Phys. Rev. B **22**, 3816 (1980)

2.233 R.M. White, *Quantum Theory of Magnetism*, (Springer, Berlin, Heidelberg, 1983)

2.234 J.-C. Bruyere, O. Massenet, R. Montmory, L. Néel: C.R. Acad. Sci. **258**, 1423 (1964)

2.235 P. Grunberg, F. Saurenbach: MRS Int'l. Mtg. Adv. Mats. **10**, 255 (1989)

2.236 A. Bardasis, D.S. Falk, R.A. Ferrell, M.S. Fullenbaum, R.E. Prange, D.S. Mills: Phys. Rev. Lett. **14**, 298 (1965)

2.237 C.F. Majkrzak, J.W. Cable, J. Kwo, M. Hong, D.B. McWhan, Y. Yafet, J.V. Waszczak, C. Vettier: Phys. Rev. Lett. **56**, 2700 (1986)

2.238 S.S.P. Parkin, N. More, K.P. Roche: Phys. Rev. Lett. **64**, 2304 (1990)

2.239 S.S.P. Parkin, R. Bhadra, K.P. Roche: Phys. Rev. Lett. **66**, 2152 (1991)

2.240 S.S.P. Parkin: Phys. Rev. Lett. **67**, 3598 (1991)

2.241 *Electrocrystallization*, ed. by R. Weil, R.G. Barradas (The Electrochemical Society, Pennington, 1981)

2.242 D.S. Lashmore, M.P. Dariel: J. Electrochem. Soc. **135**, 1218 (1988)

2.243 M. Ohring: *The Materials Science of Thin Films* (Academic, Boston, 1992)

2.244 R.F.C. Farrow, C.H. Lee, S.S.P. Parkin: IBM J. Res. Dev. **34**, 903 (1990)

2.245 R.F.C. Farrow, R.F. Marks, G.R. Harp, D. Weller, T.A. Rabedeau, M. Toney, S.S.P. Parkin: Mat. Res. Rep. (1993)

2.246 P. Grunberg, R. Schreiber, Y. Pang, M.B. Brodsky, H. Sowers: Phys. Rev. Lett. **57**, 2442 (1986)

2.247 G. Binasch, P. Grunberg, F. Saurenbach, W. Zinn: Phys. Rev. B **39**, 4828 (1989)

2.248 M.N. Baibich, J.M. Broto, A. Fert, F. Nguyen van Dau, F. Petroff, P. Etienne, G. Creuzet, A. Friederich, J. Chazelas: Phys. Rev. Lett. **61**, 2472 (1988)

2.249 W. Folkerts: J. Magn. Magn. Mat. **94**, 302 (1991)

2.250 S.S.P. Parkin, A. Mansour, G.P. Felcher: Appl. Phys. Lett. **58**, 1473 (1991)

2.251 T.R. McGuire, R.I. Potter: IEEE Trans. Mag. **MAG-11**, 1018 (1975)

2.252 P. Ciureanu: "Magnetoresistive Sensors" in *Thin Film Resistive Sensors*, ed. by P. Ciureanu, S. Middelhoek (Institute of Physics Publishing, Bristol, 1992) p. 253

2.253 J. Smit: Physica (Utrecht) **XVI**, 612 (1951)

2.254 J. Mathon: Contemp. Phys. **32**, 143 (1991)

2.255 N.F. Mott: Adv. Phys. **13**, 325 (1964)

2.256 P.L. Rossiter: *The Electrical Resistivity of Metals and Alloys* (Cambridge University Press, Cambridge, 1987)

2.257 D.C. Mattis: *The Theory of Magnetism I: Statics and Dynamics* (Springer, Berlin, Heidelberg, 1981)

2.258 R.E. Camley, J. Barnas: Phys. Rev. Lett. **63**, 664 (1989)

2.259 P.M. Levy, K. Ounadjela, S. Zhang, Y. Wang, C.B. Sommers, A. Fert: J. Appl. Phys. **67**, 5914 (1990)

2.260 J. Barnas, A. Fuss, R.E. Camley, P. Grunberg, W. Zinn: Phys. Rev. B **42**, 8110 (1990)

2.261 D.M. Edwards, J. Mathon, R.B. Muniz, M.S. Phan: Phys. Rev. Lett. **67**, 493 (1991)

2.262 J.-I. Inoue, A. Oguri, S. Maekawa: J. Phys. Soc. Jpn. **60**, 376 (1991)

2.263 D.M. Edwards, R.B. Muniz, J. Mathon: IEEE Trans. Magn. **27**, 3548 (1991)

2.264 F. Trigui, E. Velu, C. Dupas: J. Magn. Magn. Mat. **93**, 421 (1991)

2.265 S. Zhang, P.M. Levy, A. Fert: Phys. Rev. B **45**, 8689 (1992)

2.266 J.W.F. Dorleijn: in *Philips Res. Rep.* Vol. **31** (1976) p. 287

2.267 A. Barthélémy, A. Fert, M.N. Baibich, S. Hadjoudj, F. Petroff, P. Etienne, R. Cabanel, S. Lequien, F. Nguyen van Dau, G. Creuzet: J. Appl. Phys. **67**, 5908 (1990)

2.268 J. Unguris, R.J. Celotta, D.T. Pierce: Phys. Rev. Lett. **67**, 140 (1991)

2.269 S.T. Purcell, S.T. Purcell, W. Folkerts, M.T. Johnson, N.W.E. McGee, K. Jager, J. aan de Stegge, W.B. Zeper, W. Hoving, P. Grunberg: Phys. Rev. Lett. **67**, 903 (1991)

2.270 S. Demokritov, J.A. Wolf, P. Grunberg: Euro. Phys. Lett. **15**, 881 (1991)

2.271 S.S.P. Parkin, Z.G. Li, D.J. Smith: Appl. Phys. Lett. **58**, 2710 (1991)

2.272 W.R. Bennett, W. Schwarzacher, W.F. Egelhoff: Phys. Rev. Lett. **65**, 3169 (1990)

2.273 A. Cebollada, R. Miranda, C.M. Schneider, P. Schuster, J. Kirschner: J. Magn. Magn. Mat. **102**, 25 (1991)

2.274 B. Heinrich, Z. Celinski, J.F. Cochran, W.B. Muir, J. Rudd, Q.M. Zhong, A.S. Arrott, K. Myrtle, J. Kirschner: Phys. Rev. Lett. **64**, 673 (1990)

2.275 S.S.P. Parkin, D. Mauri: Phys. Rev. B **44**, 7131 (1991)

2.276 S.S.P. Parkin, unpublished

2.277 Z. Celinski, B. Heinrich: J. Magn. Magn. Mat. **99**, L25 (1991)
2.278 P. Bruno, C. Chappert: Phys. Rev. Lett. **67**, 1602 (1991)
2.279 H. Hasegawa: Phys. Rev. B **42**, 2368 (1990)
2.280 Y. Wang, P.M. Levy, J.L Fry: Phys. Rev. Lett. **65**, 2732 (1990)
2.281 D.M. Edwards, J. Mathon, R.B. Muniz, M.S. Phan: J. Phys.: Cond. Mat. **3**, 4941 (1991)
2.282 J.L. Fry, E.C. Ethridge, P.M. Levy, Y. Wang: J. Appl. Phys. **69**, 4780 (1991)
2.283 C. Chappert, J.P. Renard: Europhys. Lett. **15**, 553 (1991)
2.284 D.M. Deaven, D.S. Rokhsar, M. Johnson: Phys. Rev. B **44**, 5977 (1991)
2.285 R. Coehoorn, unpublished
2.286 P. Bruno, C. Chappert: Phys. Rev. B **46**, 261 (1992)
2.287 P. Grunberg: "Light Scattering From Spin Waves in Thin Films and Layered Magnetic Structures" in *Light Scattering in Solids V*, ed. by M. Cardona, G. Guntherodt, Topics in Applied Physics, Vol. 66 (Springer, Berlin, Heidelberg, 1989) p. 303
2.288 J.F. Cochran, J. Rudd, W.B. Muir, B. Heinrich, Z. Celinski: Phys. Rev. B **42**, 508 (1990)
2.289 J. Fassbender, F. Nortemann, R.L. Stamps, R.E. Camley, B. Hillebrands, G. Guntherodt, S.S.P. Parkin: Phys. Rev. B **46**, RC5810 (1992)
2.290 C. Carbone, S.F. Alvarado: Phys. Rev. B **36**, 2443 (1987)
2.291 D. Pescia, D. Kerkmann, F. Schumann, W. Gudat: Z. Phys. B **78**, 475 (1990)
2.292 L.M. Roth, H.J. Zeiger, T.A. Kaplan: Phys. Rev. **149**, 519 (1966)
2.293 I.R. McFadyen, P.S. Alexopoulous: "Temperature Dependence of Micromagnetic Domain Structure in Cobalt Films" in *Science and Technology of Nanostructured Magnetic Materials*, ed. by G.C. Hadjipanayis, G.A. Prinz (Plenum, New York, 1991) p. 99
2.294 E. Velu, C. Dupas, D. Renard, J.P. Renard, J. Seiden: Phys. Rev. B **37**, 668 (1988)
2.295 C. Dupas, P. Beauvillain, C. Chappert, J.P. Renard, F. Trigui, P. Veillet, E. Velu, D. Renard: J. Appl. Phys. **67**, 5680 (1990)
2.296 S.S.P. Parkin, A. Modak, D.J. Smith: Phys. Rev. B. **47**, 9136 (1993) RC (April 1, 1993)
2.297 S.S.P. Parkin: "Giant Magnetoresistance and Oscillatory Interlayer Exchange Coupling in Copper Based Multilayers", in *Magnetic Surfaces, Thin Films and Multilayers*, ed. by S.S.P. Parkin, H. Hopster, J.-P. Renard, T. Shinjo, W. Zinn, Vol, 231 (Mat. Res. Soc. Sym. Proc., 1992) p. 211
2.298 N. Hosoito, S. Araki, K. Mibu, T. Shinjo: J. Phys. Soc. Jpn. **59**, 1925 (1990)
2.299 A. Chaiken, P. Lubitz, J.J. Krebs, G.A. Prinz, M.Z. Harford: Appl. Phys. Lett. **59**, 240 (1991)
2.300 B. Dieny, V.S. Speriosu, S.S.P. Parkin, B.A. Gurney, D.R. Wilhoit, D. Mauri: Phys. Rev. B. **43**, 1297 (1991)
2.301 W.H. Meiklejohn, C.P. Bean: Phys. Rev. B **102**, 1413 (1959)
2.302 A. Yelon: "Interactions in Multilayer Magnetic Films" in *Physics of Thin Films*, ed. by M. Francombe, R. Hoffman, Vol. 6 (Academic, New York, 1971) p. 205
2.303 C. Tsang, K. Lee: J. Appl. Phys. **53**, 2605 (1982)
2.304 S.S.P. Parkin, V. Deline, R. Hilleke, G.P. Felcher: Phys. Rev. B **42**, 10583 (1990)
2.305 S.S.P. Parkin: Appl. Phys. Lett. **60**, 512 (1992)
2.306 S.S.P. Parkin: Phys. Rev. Lett. **71**, 1641 (1993)
2.307 S.S.P. Parkin: Appl. Phys. Lett. **61**, 1358 (1992)
2.308 J.Q. Xiao, J.S. Jiang, C.L. Chien: Phys. Rev. Lett. **68**, 3749 (1992)
2.309 A.E. Berkowitz, J.R. Mitchell, M.J. Carey, A.P. Young, S. Zhang, F.E. Spada, F.T. Parker, A. Hutten, G. Thomas: Phys. Rev. Lett. **68**, 3745 (1992)
2.310 S.S.P. Parkin, R.F.C. Farrow, T.A. Rabedeau, R.F. Marks, G.R. Harp, Q.H. Lam, M. Toney, R. Savoy, R. Geiss: Euro. Phys. Lett. **22**, 455 (1993)
2.311 T.A. Rabedeau, M. Toney, R.F. Marks, S.S.P. Parkin, R.F.C. Farrow, G. Harp: Phys. Rev. B. (submitted)
3.312 J.Q. Ziao, J.S. Jiang, C.L. Chien: Phys. Rev. B **46**, 9266 (1992)
2.313 R.F.C. Farrow, G.R. Harp, R.F. Marks, T.A. Rabedeau, M.F. Toney, R.J. Savoy, D. Weller, S.S.P. Parkin: J. Cryst. Growth (submitted)
2.314 J.R. Levine, J.B. Cohen, Y.W. Chung: Science **248**, 215 (1991)
2.315 S. Zhang: Appl. Phys. Lett. **61**, 1855 (1992)

3. Radio Frequency Techniques

Molecular beam epitaxy (MBE) and sputtering systems have been used in recent years to create a variety of epitaxial ultrathin films. It is of the utmost importance to employ experimental techniques which can determine their basic magnetic properties straightforwardly and quantitatively. In the following four sections of this chapter ferromagnetic resonance (FMR), Brillouin light scattering (BLS) and nuclear magnetic resonance (NMR) will be described. It will be shown that these techniques are uniquely suited for the study of ultrathin films. They provide us with all the essential parameters describing the magnetic properties of ultrathin films. FMR and BLS are very sensitive techniques enabling films one monolayer (ML) thick to be readily investigated.

3.1 Ferromagnetic Resonance in Ultrathin Film Structures

B. HEINRICH

This section will be devoted to FMR and is subdivided as follows: In Sect. 3.1.1 the main parameters describing the magnetic properties of a single magnetic layer will be introduced. Solutions of the Landau–Lifshitz equation of motion will provide a convenient link between the studied magnetic parameters and the measured rf properties. Section 3.1.2 will describe the FMR technique. A detailed description of the resonant cavities will be provided and experimental procedures allowing one to extract magnetic parameters will be introduced.

In Sect. 3.1.3 selected FMR studies of magnetic anisotropies in stable and metastable structures will be discussed. Finally, in Sect. 3.1.4, the magnetic coupling between ferromagnetic layers will be introduced and the role of the ferro- and antiferromagnetic coupling on the FMR signal will be demonstrated for the case of a simple trilayer in which two magnetic layers are coupled by a non-ferromagnetic interlayer.

B. Heinrich and J.A.C. Bland (Eds.)
Ultrathin Magnetic Structures II
© Springer-Verlag Berlin Heidelberg 1994

3.1.1 Magnetic Properties of Ultrathin Magnetic Layers and the Landau–Lifshitz Equations of Motion

3.1.1.1 Magnetic Anisotropies

The concept of ultrathin structures is explained in several parts of this book, e.g., see the Introduction and Chap. 3 by Mills in Vol. I. For the time being we assume that the exchange coupling within the layer is strong enough to maintain nearly parallel atomic magnetic moments across the film thickness. It will be shown at the end of this section how this concept can be refined for the case of rf measurements.

The description of magnetic properties associated with the behavior of ultrathin structures can be simplified significantly compared to that in bulk materials. Ultrathin layers lose their internal magnetic degree of freedom. All atomic magnetic moments across the film thickness are parallel and consequently the total magnetic moment is given by a simple algebraic sum of all atomic moments across the film thickness. The ultrathin films are essentially giant magnetic molecules which can have magnetic properties different from those in the bulk. The Landau–Lifshitz (L–L) equations of motion provide a simple means of introducing the magnetic properties of ultrathin films.

The response of the atomic magnetic moment is described by a torque equation

$$\frac{1}{\gamma}\frac{d\boldsymbol{\mu}}{dt} = -\,[\boldsymbol{\mu} \times \boldsymbol{H}^{\mathrm{a}}_{\mathrm{eff}}], \tag{3.1}$$

where $\boldsymbol{\mu}$ is the atomic magnetic moment, $\boldsymbol{H}^{\mathrm{a}}_{\mathrm{eff}}$ is the effective field acting on the atomic moment $\boldsymbol{\mu}$ and $\gamma = g|e|/(2mc)$ is the gyromagnetic ratio. The spectroscopic splitting factor is g, and for a free electron $g = 2$. The left hand side describes the time evolution of the atomic mechanical momentum and the right hand side represents a total torque acting on the atomic magnetic moment. In ultrathin films all atomic magnetic moments are parallel and therefore one can sum up all moments across the film thickness. After simple algebraic steps the L–L equations of motion can be written in the form

$$\frac{d\mathcal{M}}{dt} = -\gamma\left[\mathcal{M} \times \sum_i \frac{\mu_i}{\mathcal{M}}\, \boldsymbol{H}^{\mathrm{a}}_{i,\,\mathrm{eff}}\right]$$

$$\boldsymbol{H}_{\mathrm{eff}} = \sum_i \frac{\mu_i}{\mathcal{M}}\, \boldsymbol{H}^{\mathrm{a}}_{i,\,\mathrm{eff}}, \tag{3.2}$$

where the effective field, $\boldsymbol{H}_{\mathrm{eff}}$, acts on the total moment $\mathcal{M}$. The magnitude $|\mathcal{M}| = \mathcal{M} = \Sigma\mu_i$ is given by the algebraical sum of the magnetic moments of all of the atomic layers which constitute the ultrathin film. The magnetic moment μ_i (corresponding to the atomic layer i) is not precisely defined. The lateral sum has to be carried out across the area which is small enough that the atomic magnetic moments within this area remain parallel; the sum across the lateral unit chemical cell usually satisfies such a condition. In order to give $\mathcal{M}$ and μ_i a precise meaning *the magnetic moments $\mathcal{M}$ and μ_i will be defined per unit lateral area*. The effective field, H_{eff} is given by a sum of effective fields which are scaled by a factor $\mu_i/\mathcal{M}$. The factor $\mu_i/\mathcal{M}$ represents a weighting factor for each individual atomic layer.

In several chapters of this book it is shown that the presence of interfaces in ultrathin structures makes the individual atomic layers non-equivalent. The atomic layers in the vicinity of the interfaces have generally different magnetic properties from those which are further away. The atomic layers which do not form the interfaces directly possess bulk magnetic properties. On the other hand, due to the broken symmetry at the interfaces, the interface atomic layers can acquire magnetic properties which are very different from those of the bulk. This is a somewhat simplified picture but yet, as several chapters of this book demonstrate, it is quite applicable to epitaxial systems having sharp interfaces. *Equation (3.2) shows that in ultrathin films the effective fields are given by an admixture of the interface and bulk effective fields which are scaled with appropriate scaling factors*. The scaling factor, $\mu_i/\mathcal{M}$, can be thought of as a dilution parameter. In fact, this simple conclusion is the main point of ultrathin magnetic films. One can engineer new magnetic materials by adding to the bulk magnetic properties additional properties which originate from the interfaces.

Equation (3.2) describes both the static and dynamic response of the ultrathin film. In the static case the total magnetic moment, $\mathcal{M}$, has to be parallel with the total effective field, H_{eff}. The relationship between effective fields and magnetic energies can be derived from variational calculations [3.1] using the macroscopic concept of the energy density function $\mathscr{E}$,

$$H_{\mathrm{eff}} = -\frac{\partial \mathscr{E}}{\partial M}, \tag{3.3}$$

where M is the saturation magnetization. Since the first-principles calculations usually evaluate the energy per atom, E_{atom}, one needs to define the relationship between the macroscopic and atomic properties. Assuming that the spatial variations are slow, then

$$\mathscr{E} = \frac{E_{\mathrm{atom}}}{V_0} \quad \text{and} \quad M = \frac{\mu}{V_0}, \tag{3.4}$$

where μ is the magnetic moment per atom and V_0 is the atomic volume.

The magnetic behavior of ultrathin films (giant magnetic molecules) can be described by introducing the average saturation magnetization M_{s} ($M_{\mathrm{s}} = \Sigma\mu_i/(NV_0)$, N is the number of atoms per unit area of the film) and by the energy density function $\mathscr{E}$ which satisfies the symmetry of the film. In this article only films of materials that have bulk cubic symmetry and with the (0 0 1) surface crystallographic orientation will be considered. The treatment of other symmetries is straightforward and can be similarly accomplished.

Ultrathin films of otherwise cubic materials grown along the $[0\,0\,1]$ crystallographic direction in general have a tetragonal symmetry. The corresponding density function (magnetic anisotropy energy per unit volume) can be written in the form [3.2]

$$\mathscr{E} = -\frac{K_{1\parallel}}{2}(\alpha_x^4 + \alpha_y^4) - \frac{K_{1\perp}}{2}\alpha_z^4 - K_u\alpha_z^2, \tag{3.5}$$

where $\alpha_x, \alpha_y, \alpha_z$ are directional cosines of the saturation magnetization with respect to the $[1\,0\,0]$, $[0\,1\,0]$ and $[0\,0\,1]$ crystallographic axes, $K_{1\parallel}$ describes the strength of the four-fold in-plane anisotropy, K_u and $K_{1\perp}$ are the second and fourth order terms of the perpendicular uniaxial anisotropy. Equation (3.5) goes over to the usual expression for cubic anisotropy if $K_u = 0$ and if $K_{1\parallel} = K_{1\perp}$, in as much as $\alpha_x^4 + \alpha_y^4 + \alpha_z^4 = 1 - 2(\alpha_x^2\alpha_y^2 + \alpha_y^2\alpha_z^2 + \alpha_x^2\alpha_z^2)$.

Samples with vicinal surfaces [3.3] or with unidirectional interface chemical ordering [3.4] can exhibit in-plane uniaxial anisotropies. Let ϕ_m be the angle between the preferred axis and the $\{1\,0\,0\}$ direction and $K_{u\parallel}$ be the uniaxial anisotropy constant, then the in-plane uniaxial energy is

$$\mathscr{E} = -K_{u\parallel}(n \cdot M)^2/(M_s^2), \tag{3.6}$$

where n is the unit vector along the direction of the uniaxial axis and M (and M_s) is the average saturation magnetization of the ferromagnetic layer. The L–L equations of motion for the film described by the above energy terms can be written in a form

$$-\frac{1}{\gamma}\frac{\partial M}{\partial t} = [M \times H_{\mathrm{eff}}] - \frac{G}{\gamma^2 M_s^2}\left[M \times \frac{\partial M}{\partial t}\right], \tag{3.7}$$

where H_{eff} is given by (3.3) and the second term on the right hand side describes the Gilbert damping which accounts for the energy dissipation, as discussed by Cochran in Sect. 3.2.

In FMR the external rf magnetic field is oriented perpendicular to the dc (direct current) field and its torque leads to a precessional motion of the magnetization [3.5], also examined in sect. 3.2. In ferromagnetic samples the Gilbert damping is strong enough to allow only small deviations of the magnetization M from its static direction M_s. In this case, (3.7) can be linearized by looking for solutions in the form $M = M_s + m$, $H_{\mathrm{eff}} = H_{\mathrm{eff}}^s + h_{\mathrm{eff}}$. M_s is the static saturation magnetization and m is the rf component perpendicular to M_s. H_{eff}^s and h_{eff} (perpendicular to M_s) are the effective static and rf fields. M_s has to be parallel to H_{eff}^s to satisfy the magnetostatic condition.

FMR measurements are usually carried out in large dc applied fields which ensure that the saturation magnetization M_s is almost aligned along the applied field H_0. In general the presence of magnetic anisotropies results in an angle between the saturation magnetization M_s and the external field H_0. In this case the equilibrium orientation of M_s has to be found first. The L–L equations of motion are solved in a coordinate system which follows the direction of the saturation magnetization as the external field changes. The external field introduces the Zeeman energy

$$\mathscr{E} = - \mathbf{M} \cdot \mathbf{H}_0. \tag{3.8}$$

In films one must also include the shape anisotropy. The magnetic moment $\mathscr{M}$ tilted away from the film surface will create the magnetic charge density on surfaces which results in the restoring demagnetizing energy

$$\mathscr{E} = \frac{4\pi D M_{\perp}^2}{2}, \tag{3.9}$$

where $M_{\perp}$ is the saturation magnetization component perpendicular to the surface, and $4\pi D$ is the effective demagnetizing coefficient (for a continuum $4\pi D = 4\pi$).

In the parallel configuration the static field $\mathbf{H}_0$ and the saturation magnetization $\mathbf{M}_s$ are oriented in the film surface, and the fourth order uniaxial term $K_{1\perp}$ plays a negligible role in FMR since the corresponding effective field is proportional to the third power of the rf perpendicular magnetization component and it is therefore negligible in the linearized equations of motion.

Equations (3.3, 5, 7) then lead (for a time dependence $\exp(i\omega t)$) to the following linearized L–L equations of motion

$$\frac{i\omega}{\gamma} \mathscr{M}_{\parallel} + \left[H_0 + 4\pi D M_s - \frac{2K_u}{M_s} + \frac{K_{1\parallel}}{2M_s}(3 + \cos(4\varphi)) \right.$$

$$\left. + \frac{K_{u\parallel}}{M_s}(1 + \cos(2(\varphi - \varphi_m))) + \frac{i\omega G}{\gamma^2 M_s} \right] \mathscr{M}_{\perp} = 0$$

$$\left[H_0 + \frac{2K_{1\parallel}}{M_s}\cos(4\varphi) + \frac{2K_{u\parallel}}{M_s}\cos(2(\varphi - \varphi_m)) + \frac{i\omega G}{\gamma^2 M_s} \right]$$

$$\times \mathscr{M}_{\parallel} - \frac{i\omega}{\gamma} \mathscr{M}_{\perp} = M_s hd, \tag{3.10}$$

where $\mathscr{M}_{\parallel}$ and $\mathscr{M}_{\perp}$ are the parallel and perpendicular (with respect to the film surface) rf magnetization components per unit area integrated across the film thickness, h is the rf field parallel to $\mathscr{M}_{\parallel}$ and d is the thickness of the film. In the perpendicular configuration the static applied field H_0 and the saturation magnetization are oriented perpendicular to the sample surface. In this case the role of the in-plane four-fold anisotropy is negligible. In the perpendicular configuration the rf magnetization components $\mathscr{M}_x, \mathscr{M}_y$ in the plane of the film have identical amplitudes and can be split into right and left hand circularly polarized rf components, $\mathscr{M}_+ = \mathscr{M}_x + i\mathscr{M}_y$ and $\mathscr{M}_- = \mathscr{M}_x - i\mathscr{M}_y$. Only the $\mathscr{M}_+$ polarization undergoes FMR. The linearized L–L equation (neglecting the in-plane uniaxial anisotropy) of motion for $\mathscr{M}_+$ is

$$\left(H_0 - 4\pi D M_s + \frac{2K_u}{M_s} + \frac{2K_{1\perp}}{M_s} - \frac{\omega}{\gamma} + \frac{i\omega G}{\gamma^2 M_s} \right) \mathscr{M}_+ = dM_s h_+. \tag{3.11}$$

Note that the contributions of the dipolar interaction (demagnetizing field $4\pi D M_s$) and the uniaxial perpendicular anisotropy ($2K_u/M_s$) enter the L–L equations of motion in an additive way, and therefore one can introduce an effective demagnetizing field (effective magnetization)

$$4\pi M_{\text{eff}} = 4\pi D M_s - \frac{2K_u}{M_s}. \tag{3.12}$$

In FMR the rf magnetization components reach a maximum amplitude when the real parts of the denominators of $\mathscr{M}_{\parallel}$ and $\mathscr{M}_+$ in (3.10, 11) are equal to zero. In FMR measurements the microwave frequency, ω, is fixed and the resonance condition is reached by sweeping the external dc field H_0. The resonance field H_{res} is then given by the real part of

$$\left(\frac{\omega}{\gamma}\right)^2 = \left[H_{\text{res}} + 4\pi M_{\text{eff}} + \frac{K_{1\parallel}}{2M_s}(3 + \cos(4\varphi)) + \frac{i\omega G}{\gamma^2 M_s} \right]$$
$$\times \left[H_{\text{res}} + \frac{2K_{1\parallel}}{M_s}\cos(4\varphi) + \frac{i\omega G}{\gamma^2 M_s} \right] \tag{3.13a}$$

for the parallel configuration (neglecting the in-plane uniaxial anisotropy), and

$$\frac{\omega}{\gamma} = H_{\text{res}} - 4\pi M_{\text{eff}} + \frac{2K_{1\perp}}{M_s} \tag{3.13b}$$

for the perpendicular condition (neglecting the in-plane uniaxial anisotropy and damping). Equations (3.10, 11, 13a, 13b) are correct only in high applied magnetic fields in which the dc magnetic moment is parallel to $\boldsymbol{H}_0$.

Phenomenological effective fields in (3.10, 11) can be expressed in terms of bulk and interface magnetic properties. For simplicity let us assume again that all of the atomic magnetic moments possess the bulk magnetic properties and that the interface atomic moments have additional magnetic anisotropies which originate in the broken symmetry at the interfaces. The interface perpendicular uniaxial anisotropy, K_u^s, and the interface in-plane four-fold anisotropy, $K_{1\parallel}^s$, are usually expressed as energies per unit area. Equations (3.2, 4, 5) after simple algebraical steps, result in the well known equations

$$\frac{2K_{1\parallel}}{M_s} = \frac{2K_1}{M_s} + \left\{\frac{2K_{1\parallel}^s}{M_s d}\right\}_A + \left\{\frac{2K_{1\parallel}^s}{M_s d}\right\}_B$$

$$\frac{2K_{1\perp}}{M_s} = \frac{2K_1}{M_s} + \left\{\frac{2K_{1\perp}^s}{M_s d}\right\}_A + \left\{\frac{2K_{1\perp}^s}{M_s d}\right\}_B \tag{3.14}$$

$$\frac{2K_u}{M_s} = \left\{\frac{2K_u^s}{M_s d}\right\}_A + \left\{\frac{2K_u^s}{M_s d}\right\}_B,$$

where d is the thickness of the film and A and B represent the film interfaces. *The effective anisotropy fields in ultrathin films are given by a sum of bulk effective fields and the interface effective fields which scale inversely with the film thickness.* The reader can find a more precise treatment of the surface anisotropy fields in [3.3, 6, 7], however, in the limit of ultrathin films the above expressions provide a good description of the static and dynamic properties. One should point out that studies which are based on the magnetic torque measurements can provide only effective fields. Therefore a particular choice of M_s in data analysis is not crucial, the coefficients of the magnetic anisotropies are scaled with M_s to provide the measured values of the appropriate effective fields.

The effective demagnetizing field $4\pi M_{eff}$ strongly affects the dynamic (H_{res}) and static (orientation of M_s) response. Its value in ultrathin structures can be varied widely by changing the film thickness, as shown by (3.14). By choosing appropriate interfaces and by varying the film thickness, one can engineer materials with variable magnetic properties. The sign of $4\pi M_{eff}$ determines the orientation of the saturation magnetization with respect to the film surface. For negative values of $4\pi M_{eff}$, the saturation magnetization M_s is oriented perpendicular to the film surface. An external field larger than $|4\pi M_{eff}|$ is needed to orient the magnetization M_s parallel with the surface.

In FMR measurements the role of $4\pi M_{eff}$ is also very dramatic. For positive values of $4\pi M_{eff}$ the resonance field H_{res} is below the field ω/γ and H_{res} shifts towards higher fields with a decreasing value of $4\pi M_{eff}$. $H_{res} = \omega/\gamma$ for $4\pi M_{eff} = 0$ (neglecting the role of four-fold anisotropies). The resonance field H_{res} can be easily shifted several kOe by changing the sample thickness. The rf magnetization components are generally elliptically polarized with the larger amplitude in the plane of the film for positive values of $4\pi M_{eff}$ and with the larger amplitude perpendicular to the film surface for negative values of $4\pi M_{eff}$.

The dipolar demagnetizing field is usually described by using a magnetic continuum in which the demagnetizing field is given by $4\pi M_s$. This treatment is incorrect in samples consisting of a few atomic layers. The atomic magnetic moments are localized around their atomic sites. The discreteness of atomic moments results in a dipolar field which changes across the sample thickness and it depends on the number of atomic layers involved. The dipolar field from a given atomic layer decreases exponentially away from its surface with a decay length corresponding to the in-plane lattice spacing. Consequently the dipolar field decreases when approaching the sample surface from inside of the film. The value of the dipolar field inside of the film decreases appreciably as the thickness approaches the ML limit; the reduction is approximately 50% in a bcc lattice. The average demagnetizing factor D for a layer containing N atomic planes is

$$D = 1 - \frac{0.4245}{N} \quad \text{for bcc } (0\,0\,1) \text{ layers,}$$

$$D = 1 - \frac{0.2338}{N} \quad \text{for fcc } (0\,0\,1) \text{ layers.}$$

This is more completely discussed in [3.3, 8], and a more general treatment is given in [3.9].

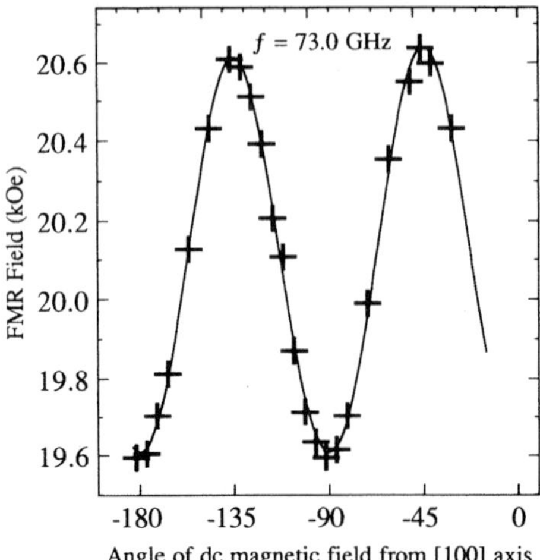

Fig. 3.1. The in-plane angular dependence of measured FMR fields carried out at $f = 73.01$ GHz for the [Ag(0 0 1) bulk substrate/5.7 ML Fe(0 0 1)/9.4 ML Ni(0 0 1)/15 ML Au(0 0 1)] at room temperature (RT). The solid line is the calculated angular dependence of the FMR field. Calculations were carried out with magnetic parameters obtained from a multiparameter χ^2 function minimization routine which gave $2K_{1\parallel}/M_s = 0.723$ kOe, $4\pi M_{eff} = 9.816$ kG, $g = 2.107$. The $\{1\,0\,0\}$ and $\{1\,1\,0\}$ crystallographic directions correspond to the in-plane magnetic easy and hard axes respectively

In-plane anisotropies further shift the resonance field. With H_0 along the easy magnetic axis the resonance field is shifted towards lower fields. This trend reverses when H_0 is oriented along the hard magnetic axis. The resonance field in films with an in-plane four-fold anisotropy exhibits a typical four-fold quasi-sinusoidal angular dependence, as seen in Fig. 3.1. The simple four-fold dependence can be affected by the presence of in-plane uniaxial anisotropy. In this case the four-fold symmetry is perturbed. Usually either hard or easy axes are not equivalent and one has to invoke the full resonance condition involving both the four-fold and uniaxial in-plane anisotropies, as in (3.10) and Fig. 3.2. The shift of the resonance field due to in-plane anisotropies usually does not exceed 1 kOe, however notable exceptions exist, as discussed in Sect. 3.1.3.

So far the ultrathin films were treated using the strict constraint of fully parallel magnetic moments, including their rf components. As for any extreme, this model is only partly true and reasonably justified only in a certain limit. At this point it is important to determine the length scale which allows us to characterize the film as ultrathin. The static case was addressed in the Introduction and was further discussed in more detail by Mills in Chap. 3, Volume, 1. Here we address the dynamic case. To keep the algebra simple let us consider perpendicular FMR. In (3.11) the exchange interaction torque has to be added in a correct manner by introducing the spatially varying exchange field, $(2A/M_s^2)\mathrm{d}^2\boldsymbol{m}/\mathrm{d}z^2$ [3.6, 7], where A is the coefficient of the exchange coupling and z is perpendicular to the film surface. Note also that the contribution of the effective surface field in (3.11) has to be removed. The L–L equation of motion is

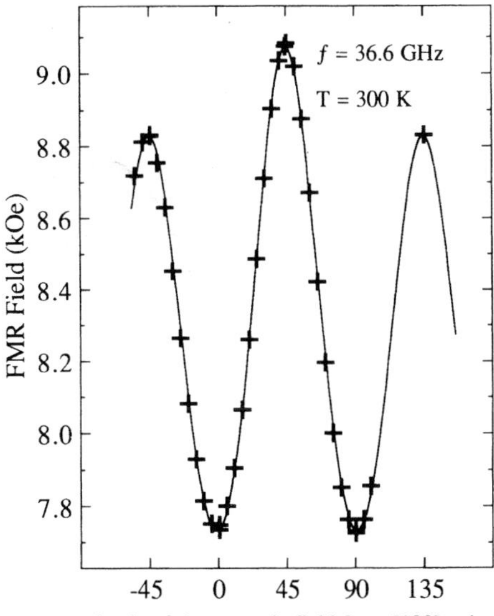

Angle of dc magnetic field from [100] axis

Fig. 3.2. The in-plane angular dependence of measured FMR fields ($f = 36.6$ GHz) for the [Ag(001) bulk vicinal substrate/6 ML Fe(001)/10 ML Ni(001)/15 ML Au(001)] at RT. The solid line is calculated angular dependence of the FMR field using following parameters: $2K_{1\parallel}/M_s = 0.789$ kOe, $2K_{u\parallel}/M_s = -0.143$ kOe, $\phi_m = 45°$ (along Fe[110]), $4\pi M_{eff} = 9.75$ kG, $g = 2.1$. Note an obvious contribution of the in-plane uniaxial anisotropy. The in-plane uniaxial anisotropy is caused by ordered atomic steps which are present in vicinal cuts. The sample in Fig. 3.1 was grown on the singular Ag(001) substrate (the surface normal within 0.25 deg of [001] crystallographic direction) and shows no noticeable contribution of the in-plane uniaxial anisotropy

(not including damping)

$$-\frac{2A}{M_s}\frac{d^2 m_+}{dz^2} + \left[H_0 - 4\pi M_s - \frac{\omega}{\gamma} \right] m_+ = M_s h_+. \tag{3.15}$$

The skin depth (~ 1000 Å) is significantly larger than the film thickness and therefore the driving field h can be assumed to be constant across the film thickness. Equation (3.15) represents a differential equation of the second order with a constant right side. Its solution with proper boundary conditions can be found by standard mathematical procedures [3.7]. However, here we want to identify only the length scale on which the rf magnetization is required to change throughout the film. Assuming that the magnetic component varies as $\exp(\pm kz)$ then the rf exchange dynamic length $\lambda_{dyn} = 1/k_{dyn} = ((H_0 - \omega/\gamma - 4\pi M_s)M_s/(2A))^{-1/2}$. However, in ultrathin films $H_{res} \sim 4\pi M_{eff} + \omega/\gamma$ and therefore $\lambda_{dyn} = i((2K_u/M_s) \cdot M_s)/2A)^{-1/2}$. This expression is similar to the exchange length introduced for the static case where $\lambda_{stat} = (4\pi M_s \cdot (M_s/2A))^{-1/2}$. The demagnetizing field $4\pi M_s$ is replaced by the uniaxial perpendicular field. The spatial variations within the film are negligible if λ_{dyn} is larger than the film thickness. Since the perpendicular uniaxial field does not usually exceed $4\pi M_s$, the dynamic length scale is often even larger than that required by the magnetostatic requirement. An upper limit of the film thickness which still reasonably satisfies the ultrathin film resonance equations (3.13a, 13b) can be found by carrying out a full calculation which should also include the treatment of electromagnetic waves by Maxwell's equations [3.7]. In Fe(0 0 1) the error in the surface uniaxial perpendicular anisotropy (which is determined from (3.13a)) is less than 7% for films thinner than $\frac{3}{4}$ of λ_{stat} (33 Å or ~ 20 ML).

3.1.2 FMR Technique and Experimental Procedures

3.1.2.1 Experimental Apparatus

The microwave range of frequencies is required to study FMR. The FMR signal is measured by monitoring the microwave losses in the studied film as a function of the applied dc field H_0. A simple diagram of a microwave spectrometer is shown in Fig. 3.3. The microwave losses in this case are measured by monitoring the amplitude of the reflected microwave electric field e which is detected by a microwave diode detector (Fig. 3.3). In FMR studies of ultrathin films, the sample is inserted in a microwave cavity and the external field is modulated with a low frequency field component. The microwave cavity enhances the role of FMR losses and the external low frequency modulation allows one to use lock-in amplifier detection. Modulation frequencies in the range of 100–200 Hz are sufficient to significantly improve the signal-to-nosie ratio and at the same time spurious field dependent signals associated with high modulation frequencies are avoided. The signal-to-noise is independent of microwave power for power greater than ~ 1 mW, but it depends on the diode dc current. The performance of a microwave diode detector can be markedly improved by operating it in an appropriate dc biased current. Usually a resistor of 1–3 kΩ in parallel to the microwave diode (resulting in a dc biased current of several μA) greatly improves the noise performance [3.10].

The choice of microwave cavities is crucial. The microwave cylindrical cavity TE_{01n} is ideally suited for FMR studies for samples which exhibit a sufficient microwave reflection (bulk metallic substrates, magnetic metallic superlattices). The TE_{01n} mode has its electric field at the end wall along the azimuthal direction only [3.11] and therefore the internal cavity quality Q factor is not affected by a poor electrical contact between the end wall and the cylindrical body of the microwave cavity. At higher microwave frequencies the cavity diameter ($\phi \sim 11$ mm at 36 GHz) is usually smaller than the substrate diameter. In this case the sample can entirely replace the end wall of the microwave cavity. The sample is separated from the cylindrical body by covering the sample with a thin non-conducting foil (e.g., Kapton foil 0.1 mm thick) (Fig. 3.4). This is an important step in the sample mounting. A non-conducting foil allows an unobscured (by eddy currents) penetration of the low frequency modulation field and at the same time eliminates degenerate and nearly degenerate microwave cavity modes which are always present in cylindrical cavities. For

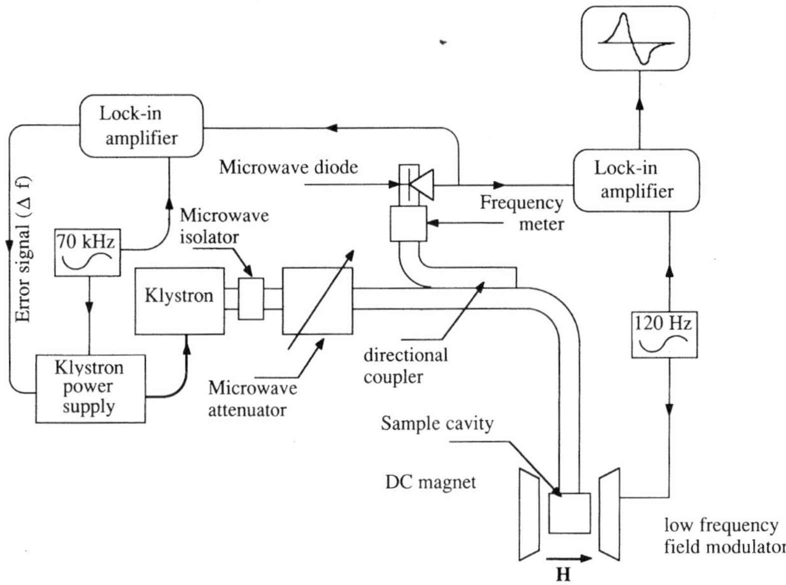

Fig. 3.3. A block diagram of a microwave spectrometer used in FMR measurements of ultrathin metallic magnetic films. The reflected microwave power from a sample resonance cavity is directed by a microwave directional coupler to a diode which is used to detect the FMR signal. A low frequency modulation (100–200 Hz) is used to monitor the field derivative of the out-of-phase microwave susceptibility, $d\chi''/dH$. The klystron repeller is modulated by a 70 kHz voltage allowing one to lock the microwave frequency to the sample resonance cavity. Lock-in amplifiers are employed for monitoring the FMR signal, $d\chi''/dH_0$, and for providing the dc error voltage which locks the klystron microwave frequency to the sample resonance cavity. The level of microwave power impinging on the sample is controlled by a microwave attenuator. The microwave frequency is measured by a microwave cavity meter. Microwave frequency meter is calibrated by Electron Spin Resonance (ESR) of a free radical (DPPH). The external field is calibrated by Nuclear Magnetic Resonance (NMR). The dc magnetic field can be usually rotated in the horizontal plane, and therefore the microwave waveguide (providing the microwave power for the sample resonant cavity) has to enter the dc magnet vertically

smaller samples, a thin Cu diaphragm with a central hole smaller than the sample is inserted between the foil and the sample.

For FMR systems in which the microwave frequency is locked to the sample cavity and for a light loading the reflected amplitude of the microwave electric field changes linearly with the absorbed microwave power in the sample. Thus the measured signal is proportional to the out-of-phase microwave susceptibility χ'', $\chi = \mathcal{M}_{\parallel}/h = \chi' - i\chi''$, where $\mathcal{M}_{\parallel}$ is the total rf magnetization component parallel to the rf magnetic field component, h, which is perpendicular to the applied dc field H_0. In this configuration the rf field lies also in the plane of the film. Equations (3.10, 11) give χ'' with a typical resonance Lorentzian lineshape. The maximum absorption occurs at the FMR field H_{res}, (3.13a, 13b). The linewidth of the resonance peak is given by the microwave

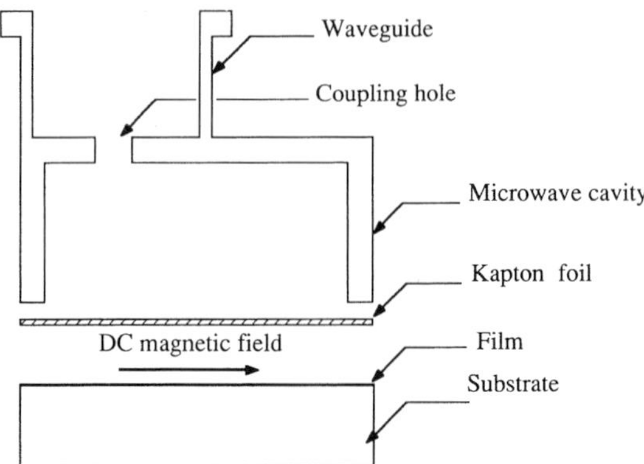

Fig. 3.4. An expanded view of the sample resonance cavity for the in-plane FMR measurements. The sample forms the end wall of a TE_{01n} cylindrical cavity. The microwave power is coupled through the coupling hole located approximately half way between the cavity axis and the cavity wall. The Kapton foil and the sample substrate are held against the cavity body by a spring loaded mechanism which allows us to obtain a gentle contact between the cavity and sample

losses. The FMR measurements are usually performed using a modulation field which is appreciably smaller than the FMR linewidth. The measured signal using a lock-in amplifier detection is then proportional to the field derivative $d\chi''/dH$. The resonance field H_{res} corresponds to the zero crossing of $d\chi''/dH$ and the FMR linewidth ΔH is given by the field interval between the extrema of $d\chi''/dH$.

In the parallel configuration the dc applied field is oriented along the sample surface and the microwave cavity is coupled to the spectrometer waveguide through its upper end wall (Fig. 3.4). The rf magnetic field is radially directed and as the external field rotates around the cavity axis the components of the microwave field perpendicular to the dc field do not change as they would do in a rectangular cavity. For the cylindrical cavity the applied field can be rotated in the sample plane without producing any major change in the magnitude of the active microwave magnetic field component. However, one should realize that changes occur in the active area of the sample which is studied by FMR. The radial rf magnetic field reaches a maximum approximately half way between the axis and the cavity wall, and has a zero value on the cavity axis and at the cylindrical wall. For a given direction of the dc field, one monitors the FMR signal in two areas shaped like kidneys which are separated by a distance which is nearly equal to the cavity radius (Fig. 3.5). This configuration is very sensitive to sample inhomogeneities. If the measured sample has a gradient in its magnetic properties, then either split or appreciably asymmetric FMR lines are observed.

In the perpendicular configuration the cylindrical cavity is mounted with its end wall perpendicular to the dc applied field. The microwave cavity is coupled to the spectrometer waveguide through its cylindrical body (Fig. 3.6). In this

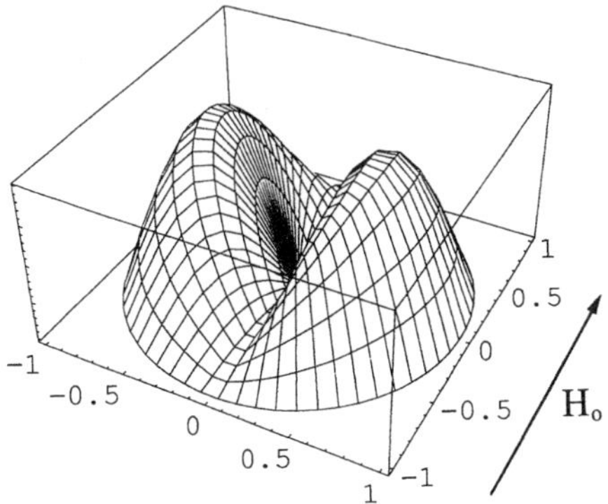

Fig. 3.5. The distribution of the rf magnetic field component $h_\perp$ which is perpendicular to the external applied dc field H_0. The external field is applied parallel to the sample surface (parallel configuration). The strength of $h_\perp$ is shown along the z-axis and the x, y coordinates determine the position along the end wall of microwave cavity. The center of the x–y plane is located at the microwave cavity axis ($h = 0$). Note that for a given field H_0 the two kidney-like shaped areas are monitored by FMR

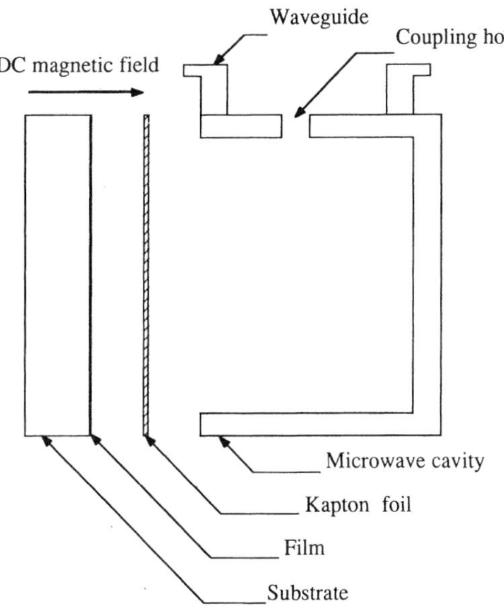

Fig. 3.6. An expanded view of the sample resonance cavity for the perpendicular FMR measurements. The coupling hole is at the midpoint of the cylindrical body (for $n = 3$, $l = (3/2)\lambda$). By rotating the dc external field in the plane which is perpendicular to the sample surface the FMR measurements can be carried out continuously from the perpendicular to the parallel configuration

configuration the dc field can be oriented from the perpendicular to the parallel configuration. Note that in the perpendicular configuration the rf magnetic field h is perpendicular to H_0 everywhere and a fully concentric area is sensitive to FMR, consequently the FMR signal in the perpendicular configuration is stronger than that in the parallel configuration.

3.1.2.2 FMR Studies

Equations (3.13a, 13b) algebraically summarize the power of FMR. Both parallel and perpendicular cavity configurations are needed to fully characterize the magnetic properties of ultrathin films. In the parallel configuration one can determine the in-plane anisotropy effective fields, $2K_{1\parallel}/M_s$ and $2K_{u\parallel}/M_s$, and the effective demagnetizing field $4\pi M_{\mathrm{eff}}$ by rotating the external field in the plane of the specimen. However, in order to make a unique interpretation one needs to find the gyromagnetic ratio which determines the effective field ω/γ. This can be done by carrying out full in-plane FMR measurements for at least two microwave frequencies. Caution must be used in the choice of microwave frequencies. One needs to use sufficiently different microwave frequencies to escape the region of strong ellipticity of the rf magnetization components. For large $4\pi M_{\mathrm{eff}}$ and low microwave frequencies, the resonance field is strongly represented only in the second bracket in (3.13a). It means that the resonance fields are primarily proportional to $(\omega/\gamma)^2/4\pi M_{\mathrm{eff}}$, which results in very inaccurate values of γ and $4\pi M_{\mathrm{eff}}$. One needs to increase the second microwave frequency to the point that the required resonance field becomes comparable to $4\pi M_{\mathrm{eff}}$. The choice of 74 GHz is usually sufficient.

The perpendicular configuration allows one to carry out FMR measurements with the magnetic moment $\mathcal{M}$ tilted away from the sample plane. The detailed angular fit of the measured resonance fields as a function of the angle between the sample and the applied field can be used to determine both the gyromagnetic ratio γ and the perpendicular effective field $2K_{1\perp}/M_s$ [3.12]. However one has to bear in mind that in the out-of-plane measurements the static magnetization is not parallel with the external field and the orientation of M_s has to be determined continuously as H_0 changes during the measurement [3.12, 13]. It is advantageous to determine ω/γ using the in-plane measurements at various microwave frequencies and then to determine the four-fold perpendicular anisotropy effective field $2K_{1\perp}/M_s$ from the perpendicular configuration. The out-of-plane resonance fields can be then used to check the magnetization process in applied magnetic fields which are oriented off of the film surface.

The FMR linewidth ΔH can be used to characterize the contribution of the intrinsic loss mechanism and determine the role of magnetic inhomogeneities. The studies of FMR linewidth in amorphous materials showed that the dependence of the FMR linewidth on the microwave frequency follows a linear dependence [3.14]

$$\Delta H = \Delta H(0) + 1.16\, \frac{\omega}{\gamma}\, \frac{G}{\gamma M_s}, \tag{3.16}$$

where the second term determines the role of viscous damping and the first term describes the frequency independent linewidth which arises from the presence of magnetic inhomogeneities. The same behavior was found in magnetic ultrathin structures [3.15, 16]. The FMR linewidth is sensitive only to those inhomogeneities which are larger on a lateral scale than the exchange length, since anything on a shorter length scale will be averaged out very effectively by strong in-plane exchange fields. MBE films can be grown with a superior quality, which exhibit lower microwave losses than well prepared bulk samples. Further discussion can be found in [3.17]. The intrinsic magnetic losses can also be effectively used in the study of critical behavior of the magnetic order parameter in ultrathin films, as was demonstrated by *Yi Li* et al. [3.18] on Ni(1 1 1) on W(1 1 0) films.

As was discussed above, the measurements of FMR fields alone cannot disentangle the dipolar demagnetizing field $4\pi DM_s$ from the perpendicular anisotropy field $2K_u/M_s$. Both fields enter the effective demagnetizing field $4\pi M_{eff}$ on an equal footing. One needs to determine the saturation magnetization by using an independent technique such as dc magnetometry or spin-polarized neutron reflection studies, as examined by *Bland*, Chap. 6, Volume 1, and in [3.19]. It should be pointed out that these techniques do not measure the saturation magnetization M_s but rather determine the total magnetic moment of the sample.

The intensity of the FMR signal (doubly integrated field derivative of the absorption χ'') is also directly proportional to the total magnetic moment of the specimen [3.13]

$$I \sim \mathcal{M}_s \left(\frac{H_{res} + 4\pi M_{eff}}{2H_{res} + 4\pi M_{eff}} \right). \tag{3.17}$$

One can therefore carry out comparision studies using a thick film as a reference sample where the total magnetic moment is known with a sufficient accuracy [3.19]. However, FMR intensity studies require careful measurements and a microwave system which can provide reproducible results. The microwave cavity coupling has to be accurately monitored and the collected data have to be rescaled to account for changing sensitivity due to variations in the internal Q of the sample cavity. FMR cavities which have their whole end wall replaced by the sample are well-suited for such studies. With appropriate care the magnetic moment comparision studies are accurate to a few percent. The above measurements can be used to quantitatively determine all the static and dynamic parameters which describe the magnetic properties of single ultrathin films.

3.1.3 Measurements of Magnetic Anisotropies

The studies of magnetic anisotropies represent one of the main trends of research on metallic magnetic ultrathin structures. Bcc Fe(0 0 1) films have been extensively studied by FMR and the results of these measurements helped to identify the presence of magnetic anisotropies created by the broken symmetry at interfaces and aided in discovering the in-plane anisotropies which are generated by ordered crystallographic defects. The results of these studies will be briefly summarized with the intention of demonstrating the power of FMR measurements.

3.1.3.1 Uniaxial Perpendicular Anisotropies

Fe(0 0 1) ultrathin films exhibit a large uniaxial anisotropy with the easy axis perpendicular to the film surface. FMR measurements [3.15] showed that the effective demagnetizing field includes a term which is inversely proportional to the film thickness. This dependence is expected in the presence of surface perpendicular uniaxial anisotropies, as presented by *Gay* and *Richter*, Chap. 2.1, Volume 1. BLS measurements on Fe whiskers by *Dutcher* et al. [3.20] and FMR measurements on bulk Fe(0 0 1) by *Purcell* et al. [3.21] showed that perpendicular interface anisotropies of the same magnitude exist at the bulk Fe(0 0 1) surface, as detailed in Table 3.1. These results clearly demonstrate that the uniaxial anisotropy in bcc Fe(0 0 1) ultrathin structures is indeed an intrinsic effect that is caused by the broken symmetry at the interfaces and it is not a consequence of some growth peculiarities.

The uniaxial interface perpendicular anisotropy is strongly dependent on the composition of interfaces. The strongest anisotropy was observed for the Fe(0 0 1)/vacuum interface (1.0 erg/cm^2), followed by Fe/Ag (0.81 erg/cm^2) and the Fe/Au (0.47 erg/cm^2) interfaces, all at room temperature (RT). Lattice expanded Pd exhibits a significantly decreased uniaxial anisotropy (0.17 erg/cm^2), the Fe/(bcc Cu(0 0 1)) interface is stronger (0.62 erg/cm^2) than in the Fe/Au interface. A summary of these results is shown in Table 3.1 [3.22]. Fe(0 0 1) films can exhibit a large perpendicular anisotropy that can overcome the perpendicular demagnetizing field $4\pi D M_s$. A 5 ML thick Fe(0 0 1) film

Table 3.1. The surface uniaxial anisotropies, K_u^s, the in-plane four-fold effective fields, $2K_{1\parallel}/M_s$, and the effective demagnetizing fields, $4\pi M_{\mathrm{eff}}$, in ultrathin Fe(0 0 1) films covered with Ag, Au, Cu, Pd epitaxial (0 0 1) layers. The values of K_u^s in individual interfaces were calculated using: $2K_u^s/dM_s = 4\pi D M_s - 4\pi M_{\mathrm{eff}} - 2K_u^s/dM_s$ (substrate). The Ag/Fe substrate interface anisotropy field, $2K_u^s/dM_s$ (substrate), was determined from the symmetric samples indicated by*. $4\pi D M_s$ (at 295 and 77 K) $= 21.55$ kG

Sample	K_u^s 295 K	K_u^s 77 K	$2K_{1\parallel}/M_s$ 295 K	$2K_{1\parallel}/M_s$ 77 K	$4\pi M_{\mathrm{eff}}$ 295 K	$4\pi M_{\mathrm{eff}}$ 77 K
*5.7 Fe/7 Ag/Au	0.69	0.72	0.092	0.204	1.987	1.214
5.7 Fe/20 Au	0.40	0.32	0.099	0.229	6.278	6.926
5.7 Fe/8 Cu/Au	0.62	0.69	0.110	0.279	3.076	1.716
5.7 Fe/8 Pd/Au	0.17	0.05	0.116	0.238	9.402	10.71
*10.7 Fe/8 Ag/Au	0.81	0.98	0.353	0.536	9.345	6.842
10.2 Fe/20 Au	0.47	0.52	0.328	0.460	11.44	9.714
7.5 Fe/vacuum[a]	1.0	—	0.172	—	2.81	—
Bulk bcc Fe/Ag[b]	0.79	—	0.547	0.617		
Bulk bcc Fe/Au[b]	0.54	—	0.547	0.617		

K_u^s in erg/cm^2; effective fields in kOe. The numbers used in the sample notation represent the number of atomic layers (ML).

[a] determined from *in situ* FMR measurements [3.23]

[b] K_u^s is determined in BLS studies using (0 0 1) facet of Fe whisker [3.20].

grown on Ag(0 0 1) having a vacuum interface was magnetized perpendicularly to the film surface [3.23] even at room temperature (RT). Fe(0 0 1) films grown on a bulk Ag(0 0 1) substrate [3.2, 13, 15] thinner than 5 ML exhibited a noticeable decrease in the surface anisotropy. Even 5–6 ML thick samples have a somewhat decreased surface anisotropy (0.7 erg/cm^2) when compared with that observed for thicker samples (0.8 erg/cm^2). The surfaces of Fe(0 0 1) layers grown on Ag(0 0 1) substrates thinner than 5 ML are affected by a vertical mismatch at the Ag(0 0 1) substrate atomic steps. Films 3–4 ML thick grown on Ag(0 0 1) and covered by Ag(0 0 1) showed a negative $4\pi M_{\text{eff}}$ only when cooled below 100 K. However, when films were grown on the (0 0 1) facet of a Fe(0 0 1) whisker the improved interface smoothness resulted in a negative $4\pi M_{\text{eff}}$ (-2.0 kOe) for the sample Ag/3.5 ML Fe(0 0 1)/Au even at RT [3.24]. Superlattices of [Ag/Fe/Ag(0 0 1)]$_n$ prepared on GaAs(0 0 1) wafers also exhibited negative $4\pi M_{\text{eff}}$ at RT and were magnetized perpendicularly for Fe layers thinner than 6 ML ($K_u^s = 0.8$ erg/cm^2), as presented by *Cabanel* et al. [3.25].

The uniaxial interface anisotropies are temperature dependent. With decreasing temperature they increase faster than the demagnetizing field $4\pi DM_s$ for Fe/Ag and Fe/Cu interfaces but decrease more slowly than $4\pi DM_s$ in Fe/Pd and Fe/Au interfaces (Table 3.1). The former behavior is expected from the two-dimensional thermodynamic behavior of ultrathin structures, also discussed by *Mills* in Chap. 3 of Vol. 1.

An atomic phenomenological description of magnetic anisotropies was introduced by *Néel* [3.26]. The spin-orbit interaction introduces an angle-dependent magnetic interaction that for two atomic magnetic moments separated by a vector r can be written as

$$\mathcal{E}(r) = \ell(r) \cos^2 \phi + q(r) \cos^4 \phi, \tag{3.18}$$

where $\ell(r)$ and $q(r)$ are expansion coefficients and ϕ is the angle between the interatomic distance r and parallel atomic magnetic moments μ. In cubic materials the first term contributes to magnetic anisotropies only in systems with broken symmetry. In a pure bcc structure with the surface oriented along [0 0 1] and assuming a nearest neighbour interaction the first term in (3.18) does not create a surface anisotropy. A possible surface relaxation results in a tetragonal distortion and leads to a surface anisotropy in Fe(0 0 1) of $\sim 0.5 e_s$ (erg/cm^2) [3.22], where e_s is the surface vertical strain component. This surface anisotropy is very small even for a large surface relaxation. Fe(0 0 1) grows on Ag(0 0 1) slightly stretched, appr. $\sim 0.7\%$. The uniaxial perpendicular anisotropy caused by the magnetoelastic effect is then given by [3.22]

$$\mathcal{E} = B_1(e_\perp - e_\parallel)\alpha_z^2, \quad B_1 = \tfrac{8}{3} N\ell, \tag{3.19}$$

where $e_\parallel$ and $e_\perp$ are in-plane and perpendicular strain tensor components, N is the number of atoms per unit volume and B_1 is the magnetoelastic coefficient [3.27]. Equation (3.19) gives a weak easy plane uniaxial anisotropy field $2K_u/M_s = -0.8$ kOe, and therefore cannot be considered as an important source of perpendicular anisotropy in Fe(0 0 1) films. The above estimates show that the experimentally observed uniaxial perpendicular fields can be explained neither by a simple *Néel* theory nor by a thickness-dependent magnetoelastic energy [3.28]. The interface uniaxial anisotropies can be evaluated only by *ab initio* calculations in which the spin-orbit interaction is included directly in the band calculations of ultrathin layers, also examined by *Gay* and *Richter*, Sect. 2.1 and *Daalderop*, Sect. 2.2, both in Vol. 1.

3.1.3.2 Four-fold In-Plane Anisotropies

The in-plane four-fold anisotropy in Fe(0 0 1) depends very strongly on the sample thickness. It is described quite well by a sum of constant and $1/d$ terms [3.22],

$$\frac{2K_{1\parallel}}{M_s} = \left[0.55 - \frac{2.5}{d} \right] (kOe) \tag{3.20}$$

at RT, where d is in ML. Note that the constant term is very close to the bulk value for Fe. The $1/d$ term most likely originates from the surface four-fold anisotropy. The change in the sign shows that the surface anisotropy has its easy axis parallel to the $\{1\,1\,0\}$ crystallographic directions. It is interesting to note that *Gay* and *Richter* [3.29] predicted the surface contribution to the four-fold in-plane anisotropy with the correct sign. However, it is very surprising that a decrease in the four-fold anisotropy is very weakly dependent on the outer interface. Samples Ag/5.7 ML Fe/Ag, Au, Ni, Cu, Pd have practically identical in-plane anisotropies at both RT and liquid-N_2 temperatures (Table 3.1). However, the in-plane four-fold anisotropies depend on the template upon which the Fe(0 0 1) film is grown [3.22].

Ni can be grown epitaxially on Fe(0 0 1). The first 3 ML follow the bcc stacking with the same lateral spacing [3.30] and the same vertical relaxation as observed for Fe(0 0 1) [3.31]. After reaching a critical thickness, 3–5 ML, the Ni overlayers transform gradually to a more complicated structure [3.2, 13]. The transformed "bcc Ni" shows that the main features of the Reflection High Energy Electron Diffraction (RHEED) patterns along the $\{1\,0\,0\}$ and $\{1\,1\,0\}$ azimuths remain, but the RHEED pattern along $\{1\,1\,0\}$ azimuths show in addition a weak diffraction streak. The additional superlattice streaks are also visible for azimuths away from the $\{1\,0\,0\}$ and $\{1\,1\,0\}$ directions. Their complicated dependence and recent Surface Extended X-ray Absorption Fine Structure (SEXAFS) [3.32] studies by *Jiang* et al. indicate that the Ni transformation is a rather severe distortion of the basic bcc Ni(0 0 1). "Bcc Ni" overlayers have a unique structure having magnetic properties which are truly different from those observed for bcc Fe(0 0 1) and pure bcc Ni(0 0 1).

FMR measurements played a crucial role in determining the magnetic properties of both the pure bcc and lattice transformed Fe/Ni(0 0 1) bilayers.

The magnetic moment in Fe/Ni bilayers measured by FMR was significantly enhanced by both the pure bcc Ni(0 0 1) and the lattice transformed "bcc Ni." The saturation magnetizations of both forms of bcc Ni, $\sim 5\,\mathrm{kG}$, are nearly as large as in fcc Ni [3.13]. This is in agreement with calculations by *Moruzzi* and *Marcus* [3.33]. The in-plane anisotropies are dramatically different. The presence of pure bcc Ni does not affect the four-fold in-plane anisotropy of the Fe(0 0 1) layer. However, it was the lattice transformed "bcc Ni" which showed remarkable magnetic properties [3.2, 3, 13]. The four-fold anisotropies observed in lattice transformed Fe/Ni bilayers far exceeded those observed in regular 3d transition metals and their alloys; e.g., $2K_{1\parallel}/M_\mathrm{s} = 2.33\,\mathrm{kOe}$ in a 6 ML Fe/15 ML Ni bilayer compared with 0.55 kOe in bulk Fe. The effective fields, $2K_{1\parallel}/M_\mathrm{s}$, in Fe/Ni bilayers are strong enough to create an appreciable drag of the saturation magnetization M_s behind the in-plane applied dc field H_0 (Fig. 3.7).

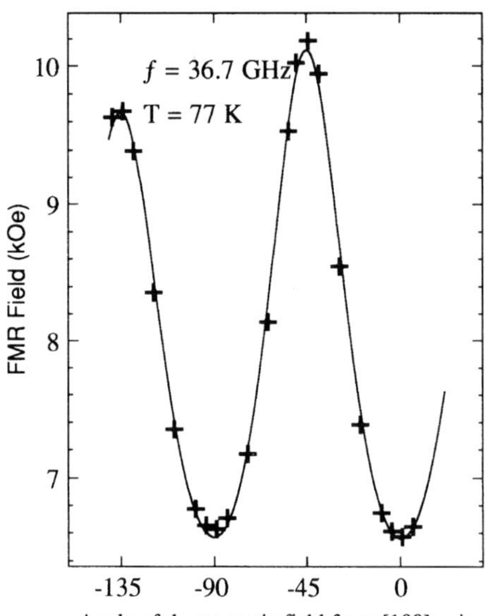

Fig. 3.7. The in-plane angular dependence of the measured FMR fields for the [Ag(0 0 1) bulk vicinal substrate/6 ML Fe(0 0 1)/10 ML Ni(0 0 1)/15 ML Au(0 0 1)] at 77 K. The solid line is the calculated angular dependence of the FMR field using the following parameters: $f = 36.76\,\mathrm{GHz}$, $2K_{1\parallel}/M_\mathrm{s} = 2.14\,\mathrm{kOe}$, $2K_{u\parallel}/M_\mathrm{s} = 0.257\,\mathrm{kOe}$, $\phi_\mathrm{m} = 41°$ (along Fe[1 1 0]), $4\pi M_\mathrm{eff} = 9.346\,\mathrm{kG}$, $g = 2.103$. Note that due to a large value of the in-plane four-fold anisotropy field $2K_{1\parallel}/M_\mathrm{s}$ the angular dependence of the resonance field, H_res, does not follow a simple sinusoidal function. The saturation magnetization was not parallel to the in-plane applied dc field when the dc field was oriented away from the magnetically easy $\{1\,0\,0\}$ and hard $\{1\,1\,0\}$ axes. The saturation magnetization has in a deeper energy minimum with the dc field along the easy magnetic axis than when oriented along the hard axis. Therefore it takes a larger angle between the saturation magnetization and the dc applied field to rotate the saturation magnetization away from the easy axis. Calculations correspond to the case where the saturation magnetization is allowed to drag behind the applied dc field

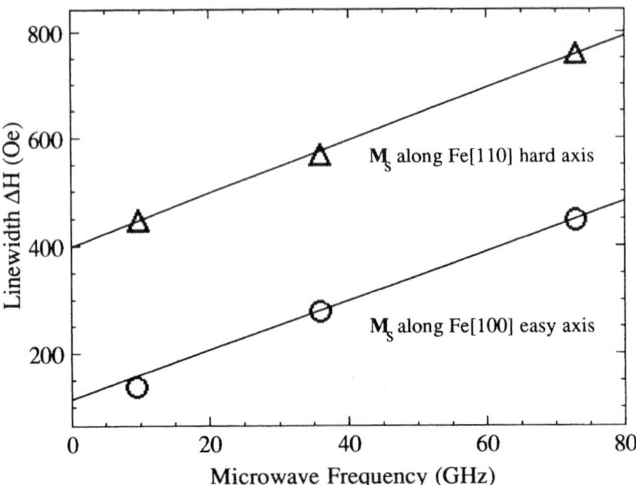

Fig. 3.8. The frequency dependence of FMR linewidth for a sample [6 ML Fe/10 ML Ni(0 0 1)] (the same sample as in Fig. 3.2) at RT. The linewidth is linear with microwave frequency. The zero frequency intersect, $\Delta H(0)$, originates in sample inhomogeneities. The slope of the solid line is caused by the intrinsic Gilbert damping. Note that the intrinsic damping is almost isotropic whereas $\Delta H(0)$ exhibits a strong angular dependence

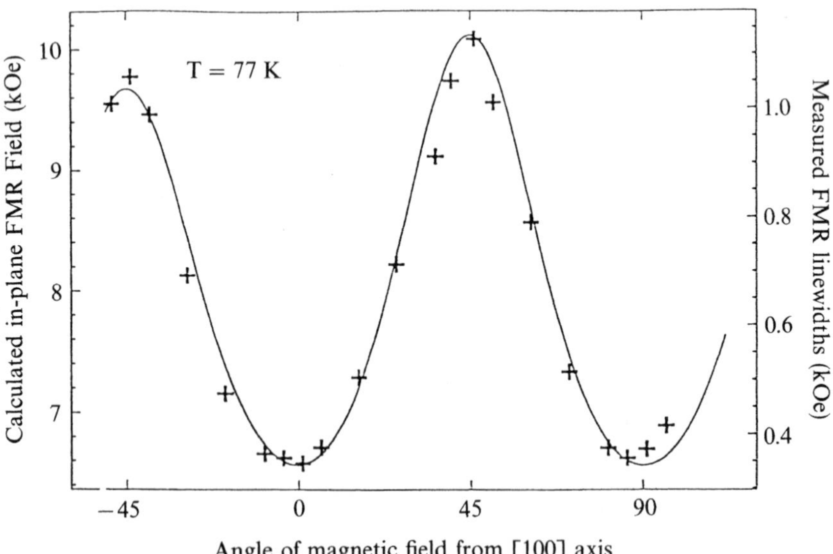

Fig. 3.9. The in-plane angular dependence of FMR linewidth for a sample [6 ML Fe/10 ML Ni(0 0 1)] using the 36.6 GHz system at 77 K. The superimposed solid curve was taken from the fit of the resonance field to demonstrate the similarity between the in-plane anisotropy of the FMR linewidth and field

A clue to the origin of these large in-plane anisotropies was found by measurements of the angular dependence of the FMR linewidth. The microwave frequency dependence of ΔH showed a typical linear dependence with a zero frequency offset $\Delta H(0)$ (Fig. 3.8). The linear slope corresponding to the intrinsic Gilbert damping is isotropic as it is in all 3d transition metals; it is the term $\Delta H(0)$ which has a strong angular dependence. The FMR linewidth $\Delta H(0)$ is caused by crystallographic faults generated in this case during the Ni overlayer transformation. The angular dependences of the in-plane resonance field and the FMR linewidth follow each other very closely (Fig. 3.9), strongly indicating that they have a common origin. The crystallographic defects created during the Ni layer transformation must form a network which satisfies the in-plane four-fold symmetry. If we assume that these defects can be described by line defects (such as a simple network of misfit dislocations) then one can use *Néel*'s expression (3.18). In the $(0\,0\,1)$ plane the first term leads to an additional perpendicular anisotropy and the second term leads to the four-fold in-plane anisotropy. In cubic structures the contribution of the spin-orbit interaction is weak. The situation changes along crystallographic defects, and along the lines of decreased symmetry the contribution of the spin-orbit interaction to magnetic anisotropies is significantly enhanced. It is remarkable that the crystallographic defects triggered by the lattice transformation of metastable bcc Ni can result in well defined four-fold anisotropies and can be used to engineer new types of magnetic materials.

The presence of four-fold perpendicular anisotropies can be easily identified by measuring FMR with the dc magnetic field inclined to the film surface (Fig. 3.10). The perpendicular four-fold anisotropies in bcc Fe$(0\,0\,1)$ films are positive

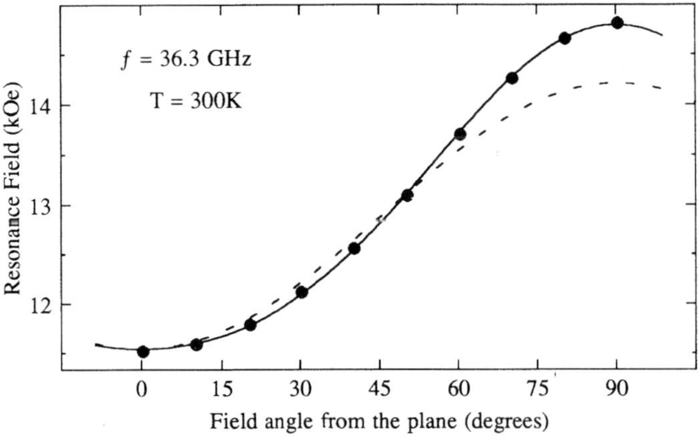

Fig. 3.10. The FMR field for a Au$(0\,0\,1)$ covered 3 ML bcc Fe$(0\,0\,1)$ film as a function of the angle between the specimen surface and the dc applied field. The solid line is the theoretical fit using the following magnetic parameters: $2K_{1\parallel}/M_s = 0.$ (found in the parallel configuration), $2K_{1\perp}/M_s = -0.62\,\text{kOe}$, $4\pi M_{\text{eff}} = 1.82\,\text{kOe}$. The dashed line omits the four-fold perpendicular anisotropy effective field $2K_{1\perp}/M_s$. The dots represent experimental points

for thicknesses greater than 9 ML and gradually approach the bulk cubic anisotropy field. For thicknesses less than 9 ML they become negative and increase their negative value with a decreasing film thickness: $2K_{1\perp}/M_s \sim -0.62$ kOe in 3 ML thick Fe(0 0 1) grown on a Ag(0 0 1) substrate and covered by Au(0 0 1). No simple linear term in $1/d$ was found in all of the samples studied. The role of the four-fold perpendicular anisotropy in the magnetostatic behavior is different from that of the perpendicular uniaxial anisotropy. The four-fold anisotropy on its own is not able to bring the saturation magnetization into the film plane (for $4\pi M_{\mathrm{eff}} < 0$). Its torque disappears when M_s is parallel to the film surface. The four-fold perpendicular anisotropy fields in Fe(0 0 1) are much weaker than the uniaxial perpendicular fields and therefore a negative value of $4\pi M_{\mathrm{eff}}$ less than -1 kOe is sufficient to orient the saturation magnetization completely along the surface normal.

3.1.4 Exchange-Coupled Ferromagnetic Layers

FMR is not a useful tool for studying very strong exchange coupling in ultrathin ferromagnetic layers or between ultrathin ferromagnetic layers which are in direct contact. The strong exchange coupling only orients the magnetic moments into a common direction. Ultrathin structures then acquire magnetic properties of their own, as was shown above. For single individual films the magnetic properties are given by (3.14). For strongly coupled layers (e.g., Fe/Ni bilayers) the torque equation (3.2) can be applied again, and after simple algebraical operations it can be shown that the effective fields are scaled by factors which are given by the partial fractions of the magnetic moments of the individual ferromagnetic layers. The overall effective fields can be written as

$$H_{\mathrm{eff}} = \alpha H_{\mathrm{eff}}^{\mathrm{A}} + (1-\alpha)H_{\mathrm{eff}}^{\mathrm{B}}, \quad \alpha = \left(\frac{\mathscr{M}^{\mathrm{A}}}{\mathscr{M}^{\mathrm{A}} + \mathscr{M}^{\mathrm{B}}}\right) \tag{3.21}$$

e.g.

$$\frac{2K_{1\|}}{M_s^{\mathrm{av}}} = \alpha\frac{2K_{1\|}^{\mathrm{A}}}{M_s^{\mathrm{A}}} + (1-\alpha)\frac{2K_{\|}^{\mathrm{B}}}{M_s^{\mathrm{B}}}; \qquad 4\pi M_{\mathrm{eff}} = \alpha 4\pi M_{\mathrm{eff}}^{\mathrm{A}} + (1-\alpha)4\pi M_{\mathrm{eff}}^{\mathrm{B}},$$

where $\mathscr{M}^{\mathrm{A}}$ and $\mathscr{M}^{\mathrm{B}}$ represent the total magnetic moments of layer A and B, respectively. A more detailed discussion is presented in [3.3]. The average saturation magnetization for each individual layer is represented by M_s^{A} and M_s^{B}, used in (3.21). The total magnetic moment is the sum $\mathscr{M}^{\mathrm{A}} + \mathscr{M}^{\mathrm{B}}$. The average saturation magnetization is given by $M_s^{\mathrm{av}} = (\mathscr{M}^{\mathrm{A}} + \mathscr{M}^{\mathrm{B}})/(d^{\mathrm{A}} + d^{\mathrm{B}})$, where d^{A} and d^{B} are the thicknesses of layer A and B, respectively. It can be shown that the total anisotropy $K_{1\|} = (d^{\mathrm{A}}K_{1\|}^{\mathrm{A}} + d^{\mathrm{B}}K_{1\|}^{\mathrm{B}})/(d^{\mathrm{A}} + d^{\mathrm{B}})$ if one uses in the effective anisotropy field the average saturation magnetization M_s^{av}, also discussed by *de Jonge* et al. in Sect. 2.3 of Vol. 1.

The situation changes when the magnetic layers are separated by a non-magnetic interlayer. The exchange coupling between layers depends upon interlayer thickness and can be decreased to the point that it can be studied by FMR. In fact, the ultrathin film limit significantly simplifies the treatment of weakly coupled layers.

3.1.4.1 Exchange-Coupled Bilayers

It is not intended to present here a rigorous treatment of the exchange coupling between two films. This can be found in previous paper by *Heinrich* et al. [3.3], *Cochran* et al. [3.34] and *Grunberg* [3.35], and recent papers which correctly treat the limit of strongly coupled layers [3.36]. Here, the description will be limited to the case of a weak coupling between ultrathin films, which has been used in a wide range of experiments investigating the exchange coupling between ultrathin films, and is also treated by *Cochran*, Sect. 3.2. The exchange coupling between two layers is usually described by

$$E_{AB} = -J^{AB}\left(\frac{M^A M^B}{M_s^A M_s^B}\right) = -J\cos\phi, \quad \text{in}[\text{ergs.cm}^{-2}] \tag{3.22}$$

where J is the interlayer exchange coupling and ϕ is the angle between the magnetic moments. The energy per interface atom is then given by

$$E_{AB}a^2 = -\boldsymbol{\mu}^A \cdot \frac{J^{AB} M^B}{(M_s^A M_s^B t_0)} \tag{3.23}$$

where a^2 is the surface area per atom, t_0 is the atomic layer separation and $\boldsymbol{\mu}^A$ is the interface atomic magnetic moment for layer A. The expression $(J^{AB}/(M_s^A M_s^B t_0))M^B$ acts like an effective field on the atomic moment $\boldsymbol{\mu}^A$. A similar expression can be derived for the atomic moment $\boldsymbol{\mu}^B$ by interchanging appropriate indexes.

The interlayer exchange fields add to the total torque acting on each individual layer. In the ultrathin film limit, this torque is shared by all of the atomic layers in a given layer. For simplicity, only the perpendicular configuration will be discussed. The effective interlayer exchange fields contribute to exchange coupling torques and result in coupled equations of motion (neglecting four-fold anisotropy)

$$\left[H_0 - 4\pi M_{\text{eff}}^A + \frac{J^{AB}}{d^A M_s^A} - \frac{\omega}{\gamma}\right]\mathcal{M}_+^A - \frac{J^{AB}}{d^B M_s^B}\mathcal{M}_+^B = d^A M_s^A h_+$$

$$-\frac{J^{AB}}{d^A M_s^A}\mathcal{M}_+^A + \left[H_0 - 4\pi M_{\text{eff}}^B + \frac{J^{AB}}{d^B M_s^B} - \frac{\omega}{\gamma}\right]\mathcal{M}_+^B = d^B M_s^B h_+. \tag{3.24}$$

The response of the sample is given by the rf susceptibility $\chi = (\mathcal{M}_+^A + \mathcal{M}_+^B)/h_+$.

The denominator of χ determines the resonance fields. The denominator has always two roots – a consequence of the two coupled systems. The precessional motions are coupled and result in an acoustic mode in which the magnetic moments in the two layers precess in phase, and in an optical mode in which the magnetic moments precess in antiphase. The character of the magnetic coupling can be determined from the relative positions of the acoustic and optical modes. In FMR, the optical mode is located at a higher field than the acoustic mode for antiferromagnetic coupling and at a lower field for ferromagnetic coupling. The positions and intensities of both modes have a complicated dependence on the strength of the exchange coupling, but they can be calculated from the coupled L–L equations of motion. The calculations become practical only using a computer program in which the general configuration can be easily considered.

It should be pointed out that the optical mode is only observable in FMR measurements if the individual layers in the absence of the exchange coupling have different resonance fields. The rf coupling to the optical mode decreases rapidly if two ferromagnetic layers converge in their magnetic properties. In the case of antiferromagnetic coupling, the acoustic mode starts from the FMR peak, which is at a lower magnetic field and then moves to higher fields as the coupling strength increases, approaching a fixed point which is given by the overall magnetic properties of two strongly coupled layers (Fig. 3.11). The acoustic peak increases its intensity with increasing coupling (Fig. 3.12a). The optical mode originates from the FMR peak, which is at a higher field. With increasing coupling it increases its resonance field and decreases its intensity (Fig. 3.12a). For ferromagnetic coupling the trend is reversed. The acoustic mode originates from the FMR peak which at a higher field, and with an increasing value of the ferromagnetic coupling it moves towards lower fields approaching a fixed point given by (3.21) (Fig. 3.11). The optical peak originates from the lower field FMR peak, and with an increasing coupling lowers its resonance field and decreases its intensity (Fig. 3.12b).

Typical resonance behavior of two coupled ultrathin films is shown in Figs. 3.13, 14. The perpendicular uniaxial surface effective field is used to split the resonance fields of the individual Fe layers. The resonance fields of individual layers are measured in separate experiments. It is important to grow films having identical interfaces as those used in the trilayer structures. Generally, one has to readjust the magnetic properties of the individual layers (in a given

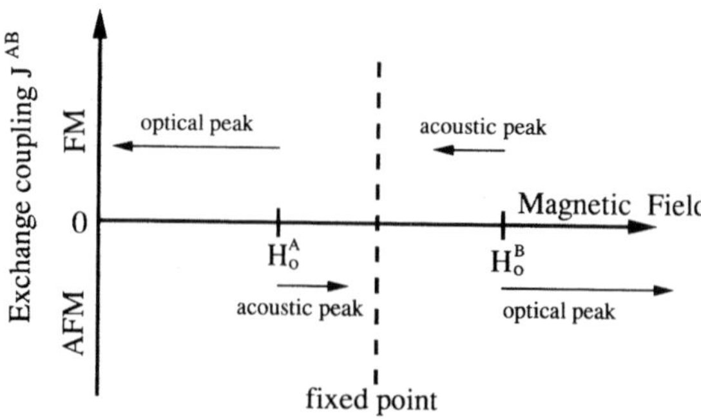

Fig. 3.11. Schematic diagram of the acoustic and optical peak positions for the ferromagnetic and the antiferromagnetic exchange coupling between two ferromagnetic layers separated by a non-magnetic interlayer

trilayer) to obtain the correct positions of the acoustic and optical modes along both the easy and hard magnetic axes. It was found that the magnetic properties of stable structures such as bcc Fe(0 0 1) are very little affected by being incorporated into the trilayer structure. The exchange coupling is the only parameter which strongly affects the position of resonance modes as illustrated in [3.37, 38] and Figs. 3.13, 14. A least squares fit of the measured positions of resonance peaks along the easy and hard axes allows one to determine the magnetic properties of individual magnetic layers and the exchange coupling between layers. It is interesting to point out that the calculated FMR intensities follow the measured signal very well (Figs. 3.13, 14) and therefore the obtained parameters represent true magnetic properties.

The results of the antiferromagnetic coupling obtained by FMR measurements were compared to those which were measured by Magneto-Optical Kerr Effect (MOKE) and they agree well. It was found that the magnetization process in ultrathin films mostly follows the path of minimum energy, however notable exceptions exists in which the magnetization process is

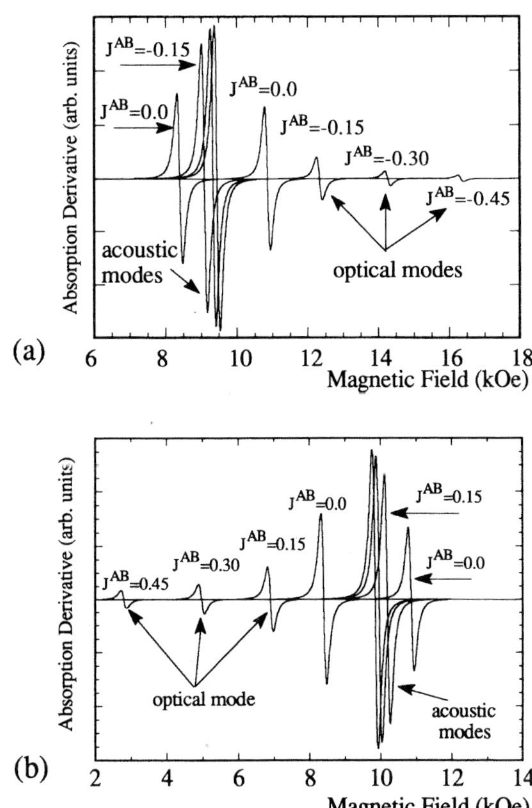

Fig. 3.12. Computer calculations of $d\chi''/dH$ as a function of the applied field at 36.3 GHz for the trilayer composed of two Fe layers A and B. The calculations were carried out using following magnetic parameters: $(4\pi M_{eff})_A = 3.5\,\text{kG}$, $(2K_{1\parallel}/M_s)_A = 0.1\,\text{kOe}$ and $(4\pi M_{eff})_B = 9.5\,\text{kG}$, $(2K_{1\parallel}/M_s)_B = 0.1\,\text{kOe}$ Figs. 12a, b show the peak positions for the antiferromagnetic and ferromagnetic coupling, respectively. Note that the intensity of the optical modes decrease rapidly with an increasing coupling. On the other hand, the intensities of acoustic peaks increase with an increasing coupling reaching a fixed point. J^{AB} in [ergs.cm^{-2}]

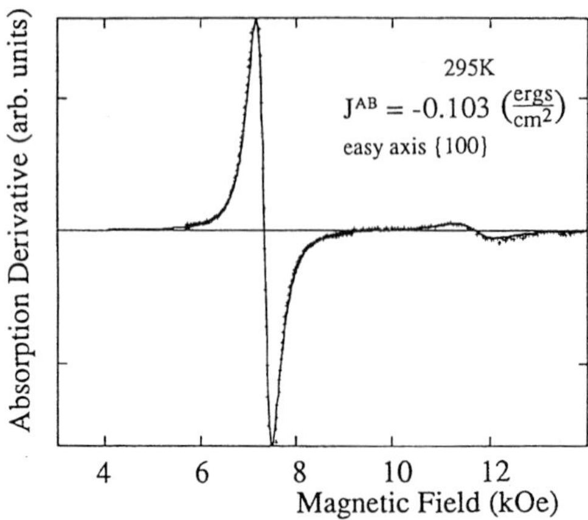

Fig. 3.13. The field dependence of $d\chi''/dH$ along the easy magnetic axis in the trilayer [5.7 ML Fe/9 ML Cu/9.7 ML Fe(0 0 1)] measured at RT. The solid line was calculated for exchange coupled Fe layers with the following magnetic parameters: $(4\pi M_{eff})_A = 13.95$ kG, $(2K_{1\parallel}/M_s)_A = 0.252$ kOe and $(4\pi M_{eff})_B = 3.16$ kG, $(2K_{1\parallel}/M_s)_B = 0.149$ kOe. The layer A and B correspond to 9.7 ML Fe, and 5.7 ML Fe, respectively. The magnetic parameters of individual layers were measured in separate experiments in structures using the same interfaces, with the following results: $(4\pi M_{eff})_A = 13.53$ kG, $(2K_{1\parallel}/M_s)_A = 0.259$ kOe and $(4\pi M_{eff})_B = 3.08$ kG, $(K_{1\parallel}//M_s)_B = 0.113$ kOe. Note the very good agreement between the magnetic properties measured in the trilayer structure and those measured in the separate individual layers. The optical peak (weak peak) is located at a higher field than the acoustic peak (strong peak) and therefore the coupling through bcc 9 ML Cu is antiferromagnetic with the exchange coupling coefficient $J^{AB} = -0.103$ erg/cm^2

closer to the rotational path. A detailed discussion of the magnetization processes involved in MOKE measurements can be found in [3.8] and it is also addressed by *Bader* and *Erskine* in Chap. 4. The situation is quite different in the case of metastable structures such as fcc Co/Cu/Co(0 0 1). The lattice strain of fcc Co(0 0 1) depends on the surrounding epitaxial layers and the magnetic properties of the Co layers change sufficiently in the trilayer structure that the exchange coupling is not the only major variable in fitting the resonance fields of the acoustic and optical modes. In this case a least squares fitting procedure has to include *a priori* the intensities of the resonance modes [3.8].

FMR has been extensively used in the study of the exchange coupling in bcc Fe/bcc Cu/Fe(0 0 1) [3.37, 40], Fe/lattice strained fcc Pd/Fe(0 0 1) [3.38, 40], Fe/fcc Ag/Fe(0 0 1) [3.40], Fe/fcc Au/Fe(0 0 1) [3.40], bcc Fe/Cr/Fe(0 0 1) [3.39] and fcc Co/Cu/Co [3.8] structures.

The BLS technique investigates rf magnetic properties across an area of 10–20 μm in diameter. This is a significantly smaller area than that in FMR measurements (~ 1 cm^2). The results of FMR and BLS measurements agree very well, as discussed by *Cochran*, Sect. 3.2, which shows that the ultrathin film structures with a large lateral homogeneity can be grown by MBE techniques.

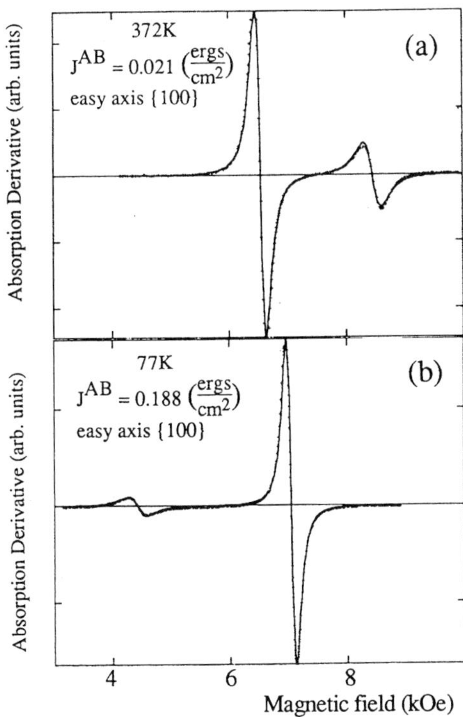

Fig. 3.14a. The upper curve shows the field dependence of $d\chi''/dH$ along the easy magnetic axis in the trilayer [5.7 ML Fe/6 ML Pd/9.8 ML Fe(0 0 1)] measured at 372 K. The solid line was calculated using the exchange coupled Fe layers using the following magnetic parameters: $(4\pi M_{eff})_A = 16.08$ kG, $(2K_{1\parallel}/M_s)_A = 0.146$ kOe and $(4\pi M_{eff})_B = 8.84$ kG, $(2K_{1\parallel}/M_s)_B = 0.072$ kOe. The magnetic parameters of individual layers were measured in separate experiments in structures using the same interfaces with the following results: $(4\pi M_{eff})_A = 16.14$ kG, $(2K_{1\parallel}/M_s)_A = 0.132$ kOe and $(4\pi M_{eff})_B = 9.07$ kG, $(2K_{1\parallel}/M_s)_B = 0.05$ kOe. Note again the very good agreement between the magnetic properties measured in the trilayer structure and those measured in the separate individual layers. The exchange coupling in this sample is weak and ferromagnetic. The FMR peaks are only slightly removed from their uncoupled positions and their intensities are also close to those corresponding to uncoupled layers

Fig. 3.14b. The lower curve shows the field dependence of $d\chi''/dH$ along the easy magnetic axis in the same trilayer measured at liquid-N_2 (LN$_2$) temperature. The solid line was calculated using the exchange coupled Fe layers with the following magnetic parameters: $(4\pi M_{eff})_A = 18.52$ kG, $(2K_{1\parallel}/M_s)_A = 0.363$ kOe and $(4\pi M_{eff})_B = 10.51$ kG, $(2K_{1\parallel}/M_s)_B = 0.239$ kOe. The layer A and B correspond to 9.8 ML Fe and 5.7 ML Fe, respectively. The magnetic parameters of individual layers were measured in separate experiments in structures using the same interfaces. The results of their magnetic parameters are as follows: $(4\pi M_{eff})_A = 18.56$ kG, $(2K_{1\parallel}/M_s)_A = 0.308$ kOe and $(4\pi M_{eff})_B = 10.71$ kG, $(2K_{1\parallel}/M_s)_B = 0.241$ kOe. The optical peak (weak peak) is located at a lower field than the acoustic peak (strong peak) and therefore the coupling through 6 ML Pd(0 0 1) is ferromagnetic with the exchange coupling coefficient $J^{AB} = 0.188$ erg/cm^2. Note that the exchange coupling increased significantly by cooling the sample to LN$_2$ temperatures. A significant increase in the coupling resulted in a reversal of peak intensities and large shifts of resonance peaks. All of the above changes, including peak intensities, are fully accountable by an increased value of the exchange coupling

3.1.5 Conclusion

This section certainly does not include, and was not intended to provide a list of all available results which employ the FMR technique in the study of ultrathin structures. Remarkable results employing this technique were also obtained by other groups and the author would like to apologize to all those who were not included in this article. Their omission should not detract from their importance. The goal of this book is mainly educational, and the main purpose of this article is to demonstrate the crucial aspects of the FMR technique. In order to achieve these aims, the author has concentrated on the work done by the Simon Fraser University group. The above discussions and the results presented demonstrate well the power of the FMR technique.

The FMR technique allows one to determine, in a straightforward way, a wide range of static and dynamic magnetic properties of simple and complex ultrathin structures. The measured FMR linewidth, particularly when combined with the measured magnetic anisotropies, can also be used to address many important structural aspects and to determine their role on the overall magnetic behavior.

Acknowledgement. The author would like to thank his colleagues J.F. Cochran, A.S. Arrott, Z. Celinski, K. Myrtle and D. Atlan for stimulating discussions and he would like to express his thanks for their help during the preparation of this manuscript.

3.2 Light Scattering from Ultrathin Magnetic Layers and Bilayers

J.F. COCHRAN

The physics of the scattering of visible light from low frequency spin waves (Brillouin light scattering, or BLS) in ultrathin magnetic films is described in this section. It is shown that the frequency shifts observed for the scattered light are closely related to the frequencies measured using ferromagnetic resonance absorption (FMR). The BLS measurements are complimentary to FMR measurements, and, like FMR, can be used to measure magnetic anisotropies in ultrathin films, and to measure the exchange coupling between pairs of ultrathin magnetic films. It is further shown that the intensity of the light scattered from ultrathin magnetic films is surprisingly strong: for a 10 Å thick iron film at 300 K in a 1 kOe applied field, for 100 mW of incident 0.5145 μm radiation, and using f-2 collection optics, one expects to obtain approximately 6×10^5 scattered photons per second. The intensity of the frequency shifted scattered light is proportional to the film thickness, to the temperature T, for $T \gg 1$ K, and it is proportional to the fourth power of the frequency of the incident light.

This chapter concludes with a discussion of two examples of light scattering from ultrathin films: (1) 3 monolayers of fcc Fe(0 0 1) grown on the (0 0 1) face of fcc copper and covered by 60 monolayers of fcc Cu (0 0 1), and (2) two ultrathin films of bcc Fe (0 0 1) grown on the (0 0 1) face of a silver template and separated by a spacer composed of 9 monolayers of Cu (0 0 1) and 1 monolayer of Ag (0 0 1); the thickness of the iron films were 9 and 16 monolayers. The two iron films in the bilayer exhibited a weak antiferromagnetic coupling. The case of Brillouin light scattering from magnetic superlattices is discussed by *Hillebrands* and *Güntherodt* in Sect. 3.3.

3.2.1 Introduction

This article is concerned with the application of Brillouin Light Scattering (BLS) to investigate the magnetic properties of ultrathin magnetic films and the interaction between pairs of ultrathin magnetic films. We are interested in film thicknesses which lie between 1 and 10 monolayers (ML). The experimentally observed magnetic properties of such ultrathin magnetic layers are particularly easy to interpret because the magnetization cannot vary across the thin dimension of the film as a result of the very strong exchange forces which hold the spins on adjacent atomic planes parallel (also discussed by *Heinrich*, Sect. 3.1). In very thin films the surface anisotropies and surface exchange coupling terms between two films play a particularly important role simply because of the relatively large number of surface atomic sites relative to the bulk sites. Ultrathin single crystalline films which have been prepared with the crystal structure of the bulk material, and with very nearly the same lattice spacing as their bulk counterparts, exhibit magnetocrystalline anisotropies which are usually quite different from those measured on massive crystals because of the enhanced ratio of surface area to volume. Moreover, there is some hope that band structure calculations which incorporate the effects of spin–orbit coupling can be carried through with sufficient accuracy for ultrathin films so as to make comparisons between theory and experiment meaningful. Spin–orbit coupling is responsible for the magnetic anisotropies, for the deviation of the magnetic g-factor from the free-electron value $g = 2.00$, and for the coupling between the magnetization and the lattice which results in intrinsic magnetic damping [3.41].

The magnetic properties of very thin single crystal films are also of interest partly because specimens can be prepared which have crystal structures which are not naturally occurring. Examples are provided by bcc films of nickel grown on a Ag (0 0 1) substrate [3.42] and fcc films of iron grown on a Cu (0 0 1) substrate [3.43]; the bcc Ni structure becomes unstable for thicknesses greater than 3–4 ML, and the fcc Fe structure becomes unstable for thicknesses greater than ~18 ML. The stable naturally occuring Ni structure is a fcc lattice, and the stable naturally occurring Fe structure is a bcc lattice.

It is the aim of this section to provide a simple, quantitative description of the light scattering experiment, and to describe how it can be used to measure fundamental magnetic parameters for ultrathin films, and for exchange coupled pairs of ultrathin films. It will be shown that the information which can be obtained from the Brillouin light scattering experiment is very similar to that which can be obtained from ferromagnetic resonance experiments, also examined by *Heinrich* in Sect. 3.1. The main differences between these two techniques as applied to single ultrathin films or to pairs of ultrathin films are:

(1) In the BLS experiment, the frequency is measured at a fixed magnetic field. In the FMR experiment the frequency is fixed and the magnetic field must be varied until the frequency of the magnetic excitation matches the applied frequency.

(2) The volume of sample required for the BLS measurement is small compared with the volume required in order to obtain an equivalent signal-to-noise ratio for the FMR experiment. As a rule of thumb, for a FMR linewidth of 200 Oe or less, an area of approximately 5×5 mm^2 is required in order to obtain a signal-to-noise ratio equivalent to the BLS signal-to-noise ratio for an iron film six monolayers (ML) thick. The BLS experiment probes an area whose diameter is ~ 20 μm.

(3) The resonant frequency can be determined in an FMR experiment with a precision of ~ 0.01 GHz: however, only one frequency can be measured using one particular piece of equipment. In a BLS experiment frequencies can be measured over the range 5–100 GHz using a single piece of equipment; however, the frequencies can be measured with a precision of only ~ 0.1 GHz.

3.2.2 The Light Scattering Experiment

A conceptual light scattering configuration is shown in Fig. 3.15. A thin film sample is placed between the poles of a magnet such that the applied field, H_0, lies in the plane of the film. A monochromatic, parallel, beam of light characterized by a frequency f_0 is directed onto the specimen by means of a partially transmitting mirror. The angle of incidence of the light on the specimen, θ, can be adjusted by rotating the specimen around an axis parallel with the magnetic field direction. In a typical experiment, an angle $\theta \sim 45°$ is used so that the specularly scattered beam is directed well away from the direction defined by the incident light beam. If the specimen could be characterized by a simple index of refraction, no light would be scattered in the direction of the incident beam. However, in actual fact, a very weak signal can be detected if a sensitive narrow band detector is placed behind the partially transmitting mirror, as depicted in Fig. 3.15. This geometry corresponds to the back-scattering configuration in which the light which is collected has been scattered along the direction defined by the incident beam. This is a very commonly used configuration for experiments on opaque materials. For a simple film this light will be found to contain a number of frequency components. The strongest intensity will usually correspond to light having the same frequency as the incident light, f_0. It is caused by dust particles or other irregularities on, or in, the specimen. In addition to the unshifted frequency component, the scattered light will contain components whose frequency has been shifted up or down from the frequency of the incident

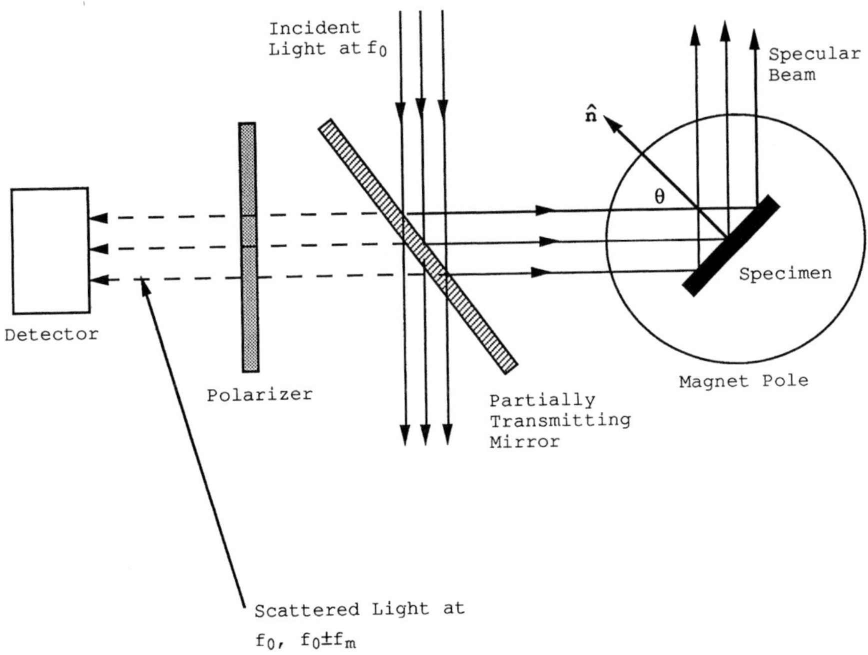

Fig. 3.15. A conceptual Brillouin light scattering experiment. A magnetic film, whose normal is specified by $\hat{n}$, is placed between the poles of a magnet such that the magnetic field is applied in the specimen plane. The scattered light is collected in the back-scattering configuration that is commonly used for opaque materials

light, $f_s = f_0 \pm f_m$, where f_m corresponds to the frequency of an excitation in the film. In the case of a magnetic film, these excitations are thermally excited mechanical oscillations (sound waves whose quanta are phonons) or thermally excited precessions of the magnetization around its equilibrium configuration (spin waves whose quanta are magnons). Light is scattered from sound waves and spin waves because the dielectric tensor which describes the interaction between the incident light and the film material contains elements which depend very slightly on the state of strain in the film and on the direction of the magnetization vector. The intensity of the scattered light is very feeble; typically it amounts to no more than 10^{-12} of the incident light intensity for visible light. The frequency shifts of interest generally range between 1 and 100 GHz. The frequency of the incident light generally lies in the visible part of the spectrum.

In a light scattering experiment, a commonly used source is the 0.5145 μm line obtained from an Argon ion laser; its frequency is $f_0 = 5.83 \times 10^{14}$ Hz. Thus a very feeble flux of light whose frequency has been shifted by a few GHz must be measured in the presence of a much more intense flux of light having a relatively large unshifted frequency. Typically, the unshifted light intensity at frequency f_0 is 10^4–10^6 times greater than the intensity of the frequency shifted light which is of interest. In order to measure the relatively weak signals which are shifted in frequency by a few GHz and which carry information about the sound wave and spin wave modes in the film it is necessary to use a spectrum analyzer which combines very high resolution with very high contrast. The instrument of choice for this purpose is a multi-pass Fabry–Perot interferometer [3.44–47]. The paper by *Mock* et al. [3.47] provides a very clear description of a modern BLS system.

We shall confine our attention to magnetic excitations in ultrathin films for the purposes of this article. The mechanical properties of thin and ultrathin films form a very interesting topic which represents a distinct sub-discipline: and are discussed in, for example, the review articles by *Grimsditch* [3.48] and by *Nizzoli* and *Sandercock* [3.49]. It is easy to discriminate between light which has been scattered from spin waves and light which has been scattered from sound waves. Light which has been scattered from spin waves is polarized at 90° to the incident light polarization, whereas light which has been back-scattered from sound waves has the same polarization as the incident light. In order to distinguish between these two cases it is only necessary to insert a suitable polarization analyzer before the detector (Fig. 3.15).

Light at the shifted frequencies $f_0 \pm f_m$ has been scattered from a thermally excited magnetic mode having a frequency f_m. As mentioned above, we are interested in normal mode frequencies which lie in the range 1–100 GHz. The energy of a quantum having a frequency of 100 GHz is $hf_m = 4.14 \times 10^{-4}$ eV, and this corresponds to a temperature $T = hf_m/k = 4.8$ K. The magnetic normal modes of the film in which we are interested are therefore very highly excited at room temperatures and may be confidently treated by means of classical methods. The interaction between these normal modes and the light is extremely weak. Therefore the amplitude of a normal mode may be calculated as if the light were not present.

Similarly, the optical electric fields in the film can be calculated as if the magnetic normal modes were absent [3.50]. The magnetization in the film processing at frequency f_m introduces a small fluctuation at frequency f_m in the components of the optical dielectric tensor through the optical analog of the Hall effect [3.51]. The product of the optical electric field oscillating at f_0 and the dielectric tensor elements oscillating at frequency f_m results in a polarization vector, P, whose components contain terms which oscillate at the sum and difference frequencies $f_0 + f_m$. The terms in the polarization which oscillate at the shifted frequencies produce radiation whose mean frequencies lie at the sum and difference frequencies $f_0 \pm f_m$. The intensity of this light is related to the average thermal amplitude of the magnetic normal mode. Even if the frequency of the incident light is perfectly sharp the intensity of the scattered light at the mean frequencies $f_s = f_0 \pm f_m$ will be distributed around f_s with a frequency spread which is inversely related to the mean lifetime of the normal mode. In principle, then, the BLS experiment can be used to investigate magnetic damping processes as well as magnetic mode frequencies. In fact, very few BLS measurements to date have been performed with sufficient resolution to provide a quantitative measure of magnetic mode lifetimes.

The calculation of the intensity of the light scattered from magnetic normal modes for a practical thin film specimen is straightforward but very complicated [3.52, 53]. Complications arise in the optical problem because one can never use an ultrathin film which is self-supporting in vacuum; the film must necessarily be supported by a substrate. One then has to deal with the optical boundary value problem in which light is incident on a film which is thin compared with the wavelength of the light so that the strength of the optical electric field components depends explicitly upon the optical properties of the substrate. Moreover, most BLS experiments reported so far have been performed outside of the ultra-high vacuum system in which the specimens have been fabricated. In such cases the specimen must be protected from oxidation by a suitable cover layer (copper or gold, for example). The presence of the cover layer provides an additional complication for the optical problem. The calculation of the magnetic normal modes in a thin film is itself a complicated problem, especially if magnetic anisotropies and magnetic damping are included [3.53–58] (and *Heinrich* (Sect. 3.1) and *Mills* Chap. 3, Vol. 1).

3.2.3 Light Scattering for a Simple Model

In order to emphasize the physics of the light scattering process, it is useful to consider a very simple model in which light falls at a normal incidence on a very thin isotropic, homogeneous, magnetic film which is somehow suspended in vacuum.

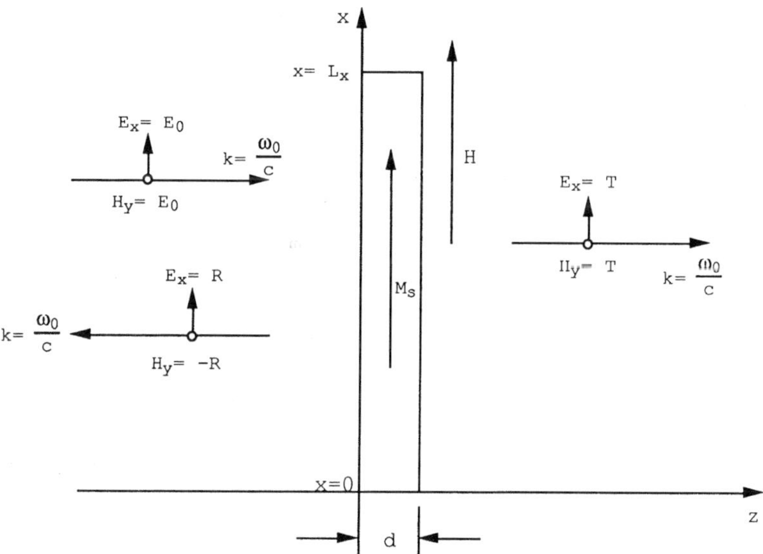

Fig. 3.16. The coordinate system used to describe light scattering from a film of thickness d and lateral dimensions L_x, L_y

Consider the simple configuration shown in Fig. 3.16 in which a plane wave of light having the circular frequency $\omega_0 = 2\pi f_0$ falls at a normal incidence upon a magnetic thin film immersed in vacuum. The thickness of the film, d, is assumed to be very small compared with its lateral dimensions L_x, L_y i.e. d/L_x, $d/L_y \ll 1$. The thickness d is also assumed to be very small compared with the wavelength of the light in vacuum. It is further assumed that the material of the film has a cubic crystal structure and that it is magnetically isotropic, i.e. the magnetocrystalline anisotropies are all zero. A static magnetic field, H_0, is applied along the x axis so that the equilibrium direction of the magnetization also lies along x. The electric vector of the incident light is also assumed to lie along the x direction for the sake of definiteness: it turns out that the intensity of the scattered light is independent of the polarization of the incident light. One has now to solve two problems: (a) the optical problem to find the electric field inside of the film, and (b) the problem of finding the magnetization distribution in the film corresponding to the magnetic normal modes.

3.2.3.1 The Optical Problem

In the linear response regime a cubic magnetic material responds to the presence of an optical electric field by developing an electric polarization per unit volume, P, which is proportional to the electric field, E. To terms linear in the magnetization, the relation between P and E can be written [3.59]

$$4\pi P = (\varepsilon_{11} - 1)\, E + K(E \times M)/M_s, \tag{3.25}$$

where the dielectric displacement vector, D, is given in the CGS system of units by

$$D = E + 4\pi P = \bar{\bar{\varepsilon}}\, E. \tag{3.26}$$

In (3.25) ε_{11} and K are complex coefficients which are frequency dependent; they are related to the band structure of the film material [3.51]. The magnetization vector, M, has the magnitude M_s which is presumed to be independent of the applied field strength for the limited range of fields considered here. The coefficients ε_{11} and K also have, in principle, a very small dependence on the strength of the applied magnetic field, H_0, but this field dependence will be neglected. Numerical values for ε_{11} and the magneto-optic coefficient, K, are listed in Table 3.2 for two commonly used laser frequencies: the red HeNe line at 0.6328 μm and the green Ar laser line at 0.5145 μm. As can be seen from the data listed in Table 3.2, the off-diagonal elements of the dielectric tensor are approximately one order of magnitude smaller than the diagonal elements for the three ferromagnetic metals Fe, Co, and Ni.

It is usual to neglect the effect of the off-diagonal dielectric tensor components when calculating the electric field distribution set up in a cubic metal film by the incident light wave, i.e. for simplicity the ferromagnetic metal film is treated as if it were an ordinary isotropic dielectric substance characterized by

Table 3.2. The electrical polarization vector P in a linear cubic ferromagnetic material can be written, to first order in the magnetization, M, as $4\pi P = (\varepsilon_{11} - 1)E + K(E \times M/M_s)$ where E is the electric field strength and $D = E + 4\pi P = \bar{\bar{\varepsilon}}E$. Values of ε_{11} from *Johnson* and *Christy* [3.60] are listed for iron, cobalt, and nickel for two common laser wavelengths. The magneto-optic coefficients, K, are taken from the article by *Krinchik* and *Artem'ev* [3.61]. A time dependence of $e^{-i\omega t}$ has been assumed

Wavelength (µm)	Material	Diagonal component of $\bar{\bar{\varepsilon}}$ ε_{11}	Off-diagonal component of $\bar{\bar{\varepsilon}}$ K
0.6328	Fe	$-1.0 + 17.8i$	$0.73 + 0.24i$
(1.96 eV)	Co	$-12.5 + 18.4i$	$0.31 + 0.14i$
	Ni	$-12.9 + 16.4i$	$0.10 + 0.023i$
0.5145	Fe	$-0.4 + 16.4i$	$0.31 + 0.24i$
(2.41 eV)	Co	$-9.5 + 14.2i$	$0.22 + 0.11i$
	Ni	$-8.0 + 12.4i$	$0.055 + 0.022i$

the dielectric constant ε_{11}. With this simplification it is easy to use Maxwell's equations plus associated boundary conditions (continuity of the tangential components of E and H across the interfaces) to calculate the optical electric field distribution within a thin film. The result, correct to terms linear in the product $kz = \omega_0 z/c$, is given by

$$\frac{E_x(z)}{E_0} = 1 + \frac{ikd(\varepsilon_{11} - 1)}{2\sqrt{\varepsilon_{11}}} + \frac{ikz}{\sqrt{\varepsilon_{11}}}.\tag{3.27}$$

We are interested in films whose thicknesses are $d \leq 20$ Å, i.e. less than 10 ML thick. The wavelength of the incident light is approximately 0.5 µm corresponding to $k = 2\pi/\lambda_0 = 1.26 \times 10^5$ cm^{-1}. The last two terms in (3.27) are therefore relatively small and may be neglected. To a good first approximation the electric field amplitude in the unsupported thin film is just equal to the amplitude of the incident wave. This conclusion makes intuitive sense since the amount of matter in the film is too small to generate an appreciable reflected wave.

The internal electric field at the circular frequency ω_0 will interact with fluctuations around equilibrium of the magnetization vector at circular frequency ω_m to produce, through (3.25), oscillations in the electric dipole density, $4\pi P$, at the sum and difference frequencies $(\omega_0 \pm \omega_m)$. The electric dipole density, $4\pi P$, oscillating at the frequencies $(\omega_0 \pm \omega_m)$ radiates light at these shifted frequencies.

Let us now turn to the problem of calculating the spin wave modes of a thin film.

3.2.3.2 The Magnetic Modes of a Magnetically Isotropic Thin Film

When the magnetization density, M, is distributed from its equilibrium orientation it precesses around that equilibrium direction under the influence of torques that act so as to restore equilibrium. The resulting motion can be described by the Landau–Lifshitz equation [3.62]:

$$-\frac{1}{\gamma}\frac{\partial M}{\partial t} = T = M \times H_{\text{eff}}. \tag{3.28}$$

In (3.28), T is a torque density, $\gamma = g|e|/2mc$ is the gyromagnetic ratio (for $g = 2.00$, $\gamma = 1.7588 \times 10^7$ radius/s/Oe.), and H_{eff} includes the applied field, H, a demagnetizing field, H_{d}, and an effective exchange field, H_{ex}, due to the exchange interaction if the magnetization density varies from place to place in the specimen. The demagnetizing field, H_{d}, is generated by div M.

It will be shown below that the magnetic modes whose frequencies can be measured by means of Brillouin light scattering experiments are characterized by little or no spatial variation of M across the thickness of the slab, and by spatial variations in the plane of the slab whose wavelengths are comparable with the wavelength of visible light. Therefore, the scale of the in-plane variations is very large compared with the slab thickness (recall that the slabs under discussion are at most 10 ML thick). Under these circumstances the demagnetizing field is almost entirely due to the discontinuity in the normal component of the magnetization at the front and rear faces of the slab. For the geometry of Fig. 3.16 the demagnetizing field is given approximately by

$$H_{\text{d}} = -4\pi M_z u_z. \tag{3.29}$$

The components of the effective exchange field are given by [3.63]

$$H_\alpha^{\text{ex}} = \frac{2A}{M_{\text{s}}^2}\nabla^2 M_\alpha, \tag{3.30}$$

where A is an exchange stiffness parameter whose magnitude is $\sim 10^{-6}$ ergs cm^{-1} for the ferromagnetic metals Fe, Co, and Ni.

The deviation of the magnetization density from its equilibrium direction is expected to be very small. It is therefore convenient to write

$$M(r, t) = M_{\text{s}} + m(r, t), \tag{3.31}$$

where M_{s} is the equilibrium magnetization density which is assumed to be everywhere the same for the simple thin slab shown in Fig. 3.16 for which the applied field, H_0, lies in the plane. The deviation of the magnetization from equilibrium at any point in the slab consists purely of a rotation since the length of the magnetization vector is fixed and equal to M_{s}. To first order in the small dispacement $m(r, t)$ one has

$$m(r, t) \cdot M_{\text{s}} = 0, \tag{3.32}$$

so that if M_{s} is oriented along the x axis, as shown in Fig. 3.16, the small vector which specifies the deviation from equilibrium, m, has only y and z components. For this case the linearized Landau–Lifshitz equations of motion become

$$-\frac{1}{\gamma}\frac{\partial m_y}{\partial t} = (H + 4\pi M_{\text{s}})m_z - \frac{2A}{M_{\text{s}}}\nabla^2 m_z, \tag{3.33a}$$

$$-\frac{1}{\gamma}\frac{\partial m_z}{\partial t} = -Hm_y + \frac{2A}{M_s}\nabla^2 m_y. \tag{3.33b}$$

The first torque term on the right-hand side of (3.33a) includes the effect of the demagnetizing field, (3.29), acting upon the x component of the magnetization density, M_s.

Plane wave solutions for the torque (3.33a, b) may be written down at once:

$$m_y(\mathbf{r}, t) = a \exp(i[k_x x + k_y y + k_z z - \omega t]), \tag{3.34a}$$

$$m_z(\mathbf{r}, t) = ib \exp(i[k_x x + k_y y + k_z z - \omega t]), \tag{3.34b}$$

where

$$\left(\frac{\omega}{\gamma}\right)^2 = \left[H + \frac{2A}{M_s}(k^2)\right]\left[B + \frac{2A}{M_s}(k^2)\right], \tag{3.35}$$

with

$$k^2 = k_x^2 + k_y^2 + k_z^2,$$

$$(b/a) = \sqrt{\frac{H + (2A/M_s)(k^2)}{B + (2A/M_s)(k^2)}}, \tag{3.36}$$

and

$$B = H_0 + 4\pi M_s. \tag{3.37}$$

The allowed values of the wavenumbers k_x, k_y, k_z are determined by boundary conditions on the magnetization density [3.64]. It can be shown that the normal derivative of the magnetization density must vanish at any surface on which there are no surface torque densities such as might be caused by the reduced symmetry at an atomic surface site. For the simple slab shown in Fig. 3.16, there are no surface torques so that these boundary conditions require that the magnetization components have the following form:

$$m_y(\mathbf{r}, t) = a_0 \cos(k_x x)\cos(k_y y)\cos(k_z z)\cos(\omega t), \tag{3.38a}$$

$$m_y(\mathbf{r}, t) = b_0 \cos(k_x x)\cos(k_y y)\cos(k_z z)\sin(\omega t), \tag{3.38b}$$

where (b_0/a_0) is given by (3.36), and

$$k_x = n\frac{\pi}{L_x}, \tag{3.39a}$$

$$k_y = m\frac{\pi}{L_y}, \tag{3.39b}$$

$$k_z = p\frac{\pi}{d}, \tag{3.39c}$$

where n, m, p are positive integers, including zero. The phase of the magnetization wave is, of course, arbitrary. For simplicity in (3.38a, b) we have measured time from a maximum in the y component of the magnetization. The allowed wavenumber values form a rectangular grid in k-space. The spacing along the k_z axis, Δk_z, will be very coarse if d is very small. For example, $\Delta k_z = 1.6 \times 10^7$ cm^{-1} for a slab whose thickness is $d = 20$ Å: this can be compared with Δk_x, $\Delta k_y \sim \pi$ for a slab whose lateral dimensions are 1 cm.

The lowest frequency occurs for $k_x = k_y = k_z = 0$; this is called the uniform mode. Its frequency (from (3.35)) is given by

$$\frac{\omega}{\gamma} = \sqrt{HB}. \tag{3.40}$$

The uniform mode corresponds to an excitation in which all of the spins in the slab precess in phase with the above frequency. The frequency of the uniform mode is measured in a ferromagnetic resonance experiment, (also examined by *Heinrich*, Sect. 3.1). Although higher order modes oscillate at a higher frequency than the uniform mode, the increase of frequency with mode number in the plane is very small. Consider a specific example. For iron at room temperatures the exchange stiffness is $A = 2 \times 10^{-6}$ ergs cm^{-1} and the saturation magnetization is $M_s = 1.72$ kOe. The exchange field correction corresponding to the next to lowest in-plane mode $(k_x = \pi/L_x, k_y = 0 = k_z)$ for $L_x = 1$ cm is $2A\pi^2/M_s = 2.3 \times 10^{-8}$ Oe. That is, the frequency increase over that of the uniform mode would be equivalent to the frequency change obtained by increasing the external field by only 2.3×10^{-8} Oe. This field increment would correspond to a frequency increase of only 0.15 Hz for an applied field of 1 kOe for which the uniform mode frequency is 13.9 GHz (calculated from (3.40) using $g = 2.09$ and $\gamma = 1.8379 \times 10^7$ per Oe.). Clearly there will be very many modes having nearly the same frequency but different wavelengths in the sample plane.

The situation is quite different for the excitation of modes which correspond to a spatial variation of the magnetization across the slab thickness. If $k_x = k_y = 0$ but $k_z = \pi/d$, the effective exchange field becomes very large. For a specimen 20 Å thick $k_z = \pi/d = 1.6 \times 10^7$ cm^{-1}; for iron this gives an effective exchange field of $2Ak_z^2/M_s = 6.0 \times 10^5$ Oe. A field of 600 kOe corresponds to a frequency of 1750 GHz which falls well outside of the range accessible to BLS experiments. The frequency spectrum in k-space therefore consists of sheets on which the frequency changes rather slowly with k_x, k_y in the long wavelength limit, but the sheets, each characterized by a particular k_z value, are very widely separated. BLS experiments are concerned with spin wave frequencies lying between 0 and $\sim$100 GHz. It follows, therefore, that the only modes of interest in an experiment on ultrathin films are those corresponding to no spatial variation of the magnetization across the film thickness, i.e. the modes for which $k_z = 0$.

The BLS experiment can be used to investigate the frequency of in-plane modes having a spatial variation in the specimen plane which is less than, or comparable to, the wavelength of the incident light. That is, only spin wave modes for which the wavenumber components in the plane are less than $\sim 2 \times 10^5$ cm^{-1} can be measured. The effective exchange field corresponding to

$2 \times 10^5 \, \text{cm}^{-1}$ is $2Ak^2/M_s = 93 \, \text{Oe}$ for iron at 300 K. To put this shift into perspective, note that the FMR linewidth observed in ferromagnetic resonance experiments at 36 GHz is approximately 200 Oe. Using (3.35) for an applied field of 1 kOe., this exchange field correction translates into a frequency increase of 0.66 GHz over the uniform mode frequency of 13.9 GHz. Exchange field corrections are not large and would just be observable using a conventional multi-pass plane Fabry–Perot interferometer. Moreover, the exchange field is proportional to the square of the wavenumber so that exchange field corrections to the uniform mode frequency become negligibly small for spin waves characterized by in-plane wave vectors whose lengths are less than $\sim 5 \times 10^4 \, \text{cm}^{-1}$.

Let us now return briefly to the subject of the dipole–dipole field corrections generated by in-plane spatial variations of the magnetization. Consider a plane magnetization wave in an ultrathin slab corresponding to no spatial variation of the magnetization across the slab thickness, i.e. $k_z = 0$. From (3.34a, b)

$$m_y = A_y \exp(i[k_x x + k_y y - \omega t]), \tag{3.41a}$$

$$m_z = A_z \exp(i[k_x x + k_y y - \omega t]). \tag{3.41b}$$

For such a wave the divergence of $\boldsymbol{m}$ is not zero. The associated magnetic charge density, $\rho_m = - \, \text{div} \, \boldsymbol{m}$, will generate magnetic field components which can exert a torque on the magnetization and which will, therefore, alter the mode frequency. The magnetic field components generated by the magnetization wave (4.41) can be readily calculated by means of the magnetic scalar potential [3.67], Ω, where

$$\boldsymbol{h} = \text{grad} \, \Omega \tag{3.42}$$

and where inside of the slab

$$\nabla^2 \Omega = - \, 4\pi \, \text{div} \, \boldsymbol{m} \tag{3.43}$$

and outside of the slab

$$\nabla^2 \Omega = 0. \tag{3.44}$$

The potential function must be continuous across the slab boundaries, and it must vanish far from the slab surfaces. In addition, the z component of $\boldsymbol{b} = \boldsymbol{h} + 4\pi \boldsymbol{m}$ must be continuous across the slab surfaces. The solution for this classical boundary value problem has been worked out by *Damon* and *Eshbach* [3.67] for an infinite slab having an arbritrary thickness. Their result can be written as an expansion in the slab thickness, d, and to terms linear in d the relevant magnetic field components averaged across the thickess of the slab are given by

$$\langle h_y \rangle = - \frac{2\pi k_y^2 d}{\sqrt{k_x^2 + k_y^2}} m_y \tag{3.45a}$$

$$\langle h_z \rangle = - 4\pi m_z + 2\pi \left[\sqrt{k_x^2 + k_y^2} \right] d m_z. \tag{3.45b}$$

In the limit of vanishing thickness only the term $h_z = -4\pi m_z$ remains: this term has already been taken into account in obtaining (3.35) for the mode frequency. If the additional terms proportional to the thickness, d, are included in the Landau–Lifshitz equations, the expression for the mode frequency becomes

$$\left(\frac{\omega}{\gamma}\right)^2 = H_1 H_2, \tag{3.46a}$$

where

$$H_1 = H + \frac{2A}{M_s}(k_x^2 + k_y^2) + \frac{2\pi M_s k_y^2 d}{\sqrt{k_x^2 + k_y^2}}, \tag{3.46b}$$

$$H_2 = H + 4\pi M_s + \frac{2A}{M_s}(k_x^2 + k_y^2) - 2\pi M_s d \sqrt{k_x^2 + k_y^2}, \tag{3.46c}$$

Notice that the frequency does not depend upon the sign of k_x, k_y in this approximation: it therefore also applies to a standing wave of the form given by (3.38).

The magnetostatic corrections to the mode frequencies are surprisingly large. Consider, for example, the case $k_x = 0$, $k_y = 2 \times 10^5$ cm^{-1} corresponding to a spatial variation in the plane on the scale of an optical wavelength. For an iron slab 10 Å thick for which $4\pi M_s = 21.6$ kOe at room temperature, the correction to the homogeneous mode frequency for an applied field of 1 kOe is 10% or 1.4 GHz in 13.9 GHz.

In summary, BLS experiments on ultrathin films are concerned with frequencies on the lowest exchange branch for which there is no spatial variation of the magnetization across the film thickness, i.e. $k_z = 0$ in (3.38). The frequency of spin waves corresponding to this lowest exchange branch depends upon the in-plane spatial variation specified by wave number components k_x, k_y. For small $k_x d, k_y d$ the frequency increase above the homogeneous mode frequency is proportional to the wave number, (3.46). For k_x, k_y smaller than $\sim 2 \times 10^5$ cm^{-1} the corrections to the homogeneous mode frequency due to exchange can generally be ignored, but the correction due to the dipole–dipole fields may have to be taken into consideration.

3.2.4 The Intensity of the Scattered Light

The light wave incident on an ultrathin film sets up an electrical polarization density in the film which is given by (3.25). The first term in (3.25) is simply proportional to the incident electric field amplitude and it therefore oscillates at the same frequency as the incident light. This term in the polarization density generates optical fields which form the reflected wave and which modify the transmitted wave. The second term in (3.25) is of greater interest because it is proportional to the magnetization density in the film; it generates the scattered light which carries information about the spin wave modes in the film. For a light wave falling at normal incidence on an ultrathin film, as shown in Fig. 3.16, and for an optical electric field in the film of the form $E_x = E_0 \exp(-i\omega_0 t)$, the magneto-optic part of (3.25) can be written

$$4\pi P_y = -\left(\frac{K}{M_s}\right) E_0 \exp(-i\omega_0 t) m_z, \tag{3.47}$$

$$4\pi P_z = -\left(\frac{K}{M_s}\right) E_0 \exp(-i\omega_0 t) m_y. \tag{3.48}$$

Insert into these expressions for the polarization components explicit expressions for the spin wave magnetization components from (3.38a, b), taking into account that we are interested only in modes for which the magnetization is uniform across the slab, i.e. those modes for which $k_z = 0$. The result is

$$4\pi P_y = -\frac{K}{M_s} E_0 \exp(-i\omega_0 t) b_0 \cos(k_x x) \cos(k_y y) \sin(\omega t), \tag{3.49a}$$

$$4\pi P_z = -\frac{K}{M_s} E_0 \exp(-i\omega_0 t) a_0 \cos(k_x x) \cos(k_y y) \cos(\omega t), \tag{3.49b}$$

where, from (3.36) $b_0/a_0 \cong \sqrt{H/B}$. In (3.49) the magnetic mode amplitudes are to be determined from the condition that the average energy of the mode be the same as that for an oscillator having the same frequency ω.

The modes of interest for the BLS experiment have frequencies which are less than 100 GHz: the temperature equivalent to 100 GHz is $T = 10^{11}\, h\, k^{-1} = 4.8$ K. It follows that the average energy per mode at room temperature may be approximated by the classical equipartition value $U_m = kT$, where T is the temperature of the film. The phase of a particular thermally excited mode of the film varies erratically with time and is uncorrelated with the phase of the other modes. Since the light scattered from each of the spin wave modes is uncorrelated, the intensity of light scattered from each mode can be calculated as if the other modes were not present. The total intensity of scattered light having a particular frequency shift, ω, is just the sum of the light intensity scattered by each mode at the frequency ω.

It is clear from the form of (3.49) that the polarization components, P_y, P_z oscillate at the two frequencies $\omega_s = \omega_0 \pm \omega$. Each volume element of the specimen acts like a point dipole and generates a radiation electric field amplitude given by [3.68]

$$d E_s(r, t) = \frac{n \times n \times \ddot{P}\, d\tau}{c^2 r}\bigg|_{t_r = t - r/c} \tag{3.50}$$

where $r = rn$ is the position vector drawn from the volume element in question, $d\tau$, to the point of observation and t_r is the retarded time. The total electric field vector at the point of observation is obtained by integrating (3.50) over the entire volume of the specimen taking into account the variation of the retarded time with position in the specimen. The required integration is easy to perform for two limiting cases: (i) the lateral dimensions L_x, L_y of the film (Fig. 3.16) are small compared with a wavelength of the emitted light so that retardation effects can be ignored; in this limit the entire specimen behaves like a point dipole. (ii) the lateral dimensions of the specimen are very large compared with the wavelength of the radiated light so that the specimen can be treated like an infinite plane sheet of thickness d.

Case (i) $L_x, L_y \ll \lambda_s$

In this case the total electric dipole moment of the specimen is $p = pL_xL_yd$ and the electric field of the scattered wave is given by

$$e_s = \frac{n \times n \times \ddot{p}}{c^2 r}. \tag{3.51}$$

But for all modes except the uniform mode the total dipole moment of the specimen is zero. In this limit the scattered radiation field amplitude is very small except for the uniform mode for which $k_x = k_y = k_z = 0$. The circular frequency of the scattered radiation will be shifted up and down from the incident light frequency by the uniform mode frequency $\omega_u = \gamma\sqrt{BH}$. This frequency, ω_u, is exactly the same frequency that would be measured in a ferromagnetic resonance experiment. This is also discussed by *Heinrich*, Sect. 3.1.

In order to estimate the intensity of the scattered light it is necessary to take into account the collection system geometry. The primitive experimental arrangement shown in Fig. 3.15 cannot be successfully used in conjunction with a multi-pass plane Fabry–Perot interferometer. The light which is analyzed by the interferometer must be very precisely collimated so that it falls on the interferometer plates at a normal incidence in order to achieve the high contrast that is required to be able to measure a weak signal against a relatively strong background of light having an unshifted frequency. In a typical case the angular divergence of the light falling upon the interferometer must be less than 10^{-3} radians. For the primitive geometry shown in Fig. 3.15, most of the scattered light of interest would be emitted at angles greater than 10^{-3} radians with respect to the Fabry–Perot analyzer axis and would therefore not be measured: in effect, one would be throwing away all of the scattered light signal except for that small fraction, $\sim 10^{-6}$, which fell within the interferometer acceptance angle. The collection efficiency of the system can be dramatically improved by using a collection lens as illustrated in Fig. 3.17. An incident laser beam whose radius r_0 (usually $r_0 \sim 1$ mm) is focused on the target by means of high quality lens (usually a camera lens) characterized by a focal length f. The illuminated spot on the target acts like a point source so that all of the scattered light collected by the lens aperture is refracted into a direction parallel with the optic axis and within the acceptance angle of the Fabry–Perot interferometer. The lens has increased the fraction of the scattered light which enters the interferometer from $\sim 10^{-6}$ for the primitive system of Fig. 3.15 to ~ 0.06 for a typical f-2 lens for which the focal length is twice the aperture diameter. The price that is paid for this 10^5-fold increase in collection efficiency is a potential overheating of the specimen. All of the power contained in the incident laser beam is directed into an area on the specimen whose radius is typically $\sim$10 µm for a laser having a Gaussian beam profile 2 mm in diameter. *Dutcher* [3.69] has shown that the local heating observed for a 2 mm thick silver single crystal substrate amounted to $\sim$26 °C for an incident laser power of 140 mW.

We wish to use the collection geometry of Fig. 3.17 to estimate the signal strength generated in a target which is made up of a large number of small islands; each island has lateral dimensions which are small compared with the wavelength of the incident light, and each island is assumed to behave like an independent magnetic unit. Each island acts like a point source of scattered radiation. The diameter of the focal spot on the target depends upon the radius of the incident beam and upon the focal length of the lens; it may be calculated using formulae appropriate for Gaussian beam optics [3.70]. The focal spot diameter is always larger than the wavelength of the incident light and therefore it is much larger than the dimensions of the magnetic islands of which the target is assumed to be composed. The scattered light collected by the lens is confined to a relatively small angle around the target normal, if the lens diameter is not

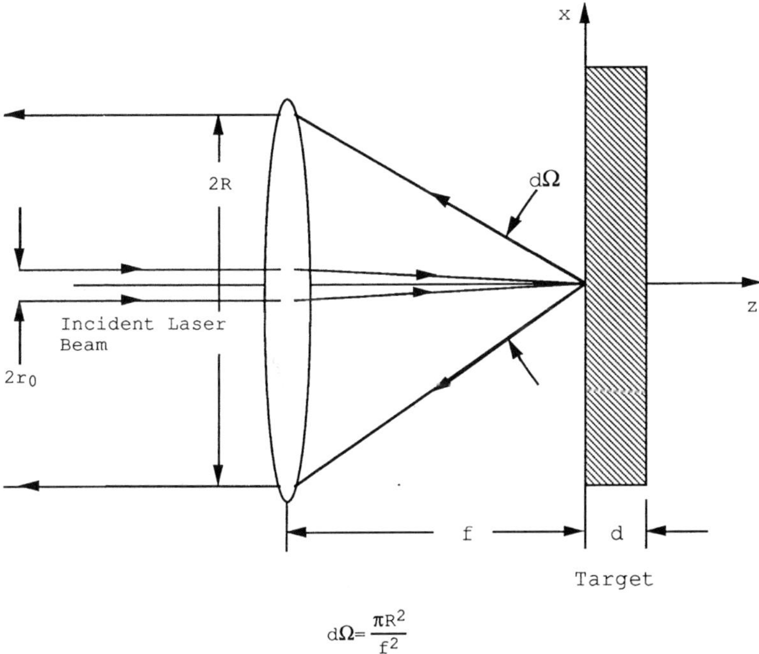

$$d\Omega = \frac{\pi R^2}{f^2}$$

Fig. 3.17. A collection lens can be used to increase the amount of scattered light which is directed parallel with the optic axis. The collected radiation must fall within the $\sim$10^{-3} radian acceptance angle of the Fabry–Perot interferometer used to analyze the light

too large. Relatively little of the scattered light will be generated by the z component of the polarization which is nearly parallel with the direction of observation; most of the scattered light intensity will be generated by the component of polarization which is transverse to the direction of observation. Thus, for incident light polarized along the x direction most of the scattered light will be due to the component $p_y = P_y L_x L_y d$ (3.49). Neglecting the small contribution from the z component of the polarization, and neglecting the slowly varying angular factor due to rays which are collected by the lens and which travel at an angle with respect to the specimen normal, one finds (3.51)

$$|E_y| = \left(\frac{|K|b_0}{8\pi M_s}\right)\frac{\omega_0^2 E_0}{c^2 r} L_x L_y d \tag{3.52}$$

for both the upshifted and downshifted frequencies: this neglects the very small differences in frequency between the incident optical frequency ω_0 and the scattered frequencies $\omega_0 \pm \omega$. The parameter b (3.52) is the amplitude of the magnetization component m_z; i.e. the average amplitude of m_z for the thermally excited uniform mode in the present case. The average rate at which scattered light energy is collected by the lens can be calculated from the area of the lens and Poynting's vector using (3.52). The result of the calculation is

$$U_s = \left(\frac{c}{8\pi}\right)\frac{|K|^2 b_0^2}{(8\pi M_s)^2}\left(\frac{\omega_0}{c}\right)^4 E_0^2 L_x^2 L_y^2 d\, d\Omega, \tag{3.53}$$

where the solid angle, $d\Omega$, is given by

$$d\Omega = \frac{\pi R^2}{f^2}, \tag{3.54}$$

and R, f are the radius and focal length of the lens. The average rate at which incident energy is transported to a small magnetic island can also be calculated from the Poynting vector

$$U_0 = \left(\frac{c}{8\pi}\right) E_0^2 L_x L_y, \tag{3.55}$$

and the fraction of this incident light which is scattered into the lens is given by

$$\frac{U_s}{U_0} = \left(\frac{|K|b_0}{8\pi M_s}\right)^2\left(\frac{\omega_0}{c}\right)^4 d\Omega L_x L_y d. \tag{3.56}$$

In order to complete the calculation, one must obtain the magnetic amplitude, b_0, by setting the average energy contained in the uniform mode equal to kT. The energy density contained in the uniform mode is composed of two terms: (1) the energy of interaction of the magnetization with the applied field,

$$F_1 = -\,\boldsymbol{M}\cdot\boldsymbol{H}, \tag{3.57}$$

and (2), the self-energy due to the demagnetizing field $-\,4\pi m_z$. The increase in the energy density of the system due to tipping the magnetization away from its equilibrium direction along x (Fig. 3.16) can be written

$$F = \frac{H}{2M_s}(m_y^2 + m_z^2) + 2\pi m_z^2. \tag{3.58}$$

The first term in (3.58) comes from $- Hm_x$ in which $m_x \approx M_s[1 - (m_y^2 + m_z^2)/2]$ as a consequence of the condition $M_s^2 = m_x^2 + m_y^2 + m_z^2$. For the uniform mode (3.38) $m_y = a_0 \cos \omega t$ and $m_z = a_0 \sqrt{H/B} \sin \omega t$, so that the time-averaged increase in energy density is given by

$$\langle F \rangle = \left(\frac{H}{2M_s}\right) a_0^2. \tag{3.59}$$

This must be multiplied by the volume to obtain the total energy increase associated with the uniform mode. When this expression for the total energy increase is set equal to kT, one obtains

$$a_0^2 = \left(\frac{2M_s}{H}\right)\left(\frac{kT}{L_x L_y d}\right). \tag{3.60}$$

However, for the uniform mode $b_0/a_0 = \sqrt{H/B}$, and therefore

$$b_0^2 = \left(\frac{2M_s}{B}\right)\left(\frac{kT}{L_x L_y d}\right). \tag{3.61}$$

Finally, inserting (3.61) into (3.56) gives the ratio of the scattered light intensity collected by the lens to the incident light intensity

$$\frac{U_s}{U_0} = \frac{d|K|^2}{32\pi^2 M_s B}\left(\frac{\omega_0}{c}\right)^4 d\Omega kT. \tag{3.62}$$

This expression does not depend upon the lateral dimensions of the elementary magnetic scatterer. If follows that if one adds together all the contributions from the illuminated magnetic areas, each one of which scatters independently of the others, (3.62) also gives the ratio of the total number of scattered photons, N_s, having a shifted frequency (either $\omega_0 + \omega$ or $\omega_0 - \omega$) collected by the lens to the total number of incident photons, N_0:

$$\frac{N_s}{N_0} = \frac{d}{32\pi}\frac{|K|^2}{M_s B}\left(\frac{\omega_0}{c}\right)^4\left(\frac{R}{f}\right)^2 kT, \tag{3.63}$$

since $d\Omega = \pi R^2/f^2$. Equation (3.63) assumes that the entire area illuminated by the laser spot is covered by magnetic scatterers (filling factor $= 1$). In any real specimen the total area of active scattering centers may be smaller than the area of the incident light spot; for that case the ratio in (3.63) must be reduced accordingly.

The above expression for the strength of the scattered light signal exhibits a number of interesting features:

(1) The signal strength is proportional to the film thickness, d. It does not depend upon the lateral dimensions of the individual magnetic patches.

(2) The signal strength is proportional to the absolute temperature, T.

(3) The signal strength is the same for the upshifted and downshifted frequencies.

(4) The scattered light is polarized at $90°$ to the polarization of the incident light.

(5) The signal strength is inversely proportional to the square of the f-number which characterizes the collection lens.

(6) The strength of the signal depends rather weakly on the static magnetic field strength, at least for fields less than $4\pi M_s$ (recall that $B = H + 4\pi M_s$).

Let us estimate the signal strength for $0.5145 \, \mu m$ light incident on a room temperature iron film $d = 10 \, \text{Å}$ thick (7 ML), and for an applied field of 1 kOe. For this light $\omega_0/c = 1.22 \times 10^5 \, \text{cm}^{-1}$ and $|K|^2 = 0.15$ from Table 3.2. For iron at 300 K $4\pi M_s = 21.5 \, \text{kOe}$ and therefore for $H = 1 \, \text{kOe}$ one has $B = 22.5 \, \text{kG}$. For a lens such that $R = 1.25 \, \text{cm}$ and $f = 5.0 \, \text{cm}$, (3.63) gives $N_s/N_0 = 2.2 \times 10^{-12}$. The scattering cross section is very small! However, an incident beam intensity of $100 \, \text{mW}$ corresponds to a photon flux of $2.59 \times 10^{17} \, \text{s}^{-1}$. Accordingly, the signal flux would correspond to a rate of $\sim 6 \times 10^5 \, \text{photons s}^{-1}$. Modern photomultiplier tubes are characterized by a quantum efficiency of $\sim 10\%$ and a dark counting rate at room temperature of ~ 1 count per second. In principle, it should be easy to measure the frequency of the uniform mode on films as thin as 1 ML.

Although the above expression for the scattering intensity, (3.63), was derived for normal incidence, a very similar expression would be obtained for light incident on the target at an oblique angle. The ratio (3.63) would simply be multiplied by an angular factor which approaches zero in the limit of grazing incidence.

Equation (3.63) was derived using unpinned boundary conditions for the magnetization components. Pinning on the broad surfaces of an ultrathin crystal can be described by surface anisotropy energies and torques, [3.55, 58]. These enter the Landau–Lifshitz equations of motion primarily as an apparent shift in the magnetization density of the specimen. Pinning torques on the lateral edges of a magnetic platelet require (3.38) to be replaced by linear combinations of sines and cosines along the x and y directions: in the limit of completely pinned spins at the platelet edges, m_y and m_z become proportional to the product functions $\sin(m\pi x/L)\sin(n\pi y/L)$ where m, n are positive integers. This change has a relatively small effect on the scattering intensity. The dipole moment of the platelet is non-zero for odd values of m, n in the fully pinned limit so that many modes can contribute to the scattered light intensity. However, the intensity contributed by each mode decreases like $1/m^2n^2$ so that only the lowest modes are important. The same sort of consideration apply to FMR: the absorption for a uniform driving field is proportional to the square of the average magnetization in the film, i.e. to $1/m^2 n^2$ for strong pinning [3.65]. It follows that the frequency measured by means of BLS should be the same as the frequency measured using FMR absorption regardless of the pinning in the limit for which the lateral dimensions of the independently scattering magnetic regions are small compared with the wavelength of the incident light.

Glancing angle electron diffraction measurements [3.55] (also discussed by *Arrott*, Sect. 5.1, Vol. 1), as well as the scanning tunneling microscope studies of *Schmid* et al. [3.71] indicate that ultrathin magnetic films consist of isolated terraces of uniformly thick material separated by regions having an irregular and uneven height profile. The terraces on silver are characterized by lateral dimensions of approximately 100–200 Å; those on properly prepared $Cu(0\,0\,1)$ substrates are as large as 1000 Å [3.71]. The magnetic films on these terraces may very likely be only weakly exchange coupled to one another and, apart from a small dipole–dipole coupling, they may behave like independent magnetic units. The dimensions of these terraces are smaller than the wavelength of visible light; they therefore satisfy the condition $L_x, L_y < \lambda_s$ for which each magnetic patch behaves like a point scatterer. If the magnetic patches are only weakly coupled, it follows that the intensity of the scattered light should be described by (3.63) and its generalization to properly include magnetocrystalline anisotropies. In this limit the spin wave frequencies measured using BLS would be the same as frequencies measured using FMR. A further consequence of scattering by small independent magnetic platelets is that the linewidth of the scattered light should accurately reflect the spin wave lifetime and should not be broadened due to an admixture of a broad range of in-plane wave vectors. This suggests that the lens used to collect the scattered light should have the smallest f-number available.

Case (ii) $L_x, L_y \gg \lambda_s$

In a scattering experiment, a beam of laser light having a Gaussian intensity profile is focussed onto the specimen by means of a lens. The same lens is used to collect the scattered light, as depicted in Fig. 3.18. The beam at the focal spot incident on the specimen can also be described by a Gaussian intensity profile, but the intensity is appreciable only over a region whose radius is typically $\sim$20 μm [3.70]. The light in the neighborhood of this focused spot can be described by a coherent superposition of plane waves, all of which propagate mainly along the direction defined by the axis of the beam, but with a narrow distribution of transverse wave vector components [3.70]. In a uniform specimen whose dimensions L_x, L_y are much larger than the focal spot formed by the incident light, the incident laser beam is scattered coherently by a spin wave having well defined wave vector components in the plane, k_x and k_y. Consider a particular spin wave for which $k_x = 0$, but one for which k_y is not zero. This propagating spin wave constitutes a kind of moving mirror which will scatter

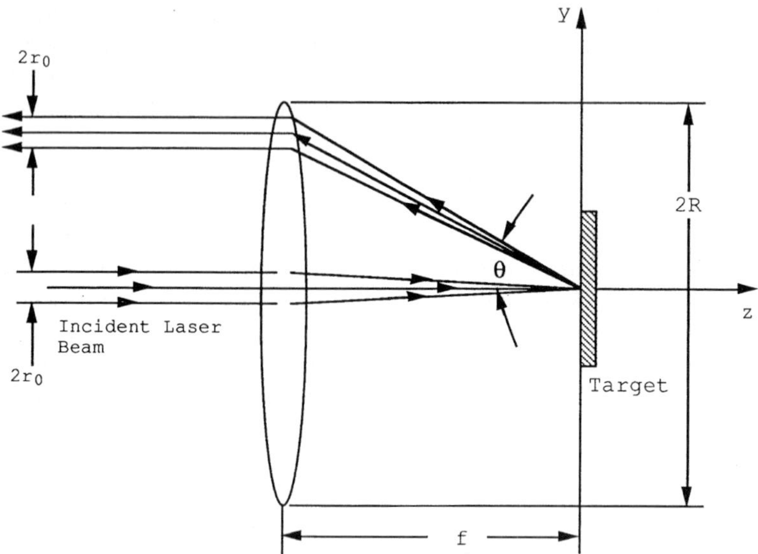

Fig. 3.18. An incident Gaussian input beam that is coherently scattered from a spin wave produces an output Gaussian beam whose diameter is the same as that of the incident beam. The angular displacement of the scattered beam is given by $\sin \theta = ck_y/\omega_0$, where k_y is the in-plane component of the spin wave wave vector, and ω_0 is the circular frequency of the light

each plane wave in the incident Gaussian beam; each wave vector component in the scattered light will be augmented by the transverse component of the spin wave wave vector, k_y, so that the scattered beam will be emitted at the angle θ given by $(\omega_0/c)\sin\theta = k_y$ (Fig. 3.18). This scattered beam will be refracted by the lens so as to produce an output beam that is characterized by a Gaussian intensity profile whose radius is the same as that of the incident beam, and one that propagates parallel with the lens axis (Fig. 3.18). The problem, then, is to calculate the ratio of the power in this scattered beam to the power in the incident beam.

In order to calculate the intensity of the light scattered by a spin wave characterized by in-plane wave vector components k_x, k_y it is useful to take advantage of the planar symmetry. It is therefore computationally easier to start from the differential equation for the optical magnetic vector potential rather than to integrate the electric dipole Green's function, (3.50), over the volume of the film. From Maxwell's equations, and using the definition $\boldsymbol{h} = \text{curl}\,\boldsymbol{A}$ for an optical medium having the same magnetic permeability as free space, one can write [3.72]

$$\nabla^2 \boldsymbol{A} - \frac{1}{c^2}\frac{\partial^2 \boldsymbol{A}}{\partial t^2} = -\frac{4\pi}{c}\boldsymbol{J}. \tag{3.64}$$

For the present problem, the driving current density is $\boldsymbol{J} = \partial\boldsymbol{P}/\partial t$, where $\boldsymbol{P}$ is the electric dipole moment per unit volume from (3.49). The magnetic slabs in which we are interested are so thin along the z direction compared with the wavelength of the light that they can be treated as δ-function current sheet sources. Furthermore, we are only interested in those modes for which $k_z = 0$ since, as shown previously, only modes for which $k_z = 0$ will give rise to light whose frequency shift lies in the range 1–100 GHz. From (3.49) it can be seen that the source term $\partial P/\partial t$ will consist of eight exponential terms: there are two time dependencies, $\exp[-i(\omega_0 - \omega)t]$ and $\exp[-i(\omega_0 + \omega)t]$, and corresponding to each time dependence there are the spatial terms $\exp[i(\pm k_x x \pm k_y y)]$. Each of these source terms will generate a scattered wave.

Now, let us consider the term $\exp[i(k_x x + k_y y - \{\omega_0 - \omega\}t]$; solutions for the other driving terms can be constructed by analogy. The problem to be solved can be stated explicitly:

$$\nabla^2 A_y - \frac{1}{c^2}\frac{\partial^2 A_y}{\partial t^2} = -\frac{(\omega_0 - \omega)KE_0 bd}{c\,8M_s}\left[\exp[i(k_x x + k_y y - \{\omega_0 - \omega\}t)]\right]\delta(z)$$

(3.65a)

$$\nabla^2 A_y - \frac{1}{c^2}\frac{\partial^2 A_z}{\partial t^2} = i\frac{(\omega_0 - \omega)KE_0 ad}{c\,8M_s}\left[\exp[i(k_x x + k_y y - \{\omega_0 - \omega\}t)]\right]\delta(z)$$

(3.65b)

The magnetic wave amplitudes a, b (see (3.34, 36)) are fixed by the requirement that the average energy associated with the magnetic mode must be equal to kT. The magnetic mode frequencies, ω, are much smaller than the optical frequency, ω_0; in the following factors like $(\omega_0 \pm \omega)$ will be replaced by ω_0 in the amplitude terms. Clearly the solutions of (3.65) which are required must be proportional to the exponential factor on the right hand side. Moreover, these solutions must satisfy the source free wave equation outside of the magnetic slab where $P = 0$. Therefore the solutions outside of the slab must have the form of a plane wave whose frequency is $(\omega_0 - \omega)$ and whose total wave vector must have the value $k_s = (\omega_0 - \omega)/c$. The required solutions outside the film have the form

for $z \geq d$,

$$A_y = A_0 \exp[i(k_x x + k_y y + qz - \{\omega_0 - \omega\}t)]$$

(3.66)

and for $z \leq 0$,

$$A_y = A_0 \exp[i(k_x x + k_y y + qz - \{\omega_0 - \omega\}t)]$$

(3.67)

with similar expressions for A_z. In (3.66, 67) $q^2 + k_x^2 + k_y^2 = ((\omega_0 - \omega)/c)^2 \cong (\omega_0/c)^2$. A_y and A_z are continuous through the film; if the film thickness can be neglected then this requires the two branches of the vector potential to be equal at $z = 0$; this requirement is satisfied by (3.66, 67). The amplitude A_0 is determined from the requirement that the discontinuities in the derivatives $\partial A_y/\partial z$ and $\partial A_z/\partial z$ at the film surfaces should generate the singular terms on the right hand side of (3.65). For example,

$$\frac{\partial A_y}{\partial z}\bigg|_d - \frac{\partial A_y}{\partial z}\bigg|_0 = -\left(\frac{\omega_0}{c}\right)\frac{KE_0bd}{8M_s}. \tag{3.68}$$

This result can be readily obtained by integrating (3.65) with respect to z through the film thickness. From (3.68) one obtains

$$A_0 = i\left(\frac{\omega_0}{c}\right)\frac{KE_0bd}{16M_s}\left(\frac{1}{q}\right), \tag{3.69}$$

where

$$q = \sqrt{\left(\frac{\omega_0}{c}\right)^2 - k_x^2 - k_y^2}.$$

Similarly, the amplitude of the component A_z can be shown to be

$$A_1 = \left(\frac{\omega_0}{c}\right)\frac{KE_0ad}{16M_s}\left(\frac{1}{q}\right). \tag{3.70}$$

The magnetic field components of the scattered waves can be calculated from the vector potential using $h_s = \mathrm{curl}\, A$. The electric field, e_s can be deduced from the fact that the electric field, the magnetic field, and the wave vector k_s form an orthogonal triad in which $|e_s| = |h_s|$ in the CGS system of units used here. The rate at which energy is radiated per unit area of film surface can be calculated from the Poynting vector, $S = c/4\pi(e_s \times h_s)$. The resulting expression for the average value of the Poynting vector is complicated by angular factors introduced because the scattered waves propagate at oblique angles with respect to the coordinate axes. One can estimate the magnitude of the radiated power by using a simplified geometry: let $k_x = 0$ so that the light is scattered by a spin wave having only an in-plane spatial variation along the y axis. In that case one obtains

$$A_y = i\left(\frac{\omega_0}{qc}\right)\left(\frac{KE_0bd}{16M_s}\right)\exp[i(k_yy \pm qz - \{\omega_0 - \omega\}t)] \tag{3.71a}$$

$$A_z = \left(\frac{\omega_0}{qc}\right)\left(\frac{KE_0ad}{16M_s}\right)\exp[i(k_yy \pm qz - \{\omega_0 - \omega\}t)]. \tag{3.71b}$$

The magnetic field has only an x component

$$h_x = \frac{\partial A_z}{\partial y} - \frac{\partial A_y}{\partial z}. \tag{3.72}$$

The radiation emitted into the back-scattered direction, i.e. $z < 0$, carries away energy at the average rate

$$\langle S \rangle = \frac{c}{8\pi} \, \text{real} \, (h_x h_x^*)$$

$$\langle S \rangle = \frac{c}{8\pi} \left(\frac{\omega_0}{c} \right)^2 \left(\frac{|K| E_0}{16 M_s} \right)^2 \left(\frac{k_y^2}{q^2} a^2 + b^2 \right) d^2 \tag{3.73}$$

per cm^2 perpendicular to k_s. The radiation emitted per unit area of magnetic film is smaller by the cosine of the angle of emission, θ. Therefore the ratio of the power scattered by the spin wave to the power incident per unit area of the film is given by

$$\frac{U_s}{U_0} = \left(\frac{\omega_0}{c} \right)^2 \frac{|K|^2}{256 M_s^2} \left(\frac{k_y^2}{q^2} a^2 + b^2 \right) d^2 \cos \theta, \tag{3.74}$$

where $k_y^2 + q^2 = (\omega_0/c)$. In this equation the spin wave amplitudes are given by (3.60, 61) providing that exchange and dipole–dipole corrections can be neglected.

The ratio (3.74) is very small: for a 10 Å iron film at 300 K, 1 cm square, and for 0.5145 μm incident radiation, one finds $U_s/U_0 \approx 4 \times 10^{-20}$. However, many beams are collected by the lens because there are many closely spaced magnetic modes in the film and each magnetic mode produces four beams corresponding to $k_x = \pm N_x (\pi/L_x)$ and $k_y = \pm N_y (\pi/L_y)$, where N_x, N_y are positive integers. Assuming that all of these magnon modes have the same frequency, i.e. assuming that exchange and dipole–dipole corrections are less than the intrinsic linewidth expressed as a magnetic field, one can add the power in each scattered beam and ascribe it to a single frequency. Crudely speaking, the maximum in-plane wave vector, K_p, for which the scattered light will be collected by a lens of radius R and focal length f, will be given by

$$\frac{c K_p}{\omega_0} = \frac{R}{f}. \tag{3.75}$$

The area in k-space spanned by these effective modes is

$$\pi K_p^2 = \pi \left(\frac{R}{f} \right)^2 \left(\frac{\omega_0}{c} \right)^2. \tag{3.76}$$

The area per mode in k-space is $\pi^2/L_x L_y$. It follows that the number of effective modes is given by

$$N \approx \frac{1}{\pi} \left(\frac{R}{f} \right)^2 \left(\frac{\omega_0}{c} \right)^2 L_x L_y. \tag{3.77}$$

If all of these modes are counted as having the same frequency the ratio of the scattered photons collected by the lens to the number of incident photons will be given by (using (3.74), assuming $(k_y/q) \sim 1$, and neglecting angular factors)

$$\frac{N_s}{N_0} \approx \left(\frac{\omega_0}{c}\right)^4 \frac{|K|^2}{256\,M_s^2}\left(\frac{a^2 d^2}{\pi}\right)\left(\frac{R}{f}\right)^2 L_x L_y. \tag{3.78}$$

Note that $(a^2/b^2) = (B/H)$, so that for $k_y/q \sim 1$ and for small fields, H, the term in a^2 in (3.74) dominates the scattering intensity. But from (3.61)

$$a^2 = \left(\frac{2M_s}{H}\right)\left(\frac{kT}{L_x L_y d}\right), \tag{3.61}$$

so that, finally,

$$\frac{N_s}{N_0} \approx \frac{d}{128\pi}\frac{|K|^2}{HM_s}\left(\frac{\omega_0}{c}\right)^4\left(\frac{R}{f}\right)^2 kT. \tag{3.79}$$

This is very similar to the result obtained for the case of a collection of very small magnetic platelets, (3.63). Clearly the scattering intensity cannot be expected to be sensitive to the magnetic coherence length in the plane of an ultrathin magnetic film. In both cases the collection rate for scattered photons is proportional to the film thickness, and is independent of the laser beam diameter, provided that the specimen dimensions are larger than the focal spot diameter on the target.

If the target is rotated around the direction of the magnetic field, the x axis as shown in Fig. 3.16, the intensity of the back-scattered light will be given by an expression similar to that for normal incidence, (3.79), but multiplied by a relatively slowly varying angular factor which goes to zero at grazing incidence, $\theta \to \pi/2$. The angular factor is mainly caused by the oblique angles which the optical electric field in the specimen, E, and the electric dipole density, P, make with respect to the specimen axes. It is tedious, but not difficult, to repeat the calculation outlined above for the non-normal incidence of the light. The main consequence of oblique incidence is that the distribution of the in-plane spin wave wave vector components k_y is centered on the value $k_y = 2(\omega_0/c)\sin\theta$ rather than on the value $k_y = 0$, as is the case for normal incidence. This comes about because the optical electric field in the film has a spatial variation when described in the film coordinate system given by $\exp[-i(\omega_0 \sin\theta/c)y]$ (this assumes a counterclockwise rotation looking along positive x). The product of this spatial variation with the spin wave y-dependence, $\cos(k_y y)$, results in terms having the form $\exp[i(k_y \pm (\omega_0/c)\sin\theta)y]$ in the source currents which generates the vector potential of the scattered radiation, (3.65). However, in order to be collected the scattered light must be emitted along a direction near that defined by the incident beam and hence its mean wave vector component along y must be given by $(\omega_0/c)\sin\theta$ (remember that the frequency of the scattered light is very nearly equal to the frequency of the incident light since $\omega/\omega_0 \ll 1$). In order to satisfy this condition, the spin wave wave vectors which produce the scattered light must be centered on $k_y = 2(\omega_0/c)\sin\theta$. The frequency shift associated with the back-scattered light is that of the spin wave whose wave number components in the plane are $k_x = 0$ and $k_y = 2(\omega_0/c)\sin\theta$. The frequency associated with this spin wave is greater than the uniform mode frequency ($k_x = 0$, $k_y = 0$) because of the exchange and dipole-dipole terms given in (3.46).

In principle, the angular variation of the scattered light frequency could be used to determine which of the two models discussed above is most appropriate to describe the results for a particular specimen: (i) the model for which the lateral dimensions of the individual magnetic patches are much smaller than the optical wavelengths, or (ii) the model for which the lateral dimensions of the magnetic patches are larger than the focal spot on the specimen and hence larger than the optical wavelength. In the former case the relevant spin wave frequency should correspond to that of the uniform mode for all angles of incidence; in the latter case, the spin wave frequency should depend upon the angle of incidence through the wave number dependent terms given in (3.46). These wave number dependent terms are small and the frequency shift with angle is expected to be ~ 1 GHz for iron at room temperatures for angles lying between 0 and 70°. No experiments on ultrathin films reported to date have been designed to measure the angular dependence of the scattered light frequencies.

Equations (3.63) and (3.79) predict a scattering rate of approximately 10^6 photons s^{-1} for a 10 ML thick iron film at room temperature and 100 mW of 0.5145 μm incident radiation. The observed counting rates for 10 ML Fe single crystals grown on silver are approximately 1000-fold smaller. A factor of 20 can be ascribed to losses in the optical train and in the interferometer. The quantum efficiency of a photomultiplier tube is typically 10% at 0.5145 μm for a bialkali photocathode, and this produces another factor of 10 reduction in sensitivity. The presence of a substrate which supports the thin film can be expected to modify the optical fields in the metal film; this can account for another factor of 2–3. Taken together, these factors lead to an overall reduction in sensitivity of approximately 500. In view of the uncertainties in the parameters which describe the system one must conclude that the calculated and observed scattering intensities are in order of magnitude agreement.

3.2.5 Magnetic Damping

Any real magnetic system contains damping mechanisms which cause the magnetization to relax to its equilibrium configuration. The effect of damping can be incorporated into the Landau–Lifshitz equations of motion for the magnetization, (3.28), by the introduction of an effective field H_{damp} where

$$H_{\mathrm{damp}} = -\frac{G}{\gamma^2 M_s^2}\left(\frac{\partial M}{\partial t}\right), \tag{3.80}$$

and where G is the Gilbert damping parameter in frequency units (also examined by *Heinrich*, Sect. 3.1). For bulk ferromagnetic metals $G \sim 10^8$ Hz. The effective field (3.80) is proportional to the rate of change of magnetization so that it corresponds to a viscous damping term of the sort which may be derived from a Rayleigh dissipation function [3.63]. For a time dependence $\exp[-i\omega t]$ the effective field (3.80) becomes

$$\boldsymbol{H}_{\text{damp}} = \left(\frac{i\omega}{\gamma}\right)\left(\frac{G}{\gamma M_s}\right)\left(\frac{m}{M_s}\right). \tag{3.81}$$

The damping field adds an imaginary component to the effective fields which occur in the linearized Landau–Lifshitz equations. The precession of the magnetization around its equilibrium configuration must be described by a complex frequency corresponding to an exponentially decreasing amplitude with time.

One finds for the uniform mode

$$\left(\frac{\omega}{\gamma}\right)^2 = \left[H + 4\pi M_s - \left(\frac{i\omega}{\gamma}\right)\left(\frac{G}{\gamma M_s}\right)\right]\left[H - \left(\frac{i\omega}{\gamma}\right)\left(\frac{G}{\gamma M_s}\right)\right]. \tag{3.82}$$

The dimensionless ratio $(G/\gamma M_s)$ is small for metals – typically of the order of 0.01. It follows that one may replace (ω/γ) on the right hand side of (3.82) by the undamped value $\sqrt{BH}$ to obtain, to first order in the small parameter $(G/\gamma M_s)$, the complex frequency expression

$$\left(\frac{\omega}{\gamma}\right) = \sqrt{BH} - i\left(\frac{B+H}{2}\right)\left(\frac{G}{\gamma M_s}\right). \tag{3.83}$$

Accordingly, if a mode is excited by an impulse it will oscillate at the frequency $f = \gamma\sqrt{BH}/2\pi$ and its amplitude will decay $\sim \exp[-t/\tau]$, where $\tau = 2M_s/(B+H)\,1/G$ seconds. This means that the electric field amplitude of the optical wave associated with scattering from this magnetic mode will also decay with the same time constant, τ, and its Fourier transform will be of the form

$$E(\Omega) \sim \frac{1}{1 - i(\Omega - \omega_s)\tau}, \tag{3.85}$$

where $\omega_s = \omega_0 \pm \omega$. The corresponding intensity distributions must therefore have the form

$$I(\Omega) = \frac{I_0}{\pi}\left(\frac{\tau}{1 + (\Omega - \omega_s)^2\tau^2}\right), \tag{3.86}$$

where I_0 is the integrated intensity of the scattered light. The intensity distributions are centered on the frequencies $\omega_s = \omega_0 \pm \omega$, and their full width at half maximum is given by

$$\Delta\Omega_s = \frac{2}{\tau} = \left(\frac{B+H}{M_s}\right)G. \tag{3.87}$$

The integrated intensity, I_0, must be identified with the intensity expressions (3.63) and (3.79) calculated for undamped magnetic modes. For small applied magnetic fields $(B+H) \approx 4\pi M_s$ so that from (3.87) $\Delta\Omega_s \cong 4\pi G$ and $\Delta f_s = \Delta\Omega_s/2\pi = 2G$. For the uniform mode of an ultrathin iron film at 300 K one finds $G \cong 3 \times 10^8$ Hz and therefore $\Delta f_s = 0.6$ GHz. No experiment on simple ultrathin films reported to date has been carried out with sufficient resolution to be able to measure linewidth broadening due to intrinsic damping; an experiment on ultrathin composite films of iron and nickel [3.54] reported linewidths which varied between 1.5 and 3.5 GHz and in consequence were broad enough to be resolved using a free spectral range of 15 GHz. The interpretation of linewidth data measured using a high resolution apparatus would have to take into account apparent line broadening due to the spread of in-plane wave vector components associated with the finite aperture of the collection lens combined with the dependence of spin wave frequencies on in-plane wave numbers (3.46). Of course, this source of inhomogeneous line broadening would not be operative for the case in which the specimen was composed of small magnetic patches whose dimensions were much smaller than the wavelength of the incident light.

3.2.6 Magnetic Bilayers

It has been observed that the magnetizations in two ferromagnetic films which are separated by a thin non-magnetic metallic spacer layer exhibit a coupling that can be either ferromagnetic or antiferromagnetic in nature depending upon the thickness of the spacer layer. The strength of the coupling varies with spacer layer thickness, and the coupling strength dependence upon thickness usually contains an oscillatory component whose period corresponds to several atomic layers (also studied by *Hathaway* and *Cullen* Sect. 2.1). The coupling between the magnetic layers can be ascribed to a surface exchange interaction. The simplest form for such an interaction was introduced many years ago by *Hoffman* et al. [3.73]:

$$F_s = - \frac{A_{AB}}{M_A M_B} M_A(0) \cdot M_B(0), \tag{3.88}$$

where M_A, M_B are the saturation magnetizations which characterize the two films, and $M_A(0)$ and $M_B(0)$ are the magnetization vectors at the two surfaces that face one another. The energy F_s is a surface term which is measured in erg/cm^2 or in Joule/m^2. Its effects are therefore most dramatic when the two magnetic films are ultrathin so that the interaction energy, F_s, becomes comparable with other magnetic energy terms such as that due to the presence of the applied magnetic field.

Just as in the case for a single magnetic film, the normal modes of the coupled bilayer system are particularly easy to understand if both films are ultrathin because the modes of interest for BLS correspond to a uniform magnetization across each film. Moreover, one is most likely concerned with the excitations that are the equivalent of the uniform mode in each film; that is because the lateral dimensions of the smooth atomic terraces are likely to be much smaller than the wavelength of the light which is used to probe the system.

The intensity of the light scattered from pairs of ultrathin films can be understood as the coherent emission of light from two coupled sources. The optical electric field due to the incident light will be the same in each magnetic film providing that the thickness of each film, including the spacer layer, is much less than the optical penetration depth, L, defined as the distance over which the electric field amplitude decreases by e^{-1} of its initial value. The optical penetration depths for 0.5145 µm light, at normal incidence, for the transition metals are $L(\text{Fe}) = 283$ Å, $L(\text{Co}) = 225$ Å and $L(\text{Ni}) = 243$ Å. Penetration depths for the noble metals are $L(\text{Cu}) = 315$ Å, and $L(\text{Ag}) = 245$ Å, and $L(\text{Au}) = 443$ Å.

Typically, the total thickness of the two magnetic layers plus the nonmagnetic spacer layer is less than 50 Å so that it is quite reasonable to assume that the optical electric field is the same in both magnetic films. Under these circumstances, the magnetic mode frequencies contained in the scattered light measured using BLS are essentially identical with the frequencies measured using FMR, and there is a one-to-one correspondence between the intensities of the BLS signals and the FMR absorption signal strengths. The optical electric field plays the same role with respect to light scattering that the rf (radio frequency) microwave magnetic field plays with respect to FMR.

The normal modes of two coupled magnetic layers have been discussed by *Heinrich* et al. [3.54], *Cochran* et al. [3.74], *Vohl* et al. [3.75] and *Hillebrands* [3.76], for the saturated case in which the static magnetizations in the two magnetic layers are parallel with the applied magnetic field (Sect. 3.1 also covers this issue). The results for two magnetic films which have identical magnetic properties, but not necessarily the same thickness, are disappointing: the coupled films can be characterized by two uniform modes, but one of those modes does not couple either to a uniform microwave field or to a uniform optical field, and in consequence is too weak to be observable. The two modes correspond to an acoustic mode in which the two magnetizations precess in phase, and an optical mode in which the two magnetizations precess 180° out of phase. The acoustic mode can be observed by means of both FMR and BLS: the magnetizations in the two films remain parallel as they precess around equilibrium and therefore the exchange coupling term (3.88) plays a minor role. The frequency observed for the acoustic mode is nearly the same as it would be for a single film. The optical mode is in principle more interesting because its frequency does depend upon the strength and sign of the exchange coupling parameter A_{AB} in (3.88): however, it is unobservable. In the case of BLS, for example, the radiated optical fields from the two films are 180° out of phase and have amplitudes that just cancel one another.

In order to investigate the exchange coupling term by means of the optical mode it is necessary to use magnetic films whose magnetic properties are different. This proves to be easy to achieve, at least for iron films, because of a uniaxial magnetic surface energy. In iron this uniaxial surface energy acts so as to turn the magnetization in the direction of the specimen normal. It behaves like a kind of anti-demagnetizing energy since it makes a contribution to the energy of the system that is proportional to the square of the normal component of magnetization density:

$$F_u = -\frac{K_u^s}{M_s^2}\, m_z^2; \tag{3.89}$$

(compare (3.89) with the demagnetization energy $2\pi m_z^2$). The torque due to this surface energy term enters the equations of motion for the magnetization in such a way that the material behaves as if it possessed a static magnetization density given by

$$4\pi M_{\mathrm{eff}} = 4\pi M_s - \frac{2K_u^s}{t M_s}, \tag{3.90}$$

where t is the film thickness. Thus, two films which have different thicknesses, but which are otherwise identical, behave like magnetic materials which are characterized by different magnetization densities. Such non-identical magnetic films can exhibit two measurable FMR absorption peaks and two BLS peaks. One peak corresponds to the acoustic mode in which the magnetizations in the two films precess in phase, and the other peak corresponds to the optical mode in which the presentation in the two films are 180° out of phase.

Let us consider the variation of the frequencies of these two modes, and the scattered light intensities associated with them, as the exchange coupling parameter A_{AB} is changed from zero at fixed magnetic field. The behavior of two coupled iron films will be used as a specific example.

3.2.6.1 Ferromagnetic Coupling ($A_{\mathrm{AB}} > 0$)

At zero coupling strength the system behaves as if it were composed of two isolated thin films. The thick film will exhibit a uniform mode frequency that is larger than that exhibited by the thin film. As a specific example, consider iron films which are 16 and 9.4 monolayers (ML) thick situated in an external magnetic field of 5 kOe applied in the film plane along an easy axis. The uniform mode frequency for the thick film will be 32 GHz; that for the thin film will be 24 GHz. The two BLS intensities will be nearly equal despite the 2:1 thickness ratio. This occurs because the BLS intensity is primarily sensitive to the component of magnetization (not the magnetization density) which is normal to the plane of the film; these normal components happen to be nearly equal for the thermally excited uniform modes in the two films.

As the ferromagnetic coupling is increased the frequencies of both modes increase. The high frequency mode becomes the optical mode and the low frequency mode becomes the acoustic mode. The frequency shift of the optical mode is proportional to the coupling parameter A_{AB}. However, the BLS intensity associated with the optical mode drops off very rapidly as A_{AB} increases; the intensity falls off approximately as A_{AB} so that for a coupling strength of $1\ \mathrm{erg\,cm}^{-2}$ ($1\ \mathrm{mJ\,m}^{-2}$) the intensity of the optical mode in our example has decreased from its zero coupling value by a factor 80.

The shift in the acoustic mode frequency saturates. In the strong coupling limit both films precess together at a frequency that corresponds to a weighted mean of the two magnetization densities. For the present example, the acoustic mode frequency approaches 28 GHz in the limit of strong coupling. As the coupling increases the intensity of the acoustic mode BLS signal increases to a value which would be expected for scattering from a single film having magnetic properties intermediate between those of the two uncoupled films.

3.2.6.2 Antiferromagnetic Coupling ($A_{AB} < 0$)

The frequencies of both modes decrease as the degree of antiferromagnetic coupling is increased. In this case, the high frequency mode becomes the acoustic mode and the low frequency mode becomes the optical mode. As $|A_{AB}|$ increases the frequency of the acoustic mode approaches a frequency corresponding to a single film having a magnetization density which is the weighted mean of the effective magnetization densities for the two films (28 GHz for the present example). The intensity of the acoustic mode also increases and saturates at a value in the large $|A_{AB}|$ limit that corresponds to the scattering from a single film having a thickness approximately equal to the combined thicknesses of the two films.

Both the frequency and the intensity corresponding to the optical mode (the low frequency mode) decrease in proportion as $|A_{AB}|$ is increased. Eventually, the frequency is driven to zero, whereupon the stable equilibrium configuration no longer corresponds to parallel magnetizations in the two films, and the dc magnetizations either rotate away from the static magnetic field direction, or the magnetization distribution becomes nonuniform [3.77, 78].

The coupling parameter, A_{AB}, can be obtained from the frequencies of the acoustic and the optical modes providing that the magnetic properties of the isolated magnetic films are known. This method can be used for both ferromagnetic and antiferromagnetic coupling. In the case of antiferromagnetic coupling, the variation of frequency with applied field exhibits cusps at those fields which correspond to a change of magnetic state. These critical magnetic fields correspond to the fields at which the variation of the static magnetization component along the applied field direction exhibits discontinuities and changes in slope as the applied field is swept from saturation in one direction through zero field to saturation in the opposite direction. Such critical fields can be used to estimate the strength of the antiferromagnetic coupling, and are particularly useful if the intensity of the optical mode at higher fields is too weak to be observable using the BLS signal.

The theory of spin wave frequencies for exchange coupled films is complicated but not difficult (3.54, 74–76). The theory for BLS frequencies and scattering intensities for a pair of saturated films (magnetizations parallel with the applied field) has been worked out by *Cochran* and *Dutcher* [3.74]. *Cochran* et al. [3.79] have carried through the calculation of frequencies and BLS intensities for pairs of ultrathin coupled films in which the magnetization in each film is permitted to rotate in the plane so as to take up the minimum energy configuration.

3.2.7 Examples

The space available for this article is too limited to be able to provide a review of BLS results for the ultrathin systems which have been measured to date; a list of references is provided in the Appendix. Two examples will have to suffice.

3.2.7.1 Single Films of fcc Fe(001) Grown on Cu(001)

In this example a 3 monolayer (ML) thick fcc iron film has been grown on the (001) surface of a copper single crystal following the procedures described by *Steigerwald* et al. [3.80]. The iron film was covered by 60 ML of epitaxial Cu before it was removed from the vacuum system. The results of BLS measurements for an incident laser power of 100 mW at the specimen are shown in Fig. 3.19. A standard Sandercock interferometer operating in the four pass plus two pass configuration was used to collect the data [3.81]; the scattered light was collected using a f-2 lens. The large off-scale peaks shown in Fig. 3.19 are Fabry–Perot resonances associated with the scattered light having the frequency of the unshifted incident $0.5145\ \mu m$ laser light. These peaks are spaced the equivalent of 30 GHz apart corresponding to a spacing of 5.00 mm between the interferometer plates. The smaller peaks displayed in Fig. 3.19 correspond to light which has been shifted up or down in frequency by 17.8 GHz from the incident frequency: these are due to the presence of thermally excited spin waves in the 3 ML thick iron film.

The frequency interval of 60 GHz illustrated in Fig. 3.19 was divided into 204 channels, and data was collected for a total of one second in each channel. The integrated strength of each spin wave peak shown in Fig. 3.19 was approximately 120 counts per second. The strength calculated from (3.63) using 1 ML = 1.81 Å, a coupling constant and a magnetization density appropriate for bcc Fe, an incident power of 100 mW, and the assumption that 1 count was registered for each 200 photons collected by the lens, is $\sim 10^4$ counts per second. The origin of the factor 10^2 discrepancy is unknown. (In (3.63) it is necessary to use $B = B_{eff} = H + 4\pi M_{eff}$ because a large surface anisotropy term acts to orient the magnetization perpendicular to the specimen plane.)

Data such as that shown in Fig. 3.19 was used to construct a plot of spin wave frequency versus applied magnetic field (Fig. 3.20). Obviously something interesting occurs at an applied field of 5.35 kOe. Thin fcc Fe(001) films exhibit a uniaxial anisotropy that acts so as to align the magnetization along the specimen normal. For applied fields less than 5.35 kOe the magnetization swings out of the plane; for an ideally homogeneous specimen with the applied

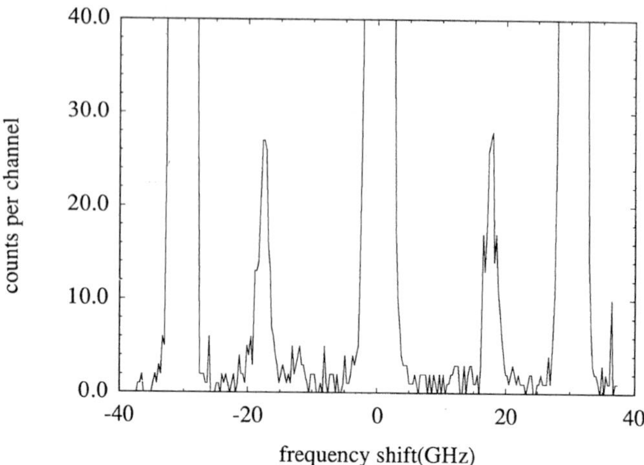

Fig. 3.19. The frequency spectrum observed for 0.5145 μm light incident at 45° and scattered from a 3 monolayer thick fcc Fe(0 0 1) film at 300 K. A 9.3 kOe magnetic field was applied parallel with the specimen plane; the incident light intensity was 100 mW. The iron film was grown on an fcc Cu(0 0 1) substrate, and it was covered by 60 monolayers of fcc Cu(0 0 1). The signals at ± 30 GHz are Rayleigh peaks due to a component of scattered light which is unshifted from the incident laser light frequency. The signals at ± 17.8 GHz are due to light which has been scattered by spin waves. The interval between -30 and $+30$ GHz was divided into 204 channels and data was collected in each channel for a total of 1 s. The scattered light was collected using an f-2 lens. The instrumental full width at half maximum is approximately eight channels; there are therefore ~ 120 counts in each peak

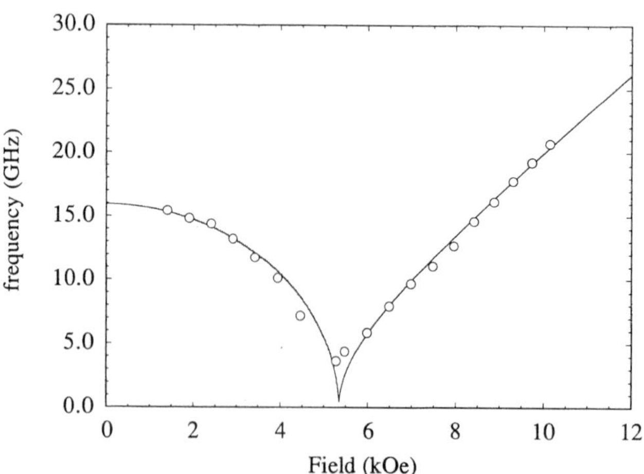

Fig. 3.20. Spin wave frequency vs. in-plane applied magnetic field for the 3 monolayer thick fcc Fe(0 0 1) film of Fig. 3.19. At a critical field of 5.35 kOe the magnetization swings out of the specimen plane because of a magnetic uniaxial anisotropy in which the easy axis is normal to the specimen surface. The solid line was calculated according to the description given by *Dutcher* et al. [3.87] using $4\pi M_{\mathrm{eff}} = -5.35$ kOe, $g = 2.09$, and $H_2 = 4K_{\mathrm{u}}^{(2)}/M_{\mathrm{s}} = 0.1$ kOe

field accurately aligned in the specimen plane the frequency should approach a very small value which depends upon the strength of the exchange parameter, A [3.82]. This topic is also covered in *Mills*, Chap. 3, Vol. 1, and in *Erickson* and *Mills* [3.58]. The intensity of the spin wave scattered light becomes relatively large for applied fields near the critical value of 5.35 kOe [3.83]. This increase can be understood on the basis of the simple theory which led to (3.63): at the critical field the restoring torque on the magnetization becomes small, and therefore the mean square thermally excited magnetization amplitude becomes large.

3.2.7.2 Ultrathin Bilayers of bcc Fe(0 0 1) Grown on Ag(0 0 1)

Two bcc iron films were grown on a single crystal Ag(0 0 1) template; the iron layers were 9.4 ML and 16 ML thick and they were separated by 9 ML of bcc Cu(0 0 1) plus 1 ML of Ag(0 0 1). This bilayer sandwich was covered by 20 ML of Au(0 0 1) in order to protect the iron layers when the specimen was removed from the vacuum system. The results of BLS measurements for an in-plane applied magnetic field of 10 kOe and for an incident light intensity at the

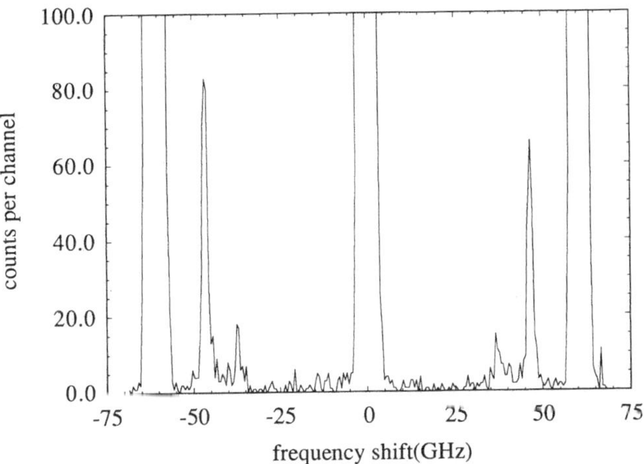

Fig. 3.21. The frequency spectrum observed for 0.5145 μm light incident at 45° and scattered at 300 K from a sandwich composed of two fcc Fe(0 0 1) layers, 9.4 and 16 monolayers thick, separated by a spacer layer composed of 9 monolayers of bcc Cu(0 0 1) and 1 monolayer of Ag(0 0 1). This sandwich was grown on a Ag(0 0 1) template and was covered by 20 monolayers of Au(0 0 1). A 9.9 kOe magnetic field was applied parallel with the specimen plane and along an easy (1 0 0) direction; the intensity of the incident light was 120 mW. The signals at ± 60 GHz are Rayleigh peaks due to a component of scattered light which is unshifted from the incident laser light frequency. The signals at ± 37.5 and at ± 47.2 GHz are due to light which has been scattered by spin waves. The interval between -60 and $+60$ GHz was divided into 221 channels and data was collected in each channel for a total of 1 s. The scattered light was collected using an f-2 lens. The instrumental full width at half maximum is approximately four channels

specimen of 120 mW are shown in Fig. 3.21. Two upshifted spin wave peaks and two downshifted spin wave peaks are clearly visible. The more intense high frequency peaks correspond to the acoustic spin wave mode in which the magnetizations in the two films precess in phase. The weak low frequency peaks correspond to the optical mode in which the two magnetizations precess in anti-phase. The magnetizations in the two films are clearly coupled; the intensities of the two peaks should be approximately the same in the limit of no coupling [3.79]. The observed intensity ratio, $\sim 5:1$, corresponds to a coupling strength $A_{AB} \sim -0.1$ erg cm^{-2} according to the calculations described in [3.79]. The coupling is anti-ferromagnetic since the acoustic mode frequency is larger than the optical mode frequency.

Data such as that illustrated in Fig. 3.21 can be used to construct plots of frequency versus applied field (Fig. 3.22). The diagram of frequency versus magnetic field clearly exhibits a cusp-like feature for fields near 1 kOe. It corresponds to a field at which the antiferromagnetic coupling between the two films has overcome the external field and the two magnetizations have swung away from the applied field direction. (The field was applied along the (1 0 0) direction for the data shown in Fig. 3.22.) The data can be compared with theory [3.79] to deduce a value $A_{AB} = -0.20$ erg cm^{-2} for the strength of the exchange coupling between the two iron films. This value is in good agreement with that deduced from FMR measurements.

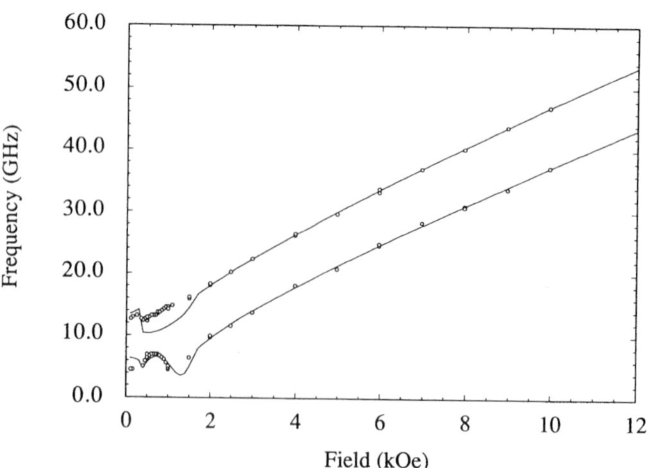

Fig. 3.22. Spin wave frequencies versus in-plane magnetic field for the bilayer specimen of Fig. 3.21. The magnetic field was applied along a (1 0 0) direction (easy axis). At a field of approximately 1.8 kOe the magnetizations in the iron films begin to swing away from the applied field direction because of the antiferromagnetic coupling between them. The solid lines were calculated according to the description given by *Cochran* et al. [3.79] using $A_{AB} = -0.20$ erg cm^{-2}, $d_A = 13.4$ Å, $d_B = 22.9$ Å, $4\pi M_{eff/A} = 8.28$ kOe, $4\pi M_{eff/B} = 16.11$ kOe, $g_A = g_B = 2.08$ and in-plane anisotropies $K_{1A} = 2.89 \times 10^5$ and $K_{1B} = 3.33 \times 10^5$ ergs cm^{-3}. No magnetic dipole correction corresponding to an in-plane spatial variation of the magnetization wave was applied

3.2.8 Conclusions

BLS measurements are complimentary to FMR measurements because they can be easily carried out over a broad frequency range, unlike FMR which is essentially a fixed frequency experiment. However, the precision of a given frequency measurement is an order of magnitude less than that for an FMR measurement, and that makes it very difficult to measure spin wave lifetimes using BLS. Light scattering experiments can be carried out using relatively small specimens since the incident light is focussed to a spot that is typically 20 μm in diameter; in order to obtain equivalent signal-to-noise using FMR the specimens must be appropriately 5 mm in diameter. BLS also offers the possibility of carrying out magnetic measurements without having to remove the specimen from the vacuum chamber [3.84]. For that purpose it is very useful to have available an instrument that can operate over a broad frequency range since it is very difficult to provide for a magnetic field in the ultrahigh vacuum chamber which is much larger than 2–3 kOe.

Acknowledgements. The author would like to thank his colleagues B. Heinrich, M. From, Z. Celinski, and K. Myrtle for the stimulating interactions which have been very important for the formulation of the matter contained in this article.

Appendix

Herewith follows a list of ultrathin magnetic film systems that have been investigated using Brillouin light scattering, and relevant references. The first metal listed is the bulk metal template on which the ultrathin films were grown. An ultrathin magnetic film has been defined as a film thinner than 30 monolayers (see Chap. 1).

Ag(0 0 1)/Fe(0 0 1)/Au(0 0 1) [3.85].

Ag(0 0 1)/Fe(0 0 1)/Cu(0 0 1)/Fe(0 0 1)/Au(0 0 1) [3.79, 86].

Cu(0 0 1)/Fe(0 0 1)/Cu(0 0 1) [3.83, 87, 88].

Ag(0 0 1)/Fe(0 0 1)/Pd(0 0 1)/Fe(0 0 1)/Au(0 0 1) [3.89, 90].

Cu(0 0 1)/Co(0 0 1)/Cu(0 0 1) [3.91].

W(1 1 0)/Fe((1 1 0)/Vacuum [3.84, 92].

Pd(1 1 1)/Fe(1 1 0)/W(1 1 0) [3.93].

Co(0 0 0 1)/W(1 1 0) [3.93].

Co(0 0 0 1)/Pd(1 1 1)/W(1 1 0) [3.93].

3.3 Brillouin Light Scattering in Magnetic Superlattices

B. HILLEBRANDS and G. GÜNTHERODT

The magnetism in superlattices, constructed from a large number of alternating magnetic and nonmagnetic layers, is not only an extension of thin film magnetism in the sense that for superlattices many thin film properties, e.g., interface anisotropies, are multiplied by the large number of magnetic layers. A new interaction is introduced by the coupling between the individual magnetic layers. This interaction may be in nature of a dipolar type as well as in particular cases of an exchange type. The magnetic properties of superlattices will therefore depend on the properties of the individual layers as well as on the details of the interlayer coupling.

Of great importance is the fact that the type of interlayer coupling as well as its sign and strength can be tailored by the appropriate choice of the spacer material and its layer thickness. For instance, the discovery of the existence of antiferromagnetic interlayer coupling between ferromagnetic layers has boosted the discovery and investigation of a large variety of new magnetic structures in superlattices. New, large research fields have become available with important areas of applications, particularly in magneto-optic recording media and magnetoresistive field sensors. Perhaps the most important consequence of these interlayer coupling mechanisms is the formation of so-called collective spin wave excitations due to the stacking periodicity (modulation wavelength), which are coherent throughout all layers of the superlattice structure. These excitations are unique to magnetic superlattice structures and not known from bulk materials.

3.3.1 Introduction

This section is intended to provide an introduction into the physics of collective spin wave excitations in superlattices and their applications to the determination of material properties. The experimental method of choice is Brillouin light scattering, and therefore much attention is paid to illustrate the basic principles with Brillouin light scattering experiments. The Brillouin light scattering method provides information on (i) the magnetic properties of magnetic layers, such as saturation magnetization and anisotropies, (ii) on the magnetic ground state of the coupled layer system, i.e. the magnetization orientation distribution, (iii) on the details of the coupling between magnetic layers (dipole or exchange), and (iv) due to the finite spin-wave wavelength, on inhomogeneities of internal fields, induced by, e.g., lateral variations in layer thicknesses or anisotropies. Due to the limited space we will restrict our attention to superlattice structures constructed from periodically layered alternating magnetic and nonmagnetic materials of many repetitions. The very interesting "building blocks" of superlattice structures, such as magnetic sandwich structures and triple layers, can only be considered very briefly.

This section is organized as follows: In Sect. 3.3.1, a short outline of the underlying theory of spin wave excitations is given, as far as it exceeds the thin film properties discussed in Sect. 3.2 by *Cochran*. Section 3.3.2 demonstrates the influence of dipolar interlayer coupling in Co/Pd superlattices. In Sect. 3.3.3, the particular properties of interlayer exchange coupling are discussed with respect to exchange coupled collective spin wave excitations as well as with a special regard to antiferromagnetically coupled layers in the case of Co/Ru superlattices. The characterization of spatial variations of anisotropies and layer thicknesses is presented in Sect. 3.3.4 for the case of Co/Pt superlattices. In the conclusion the results are summarized and an outlook is given. Due to space limitations the theory of the Brillouin light scattering cross section in superlattices could not be discussed, although some remarks are added in the conclusion. Also for instrumentation the reader is referred to [3.94, 95].

3.3.2 Theoretical Background

Although a large body of experimental and theoretical work in Brillouin light scattering exists for bulk magnetic materials and for single magnetic layers (also discussed in Sect. 3.2), the field of spin wave excitations in superlattices is still developing. The first calculations, restricted to the dipolar limit, were reported by *Camley* et al. [3.96], *Grünberg* and *Mika* [3.97], *Emtage* and *Daniel* [3.98], and, including volume anisotropy contributions, by *Rupp* et al. [3.99]. Exchange modes in multilayers have been considered by *van Stapele* et al. [3.100], *Dobrzynski* et al. [3.101], *Albuquerque* et al. [3.102], *Hinchey* and *Mills* [3.103], *Vayhinger* and *Kronmüller* [3.104, 105], and *Barnaś* [3.106–108]. A first inclusion of interface anisotropies as well as exchange interlayer coupling was performed by *Hillebrands* [3.109, 110]. Large out-of-plane anisotropies resulting in perpendicularly magnetized superlattices were included into the spin wave calculations by *Stamps* and *Hillebrands* [3.111–113].

In the following we use a model which in general contains all salient features of spin waves in superlattices including exchange contributions and anisotropy. We use a continuum model approach first used for single layers by *Rado* and *Hicken* [3.114], *Cochran* and *Dutcher* [3.115], and applied to superlattices by *Hillebrands* [3.109, 110] and *Stamps* and *Hillebrands* [3.111–113]. The full equivalence to a microscopic model starting from the spin Hamiltonian has recently been demonstrated by *Stamps* and *Hillebrands* [3.116].

The coordinate system used is shown in Fig. 3.23. The x-axis is perpendicular to the magnetic layers. For a N-layer system the positions of the interfaces are defined by d_n, $n = 1 \ldots N$, such that for the nth layer the interfaces lie at $x = d_{n-1}$ and $x = d_n$. We use the index "n" to indicate parameters of the nth layer, but when appropriate this index is omitted for better clarity of the formulae. We define the angles θ and ϕ as the angles between the direction of the

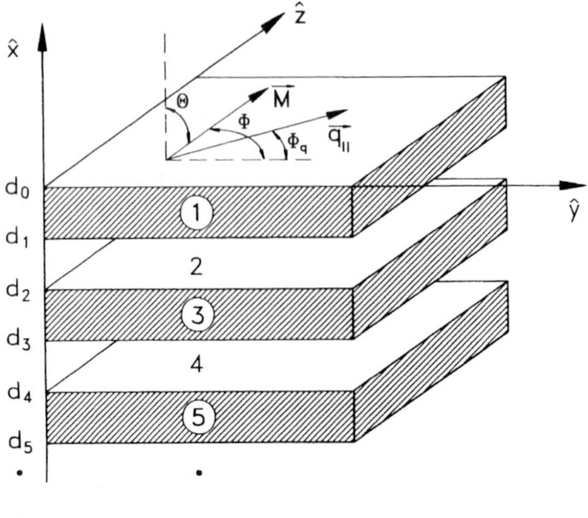

Fig. 3.23. Coordinate system used for calculating spin wave frequencies

magnetization M and, respectively, the surface normal, and a crystallographic reference direction within the film plane, which is normally $[1\,0\,0]$ (dashed line in Fig. 3.23). The direction of the spin wave propagation, defined by the mode's wave vector component parallel to the film plane, $q_{\parallel}$, is within the (y, z) plane. Its angle with the reference direction is ϕ_q.

We begin with the full Landau–Lifshitz torque equation of motion:

$$\frac{1}{\gamma}\frac{\partial M}{\partial t} = M \times H_{\text{eff}}, \tag{3.91}$$

where $\gamma = \gamma_e \cdot g/2$ is the gyromagnetic ratio, $\gamma_e = 1.759 \times 10^7\,\text{Hz/Oe}$ is the value of γ for the free electron and g is the spectroscopic splitting factor. The effective magnetic field acting on the magnetization, H_{eff}, is given by

$$H_{\text{eff}} = H - \frac{1}{M}\nabla_{\alpha_M} E_{\text{ani}} + \frac{2A}{M^2}\nabla^2 M. \tag{3.92}$$

E_{ani} is the usually defined volume anisotropy energy density and A is the exchange stiffness constant. We have omitted the layer index, n, for clarity. The first term on the right hand side is the external applied field including fluctuating fields generated by the precessing spins. The second term is an effective field due to magnetic anisotropies with ∇_{α_M} the gradient operator for which the differentiation variables are the components of the unit vector α pointing into the direction of M. The last term is the exchange field due to volume exchange interaction.

Also, the magnetostatic Maxwell equations have to be fulfilled:

$$\nabla \times \boldsymbol{H} = 0, \tag{3.93}$$

$$\nabla \cdot (\boldsymbol{H} + 4\pi \boldsymbol{M}) = 0. \tag{3.94}$$

From the equation of motion (3.91) and the Maxwell equations (3.93, 94), boundary conditions are derived. At each interface the parallel component of $\boldsymbol{H}$ and the perpendicular component of $\boldsymbol{H} + 4\pi \boldsymbol{M}$ have to be continuous. From the equation of motion (3.91) we obtain the condition that the sum of the interface torques must be zero for each interface. The general boundary condition at the $(x = d_n)$-interface is given by the so-called *Hoffman* boundary condition, which includes exchange coupling to the interface of the next magnetic layer at $x = d_n$, [3.117, 118]:

$$\boldsymbol{M}_n \times \left[\frac{1}{M_n} \nabla_{\alpha_{M_n}} E_{\text{inter},n} - \frac{2A_n}{M_n^2} \frac{\partial \boldsymbol{M}_n}{\partial n_n} \right]\Bigg|_{x=d_n}$$

$$- \boldsymbol{M}_n \times \frac{2A_{nn'}}{M_n M_{n'}} \left[\boldsymbol{M}_{n'} + a_{n'} \frac{\partial \boldsymbol{M}_{n'}}{\partial n_{n'}} \right]\Bigg|_{x=d_{n'}} = 0, \tag{3.95}$$

where E_{inter} is the interface anisotropy energy. $\partial/\partial n$ is the partial derivative with respect to the surface normal unit vector, $\boldsymbol{n}$. The latter points from the interface into the corresponding magnetic layer. The interlayer exchange constant between layers n and n' is $A_{nn'}$. Without loss of generality we call this parameter A_{12} in the following. The lattice constant is denoted by a_n. The first term in (3.98) is the so-called *Rado–Weertman* boundary condition, describing the surface torque of a single magnetic film due to anisotropies and exchange [3.119], whereas the second term describes the exchange coupling between the two layers [3.120]. We need to discuss two limiting cases of the interlayer exchange constant, A_{12}. For $A_{12} = 0$ (3.98) resembles the *Rado–Weertman* boundary conditions, i.e., the interface torque must be separately zero for $x = d_n$ and $x = d_{n'}$. For large absolute values of A_{12}, i.e., $|A_{12}| \approx A_n/a$, we obtain $\boldsymbol{M}_n \times \boldsymbol{M}_{n'} = 0$, i.e., $\boldsymbol{M}_n$ and $\boldsymbol{M}_{n'}$, are aligned either parallel or antiparallel, depending on the sign of A_{12}.

The interface anisotropy energy, E_{inter}, is expressed in lowest order by

$$E_{\text{inter}} = - K_{\text{u}}^{\text{s}} \cos^2 \theta + K_{\text{p}}^{\text{s}} \sin^2 \theta \cos^2 \phi, \tag{3.96}$$

where K_{u}^{s} is the out-of-plane interface anisotropy constant and K_{p}^{s} is the in-plane interface anisotropy constant, respectively. A positive sign of K_{u}^{s} corresponds to the surface normal being an easy axis.

We now turn to the calculation of the spin wave frequencies, which is fully analogous to the procedure for thin films described by *Cochran* in Sect. 3.2. We assume that the fluctuations in $\boldsymbol{M}(t)$ and $\boldsymbol{H}(t)$ associated with the spin waves are small compared to the static values. This condition is almost always fulfilled for thermally driven spin waves at temperatures considerably less than T_c [3.121]. We split $\boldsymbol{M}(t)$ and $\boldsymbol{H}(t)$ into time independent static parts $\boldsymbol{M}_s$ and $\boldsymbol{H}$ and dynamic parts $m(t)$ and $h(t)$:

$$M(t) = M_s + m(t), \qquad |m(t)| \ll |M_s|, \tag{3.97}$$

$$H(t) = H + h(t), \qquad |h(t)| \ll |H|. \tag{3.98}$$

In principle we have to find the static equilibrium orientations of the magnetizations for the layered system before calculating the spin wave frequencies. Due to interface anisotropies and exchange coupling effects the static equilibrium direction might differ from the bulk direction, in particular for $A_{12} < 0$, i.e., for antiferromagnetic interlayer coupling. The direction of magnetization can be obtained by solving the equations of motion (3.91) and the magnetostatic Maxwell equations (3.93, 94) together with the boundary conditions (3.93–95) for time independent M and H. It should be noted that in the general case the direction of the magnetization is a function of the position in each magnetic layer. Once we have solved the static problem all time independent terms contained in (3.91–95) cancel to zero. The further calculations are straightforward as is shown for the single film case in Sect. 3.2, albeit algebraically and numerically extensive. The reader is referred to [3.109–113] for full details [3.122]. Since there are six partial solutions for $m(t)$ and $h(t)$ for each magnetic layer and two for each nonmagnetic layer, the boundary condition determinant increases in dimension by eight for each bilayer within the superlattice stack. However, the numerical expenses can be reduced in various ways: (i) by dropping exchange energy contributions, the dimensional increase of the boundary condition determinant is reduced to four and the numerical problem can be solved using efficient numerical tools to evaluate the resulting band matrices, or by using a transfer matrix method; (ii) for systems with modes of not too large exchange energy contributions, i.e. for modes with internal fields varying only slowly from layer to layer, effective-medium models have been proposed, in which parameters are averaged across a suitably defined unit cell (e.g., a double layer within the superlattice) [3.123, 124]. So-called effective susceptibilities are calculated including contributions from interface anisotropies and interlayer coupling and the superlattice structure is treated as a homogeneous magnetic film of effective, renormalized parameters.

3.3.3 Dipolar Coupled Collective Spin Waves

In this section, we will consider collective spin wave excitations formed by dipolar coupling of modes of the individual magnetic layers.

3.3.3.1 The Coupling Scheme

Before we present experimental data let us first discuss the basic mechanisms of collective spin wave excitations in superlattices as it follows from model calcu-

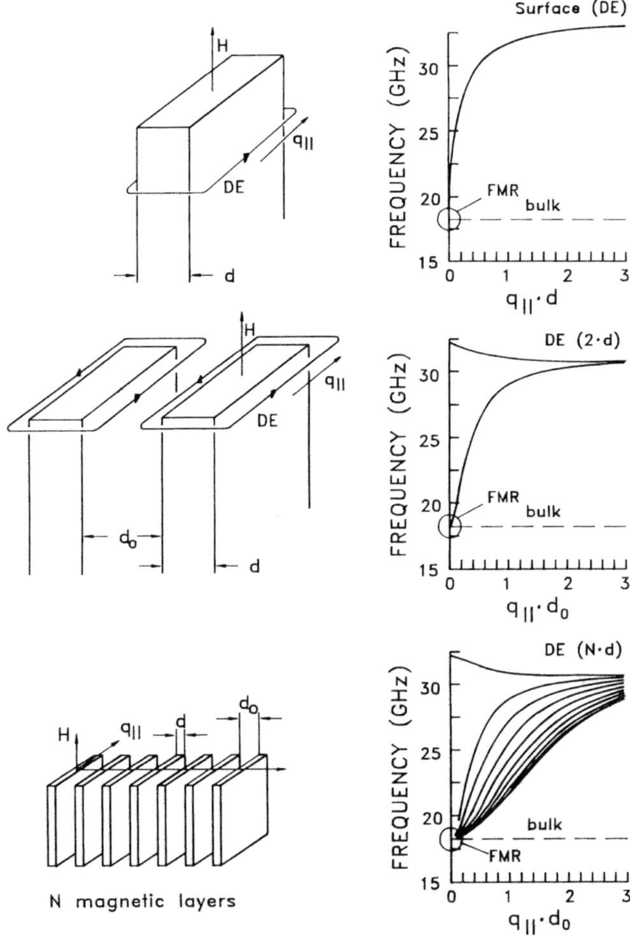

Fig. 3.24. Top: The Damon–Eshbach mode in a single layer. Middle: Coupling scheme of dipolar spin waves in a double layer. Bottom: Coupling scheme of dipolar spin waves in a multilayer. The magnetic and spacer layer thicknesses are denoted by d and d_0, respectively

lations. In Fig. 3.24 a sketch of interacting spin waves in superlattices is displayed. We will restrict our attention to the simple case that only dipolar interactions are considered and that any anisotropy contributions apart from the shape anisotropy are zero. Consider first a single magnetic layer with a magnetic field applied parallel to the layer as displayed in the upper part of Fig. 3.24. In the film there exists the so-called Damon–Eshbach mode, which propagates perpendicular to the applied field with a defined sense of revolution about the film. As discussed in Sect. 3.2, the frequency ω of this mode decreases with decreasing film thickness, d, and decreasing wave vector parallel to the surface, $q_\parallel$, as shown on the right hand side. In the limit of $qd \rightarrow 0$ the mode frequency approaches that of the so-called uniform mode, characterized by a constant precession phase throughout the entire film, as it can be measured by ferromagnetic resonance (FMR). If there are now two magnetic films of the same thickness at a distance, d_0, small enough to couple the films via their dipolar stray fields, the two Damon–Eshbach modes, each existing on each film, will change in frequency due to the coupling. The frequency splitting increases with increasing coupling, i.e. with decreasing spacer thickness, d_0. In the limit of vanishing d_0, the frequency of one mode increases to the frequency of a film of thickness $2d$ and that of the other mode decreases to the frequency of the uniform mode. Now in the case of a superlattice consisting of N magnetic layers separated by $N - 1$ spacer layers (Fig. 3.24, bottom) the frequency degeneration of the N Damon–Eshbach modes is lifted, and, in the case of large N, a band of so-called collective spin wave modes is formed. Out of the N modes of the band, one mode (highest frequency mode in Fig. 3.24 bottom) is characterized as a surface mode of the total multilayer stack, with the mode energy (precession amplitude) localized near the surface of the stack, and travelling about the total stack with a well defined sense of revolution. The remaining modes have, depending on the wavevector component perpendicular to the stack, both surface-mode- and bulk-mode-like characteristics to a greater or lesser degree.

We now demonstrate the properties of the collective spin wave band with some sample spectra of Fe/Pd superlattices [3.125]. The samples were prepared on single crystal sapphire substrates using a radio frequency (rf) sputtering technique [3.126–128]. In Figs. 3.25, 26 typical Brillouin spectra of spin wave excitations in Fe/Pd superlattices are displayed. The measured scattering intensities are plotted as a function of frequency shift, v, with respect to the laser frequency. The central peak near $v = 0$ is due to elastically scattered laser light. The magnetic field applied parallel to the layers is 1 kG. Figure 3.25a shows the spectrum of a Fe/Pd superlattice with a stacking periodicity of $\lambda_{\mathrm{SL}} = d + d_0 = 46.2$ Å. The thickness $t = 21.9$ Å of the magnetic material is close to that of the

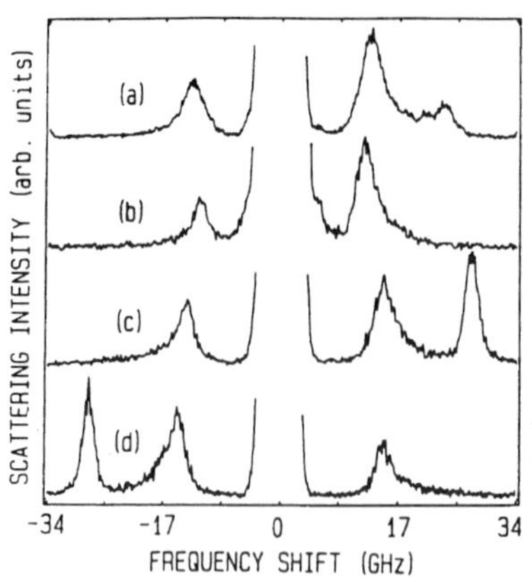

Fig. 3.25. Room temperature Brillouin spectra of Fe/Pd superlattices in an applied magnetic field of 1 kG: (**a**) $d = 21.9$ Å and $d_0 = 24.3$ Å, (**b**) $d = 41.7$ Å and $d_0 = 138.7$ Å, (**c**), (**d**) $d = 41.0$ Å and $d_0 = 9.1$ Å. In (**d**) the direction of the applied field has been reversed compared to (**c**). The magnetic and spacer layer thicknesses are denoted by d and d_0, respectively

spacer material $d_0 = 24.3$ Å. The band of collective spin wave excitations can clearly be identified in the right hand part of the spectrum by its specific asymmetric shape: the density of states is largest at small frequency shifts and decreases asymmetrically toward the upper band edge. At the latter a few discrete spin wave modes can still be resolved due to the small thickness and the still finite number of bilayers, which is 90 [3.127, 128]. The large Stokes/anti-Stokes asymmetry identifies them as surface-mode-like spin waves. On the other hand, the modes near the lower edge of the spin wave band are found to be bulk-mode-like from the small Stokes/anti-Stokes asymmetry. If we neglect the discrete modes near the upper band edge the shape of the spin wave excitation band is qualitatively very similar to the calculated Brillouin scattering cross section for the semi-infinite superlattice system Mo/Ni [3.96].

In Fig. 3.25b we show the Brillouin spectrum of an Fe/Pd superlattice with $d = 41.7$ Å and a much larger spacer thickness $d_0 = 138.7$ Å. In this case the spin wave band becomes narrower due to the reduced coupling across the spacer layers. A very different spectrum is found for the case of d_0 (9.1 Å) much smaller than d (41.1 Å), as shown in Fig. 3.25c. Here a very intense discrete mode is found near 27.7 GHz in the anti-Stokes spectrum apart from the band of collective modes near ± 15 GHz. This superlattice surface spin wave mode, which travels about the total superlattice stack, is allowed to exist besides the collective spin wave band. It would merge with the latter for $d = d_0$ [3.129].

The effect of inverting the direction of the applied magnetic field is demonstrated in Fig. 3.25d. Since the direction of the applied field defines the sense of revolution of each surface spin wave mode about each magnetic layer, an inverted field causes the Stokes and anti-Stokes parts of the spectrum to be exchanged. Figure 3.26 shows Brillouin spectra of an Fe/Pd superlattice with

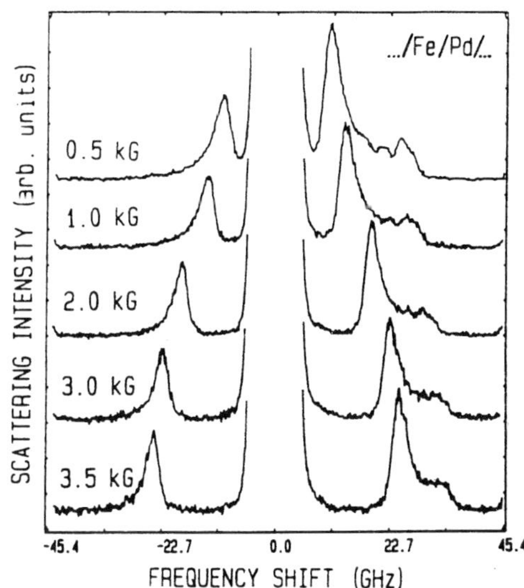

Fig. 3.26. Room temperature Brillouin spectra of a Fe/Pd superlattice consisting of 49 bilayers with $d = 89.4$ Å and $d_0 = 99.0$ Å for different applied magnetic fields, as indicated in the figure

$d = 89.4$ Å and $d_0 = 99.0$ Å for different applied magnetic fields. With increasing field the spin wave frequencies increase in a quasi-linear fashion, accompanied by a slight bandwidth narrowing.

3.3.3.2 Spin Waves in Systems with Large Perpendicular Anisotropies

Of particular technological interest are superlattice structures with anisotropies large enough to compensate for the shape anisotropy and to turn the direction of magnetization out of plane. This is achieved by choosing systems with a large perpendicular interface anisotropy constant, K_u^s, and with a small magnetic layer thickness, d, such that the effective anisotropy contribution, $K_{eff} = K_v + 2K_u^s/d$, is larger than the shape anisotropy contribution, with K_v being the volume anisotropy contribution of the same symmetry axis.

We will consider the case where the external magnetic field is applied parallel to the layers. With increasing field strength the direction of magnetization M is increasingly tilted into the layer planes until a critical field strength, $H_{crit} = 2(K_{eff} - 2\pi M_s^2)/M_s$, is reached, above which the directions of magnetization and external field are co-linear.

The spin wave frequencies are very dependent on the out-of-plane angle θ between M and the x axis (stacking axis) [3.112]. With increasing angle θ caused by an increasing applied field the spin wave frequencies decrease and some modes may even go soft in the vicinity of the critical field strength, H_{crit}. For $H > H_{crit}$, i.e. in the regime of H parallel to M, the spin wave frequencies increase quasi-linearly with further increasing field. In Fig. 3.27 we show the calculated spin wave frequencies as a function of the applied field for a six-bilayer stack. The magnetic parameters were taken from experiments on Co/Pt superlattices [3.130], and are $A = 2.85 \times 10^{-6}$ erg/cm, $4\pi M_s = 14.5$ kG, and $g = 2$. The hcp-Co bulk anisotropy constants K_1 and K_2 are both set to zero for simplicity. The thickness of the Co layers is 8.8 Å and of the spacer layers is

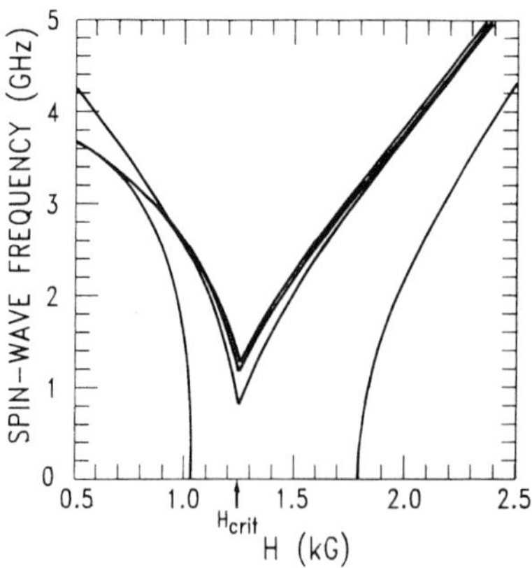

Fig. 3.27. Frequencies of spin wave modes for a six-bilayer stack as a function of the in-plane applied field H. The parameters are appropriate for Co and are given in the text. The Co layers are 8.8 Å thick and the spacer layers are 7.6 Å thick

7.6 Å. The interface anisotropy constant is chosen to be $K_u^s = 0.4\,\text{erg/cm}^2$ resulting in $H_{crit} = 1.26\,\text{kG}$.

For applied fields less than ≈ 500 G, the magnetization is almost completely normal to the film plane. In this case a dipolar-dominated surface mode cannot exist in a single thin film or in a superlattice, and the modes are mostly exchange-type in character [3.131]. We will only consider modes of dominantly dipolar character, which exist for $H \gtrsim 500$ G. For fields larger than H_{crit} the magnetization is forced into the layer planes and the spectrum consists of five nearly degenerate bulk-like modes and a lower frequency surface mode. The fact that the surface mode lies below the bulk-like modes is an indicator for large perpendicular anisotropies. For fields below H_{crit} the magnetization has an out-of-plane component and the surface mode appears to cross through and rise above the bulk band as the field is lowered. Near H_{crit} the bulk modes take their minimum values while the surface mode goes completely soft. Such a softening can often be associated with a surface magnetic phase transition in the spin structure [3.132, 133]. In the example of Fig. 3.27 there is a strong possibility for the direction of magnetization to vary across the stack in order to minimize the net demagnetization energy of the structure.

3.3.4 Interlayer-Exchange Coupled Collective Spin Waves

So far only dipolar interactions between magnetic layers within the superlattice stack have been considered. In this section we will now discuss the additional influence of interlayer exchange interactions on the spin wave properties.

3.3.4.1 Exchange Dominated Collective Spin Waves

A considerable influence of interlayer exchange interaction on the spin wave frequencies exists if the spacer layers are thin enough ($\lesssim 10$ Å), since the interaction decays rather rapidly with increasing spacer thickness. For magnetic sandwich structures interlayer exchange interaction has been extensively studied by *Grünberg* and coworkers with Brillouin light scattering [3.134–136], *Heinrich* et al. [3.137] and *Cochran* and *Dutcher* [3.138]. Depending on the spacer material the interlayer coupling is ferro- or antiferromagnetic or it even oscillates as a function of the spacer thickness, as demonstrated below in this section.

In the presence of interlayer exchange coupling in superlattices, all but the stack surface mode of the dipolar collective modes are converted into exchange modes, which in the full coupling limit become the so-called standing spin waves of the total superlattice stack [3.109, 110]. This new type of collective exchange-dominated modes was predicted by the model outlined in Sect. 3.3.2 [3.109, 110]. Figure 3.28 shows the calculated frequency dependence of the modes for Co/Pd multilayers of nine periods as a function of the individual layer thickness, assuming that the thickness of all magnetic (d_{Co}) and nonmagnetic (d_{Pd})

layers are the same. For this calculation the parameters of Co listed in the figure caption have been used, which were obtained by Brillouin light scattering and Superconducting Quantum Interference Device (SQUID) magnetometry measurements on Co/Pd multilayer samples prepared with the same specifications [3.139, 140]. For $d_{Co} = d_{Pd} > 70$ Å dipolar collective modes are seen to exist in the frequency region between 22 and 28 GHz (Fig. 3.28). The stack surface mode is well-separated from the remaining eight modes, which form a narrow band of collective excitations. For $d_{Co} = d_{Pd} > 130$ Å the first standing spin wave, which is an exchange mode of each single layer, is obtained (decreasing from 100 to 59 GHz in Fig. 3.28) with its characteristic $1/d_{Co}^2$ dependence. For thinner layers, $d_{Co} = d_{Pd} < 50$ Å, all collective dipolar modes except the highest frequency dipolar mode (stack surface mode) increase in frequency and cross the stack surface mode due to the onset of the interlayer exchange interaction.

An analysis of the mode properties shows that these modes are dominated by the exchange energy. In order to model the interlayer exchange coupling strength as a function of the Pd spacer layer thickness, d_{Pd}, we assume in Fig. 3.28 that A_{12} decreases exponentially with increasing d_{Pd}. That is $A_{12}(d_{Pd}) = A_{12}^0 \exp(-d_{Pd}/d_C)$ with $A_{12}^0 = 10$ erg/cm^2 and the decay constant $d_C = 10$ Å. In the crossing regime the modes are hybridized, exhibiting a very small mode repulsion which can only be resolved on the scale of Fig. 3.28 for the crossing of the stack surface mode and the lowest-frequency bulk mode.

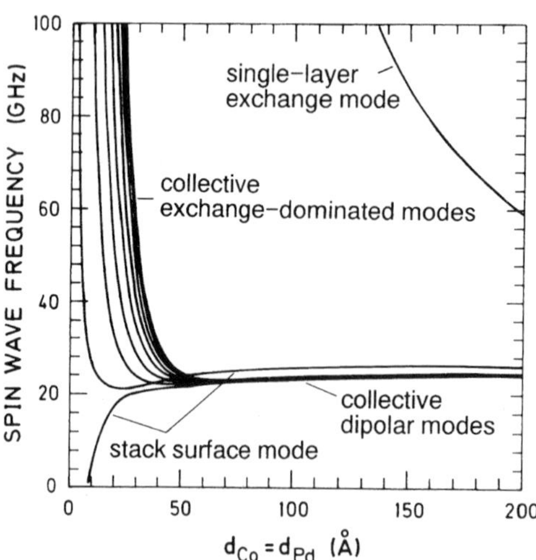

Fig. 3.28. Calculated spin wave excitations in a superlattice of nine periods. For the magnetic layer the parameters of sputtered Co films are used [3.139]: The saturation magnetization is $4\pi M_s = 14.5$ kG, the exchange constant is $A = 2.85 \times 10^{-6}$ erg/cm and the g-factor is $g = 2.03$. For the sum of the two volume anisotropy constants of hcp symmetry, K_1 and K_2, we used the value $K_1 + 2K_2 = 3.05 \times 10^6$ erg/cm^3 and for the interface anisotropy constant we used $K_u^s = 0.4$ erg/cm^2

We now turn to the experimental demonstration of the numerical findings of the collective exchange modes [3.141, 142]. The sample series used were magnetically enhanced triode sputtered [1 1 1] c-axis textured Co/Pd superlattices of 30 bilayers. The samples were grown with the number of atomic layers, n, in each Co and Pd layer the same, and which varies for different samples between 1 and 32. The sample preparation and characterization is reported in [3.139].

Figure 3.29 shows the Brillouin light scattering spectra of a series of Co/Pd multilayer samples in an external magnetic field of $H_0 = 5$ kG, which is applied parallel to the layer planes and perpendicular to the scattering plane. The applied field of 5 kG is large enough to generate a single magnetic domain, with the direction of magnetization lying in-plane. The number of bilayers is $N = 30$. For the sample with thickest layers, $n = 32$, we obtain a spectrum typical for collective dipolar spin wave excitations. The mode character changes from surface-mode-like at the upper band edge (indicated by the open arrows in Fig. 3.29) to bulk-mode-like at the lower band edge (full arrows in Fig. 3.29) as determined by the characteristic Stokes/anti-Stokes asymmetry [3.125]. With

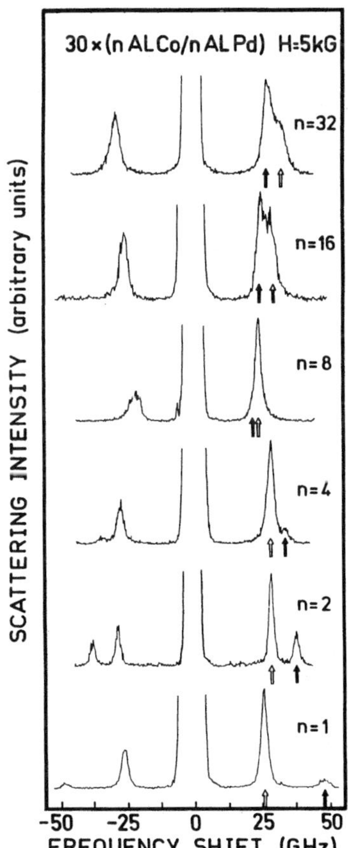

Fig. 3.29. Brillouin spectra of Co/Pd multilayers at $H = 5$ kG applied parallel to the layer planes and perpendicular to the scattering plane. The number of atomic layers per magnetic and nonmagnetic layer is indicated by n. The number of bilayers is $N = 30$. The open arrows denote modes which are predominantly surface-mode-like in character, the full arrows denote those which are mainly bulk-mode-like in character

decreasing layer thickness (decreasing n) the width of the band of collective excitations becomes smaller at first, until for $n = 4$ the observed bandwidth comes close to the experimental resolution. The surface mode then remains essentially unchanged in frequency as the thickness is decreased further. However, the frequencies of bulk modes move above the stack surface mode, and show a large increase in frequency with decreasing n. Here the bulk modes contain to a large degree interlayer exchange energy and thus are identified as collective exchange-dominated spin wave modes. From the frequency positions of these modes the interlayer coupling constant, A_{12}, can be determined [3.143].

3.3.4.2 Spin Waves in Antiferromagnetically Coupled Superlattices

Antiferromagnetic interlayer exchange coupling between ferromagnetic layers in sandwich and superlattice structures has recently become one of the most discussed phenomena in the magnetism of layered structures. Although the existence of antiferromagnetic coupling is already very surprising in itself, the observation of exchange coupling oscillating in strength and sign as a function of spacer layer thickness was even more exciting. *Grünberg* et al. [3.134] first discussed experimental evidence for antiferromagnetic interlayer exchange coupling in Fe/Cr/Fe sandwich structures, followed by the discovery of oscillatory interlayer exchange as a function of the spacer material thickness in different multilayered structures by *Parkin* et al. [3.144]. Since then many other systems with antiferromagnetic or oscillatory interlayer coupling have been found [3.145–152], recently even with two oscillation periods [3.153–158]. The interest in these phenomena was boosted by the discovery of the so-called "giant magnetoresistance" effect in the antiferromagnetic coupled regimes [3.159–161], making the effect a promising subject for, e.g., designing magnetoresistive pick-up heads for magnetic storage devices.

The interlayer exchange coupling constant, A_{12}, can be obtained in the antiferromagnetic regimes from magnetometry measurements of the saturation field [3.144]. For the ferromagnetic regimes so-called exchange-biased [3.161] and spin-engineered [3.162] layered structures were investigated. An easier access to A_{12}, both in the ferro- and the antiferromagnetic regime, is provided by Brillouin light scattering, as has been demonstrated for sandwich structures [3.110, 134–138, 158].

In this section we demonstrate the influence of antiferromagnetic as well as oscillating interlayer exchange coupling on the spin wave frequencies in magnetic superlattice structures for the case of sputtered Co/Ru superlattices [3.163]. For this system A_{12} oscillates as a function of the ruthenium thickness with a periodicity of 11.5 Å [3.144, 163]. Figure 3.30 shows four spectra of Co/Ru multilayers with a Co layer thickness of 20 Å and a Ru layer thickness of (a) 20.9 Å, (b) 15.2 Å, (c) 9.5 Å and (d) 3.8 Å measured with an applied magnetic field of 1 kG parallel to the layers. The peaks at ± 8.5 GHz correspond to the surface phonon (Rayleigh mode) of the system and are not further considered here. In all of the spectra we observe a band of collective spin wave

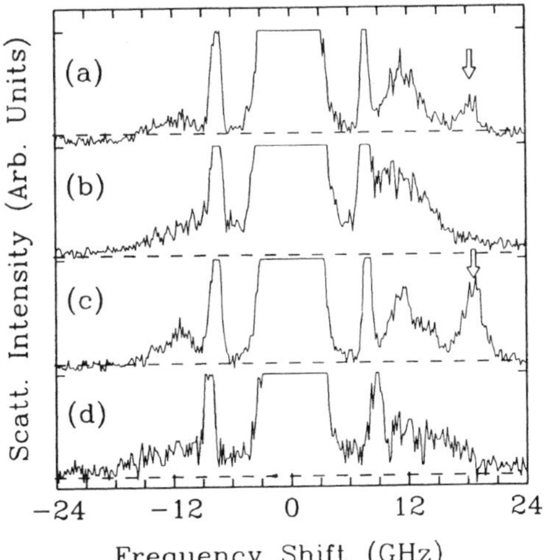

Fig. 3.30. Spin wave spectra of Co/Ru multilayers with a Co layer thickness of 20 Å and a Ru thickness of (a) 20.9 Å, (b) 15.2 Å, (c) 9.5 Å and (d) 3.8 Å. The magnetic field applied perpendicular to the spin wave propagation direction is 1 kG. The background due to the photomultiplier dark count rate is indicated by dashed lines. The stack surface mode is indicated by an arrow

excitations in the frequency range between 10 and 20 GHz. Near 19 GHz the stack surface mode (marked in Fig. 3.30 with an open arrow) is identified in Figs. 3.30a, c by its characteristic Stokes/anti-Stokes intensity asymmetry. This pronounced mode is only observable in the regimes of $d_{Co} = 10$–14 Å and 20–24 Å, which are identified as the ferromagnetic coupled regimes. Otherwise (cf. Figs. 3.30b, d) the mode is shifted to lower frequencies and merges with the other band modes.

In the upper part of Fig. 3.31 the frequency positions of the stack surface mode (squares) and the center of the bulk modes (circles) measured in an applied field of 3 kG, are plotted as a function of the Ru layer thickness. Oscillations with a period of 11.5 Å are well resolved. For comparison the spin wave frequencies have been calculated for the exchange uncoupled case ($A_{12} = 0$) using the model described in Sect. 3.3.2. They are shown as full lines in Fig. 3.31, upper part. The frequencies are adjusted to the experimental data of the stack surface mode in the ferromagnetic coupling regimes by choosing an appropriate value for the uniaxial perpendicular anisotropy constant of 4.7×10^6 erg/cm^3, which is a value typical for Co layers. The regimes of Ru thickness exhibiting reduced spin wave frequencies are identified as the antiferromagnetic coupling regimes as described further below. In particular for $d_{Ru} < 6$ Å a very large antiferromagnetic coupling is found by the large frequency decrease.

We will now discuss the spin wave properties in the antiferromagnetic coupling regimes. For not too large external fields the magnetizations of neighboring layers are canted with respect to each other. The canting angle ϕ_{12}, which is the angle between the directions of the saturation magnetizations in neighbored layers, depends on the (negative) value of A_{12} and the strength of the applied field. Calculations of the spin wave frequencies in this regime have been performed, which are based on an effective medium model [3.123, 124]. The total multilayer stack is treated as a ferromagnetic film with effective susceptibil-

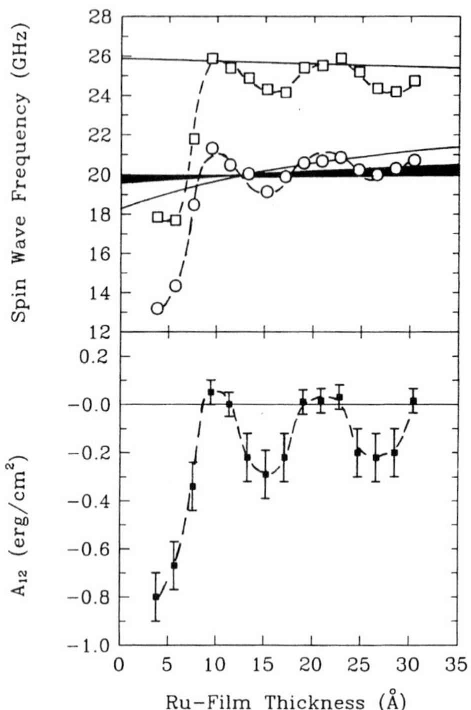

Fig. 3.31. Upper part: Spin wave frequencies of the stack surface mode (open squares) and the bulk modes (open circles) as a function of the Ru layer thickness of Co/Ru multilayers measured at an applied field of 3 kG. The Co layer thickness is 20 Å and the number of bilayers is $N = 20$. For comparison, the spin wave frequencies calculated for zero interlayer exchange coupling are shown as full lines. Lower part: Experimentally determined values of the interlayer exchange constant, A_{12}, as a function of Ru layer thickness

ities which include exchange coupling. The susceptibilities are calculated assuming that the electromagnetic fields vary only slightly across each bilayer period. Therefore this model allows one to calculate only the frequencies of the stack surface mode and the first few bulk modes, but these are the modes which contribute to the light scattering cross section. In Fig. 3.32 we have plotted the calculated spin wave frequencies as a function of the interlayer exchange constant, A_{12}. For comparison, the calculated canting angle of the saturation magnetization between neighboring magnetic layers is shown as a dashed line. For $A_{12} > -0.25$ erg/cm^2, i.e. for zero canting angle, the spin wave frequency of the stack surface mode is independent of A_{12}, since here the net magnetization is constant.

Canting of the magnetization occurs in Fig. 3.32 for $A_{12} < -0.25$ erg/cm^2. Here the spin wave frequencies display a more complicated behavior. There are now two surface modes, indicated in the figure by thick solid lines, and we see that one of these surface modes goes soft when the magnetizations lie parallel to one another at $A_{12} = -0.25$ erg/cm^2. The bulk spin wave bands are shown as shaded areas. For $d_{Co} = d_{Ru}$, the surface modes are not well defined and exist at the top of the bulk bands, thus forming the upper frequency limit of the dipolar bulk modes.

There are now two bands. One band is the continuation of the collective spin wave band of ferromagnetically coupled layers. Its frequencies decrease with

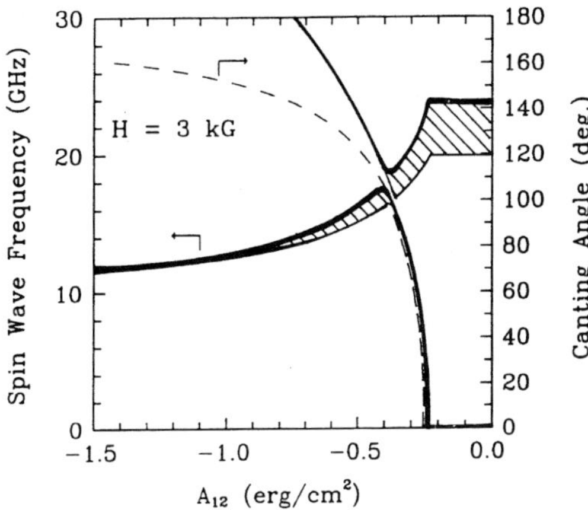

Fig. 3.32. Spin wave frequencies calculated as a function of the interlayer exchange constant, A_{12}, using an effective medium approach as described in the text. The surface modes are indicated by thick lines. The applied field is 3 kG. The canting angle ϕ_{12} between the saturation magnetizations of neighboring layers is shown as a dashed line

increasing negative value of A_{12}, i.e. increasing canting angle. This band is crossed by a new collective band, which is reminiscent of the "optic" high frequency spin wave mode of antiferromagnetic bulk material, and which goes soft for $A_{12} \geq -0.25$ erg/cm². The behavior of the former band, apart from the crossing regime, can be easily understood. The modes respond to the net magnetization in the direction of the field. As the canting angle ϕ_{12} increases the net magnetization in the direction of the field decreases approximately according to

$$M(\phi_{12}/2) = M\cos(\phi_{12}/2) = -\frac{H_0 M^2 d_{Co}}{4A_{12}}, \tag{3.99}$$

where M is the magnetization of each layer [3.124]. In the simple case of absence of anisotropy, the frequency of the surface mode for a semi-infinite multilayer is well described by

$$\frac{\omega}{\gamma} = H_0 + 2\pi M_s \cos(\phi_{12}/2). \tag{3.100}$$

Similarly, the bottom of the associated bulk band is given by

$$\frac{\omega}{\gamma} = (H_0(H_0 + 4\pi M_s \cos(\phi_{12}/2)))^{1/2}. \tag{3.101}$$

Thus measurements of the frequencies are measures of the interlayer exchange constant A_{12} and the canting angle ϕ_{12}, which are interrelated by (3.99). We emphasize that modes at these frequencies make the largest contribution to the light scattering cross section. For multilayer structures of finite thickness with anisotropies, however, the frequencies can be determined only numerically.

Fitting this model to the experimental data, values for the interlayer exchange coupling constant, A_{12}, are obtained. They are displayed in the lower part of Fig. 3.31. Although the error bars are rather large due to the experimentally observed large line-width of the modes, the oscillations between ferro- and antiferromagnetic coupling are clearly observed. However, the error bars are too large to determine the decay in oscillation amplitude with increasing d_{Ru}.

3.3.5 Superlattices with Spatial Inhomogeneities

In the preceding sections we have considered magnetic superlattice structures of "perfect" layering, i.e. any phenomena such as interdiffusion at the interfaces, and/or interface roughness and lateral or layer-to-layer variations of layer thicknesses were not considered in the models. Now we want to allow for non-homogeneous superlattices and discuss these phenomena in detail.

3.3.5.1 Spin Waves in Superlattices with Lateral Variations of Internal Fields

Certainly the spin wave mode spectrum will depend on lateral inhomogeneities. We distinguish between two limiting cases depending on the length scale Λ of lateral inhomogeneities. First, what we call "inhomogeneities" are essentially variations of internal fields caused by magnetization-, anisotropy- or thickness- (corrugation) variations, variations in the exchange constant or variations in the interlayer thickness. If Λ is much smaller than the spin wave wavelength λ (≈ 3000 Å), the spin waves depend on modified magnetic parameters which are averaged over the inhomogeneities. An example is interface anisotropies originating from small-scale interface roughness.

In the regime $\Lambda > \lambda$ the system is essentially homogeneous on the length scale of λ. If the sampling area, i.e. the laser spot in a Brillouin light scattering experiment ($\phi \approx 100$ μm), is still much larger than Λ, the observed spin wave band may also show a broadening due to sampling of a large set of spin wave spectra from different, well defined regions with different parameters and therefore with different spin wave frequencies.

This will now be demonstrated for Co/Pt superlattices [3.143]. For a superlattice with $d_{Co} = d_{Pt} = 5$ Å the collective exchange modes are well-separated in frequency from the dipolar stack surface mode due to large ferromagnetic interlayer exchange coupling [3.143]. The observed width of the stack surface mode of 14 GHz is still much larger than the experimental resolution, which is about 3 GHz. This broadening is attributed to spatially varying anisotropies caused by thickness variations of the Co layers as follows: in Fig. 3.33 the frequency of this mode is calculated as a function of the Co layer thickness, d_{Co}, for an applied magnetic field of 8 kG using for $4\pi M_s$ and the uniaxial anisotropy constant the results obtained from a fit to the experimental data, as reported in [3.164]. By varying d_{Co}, the contribution of the interface anisotropy field to the internal field varies with $1/d_{Co}$. The frequencies go to zero at $d_{Co} \simeq 3.4$ Å, indicating a perpendicular magnetized state for smaller values of d_{Co}. The

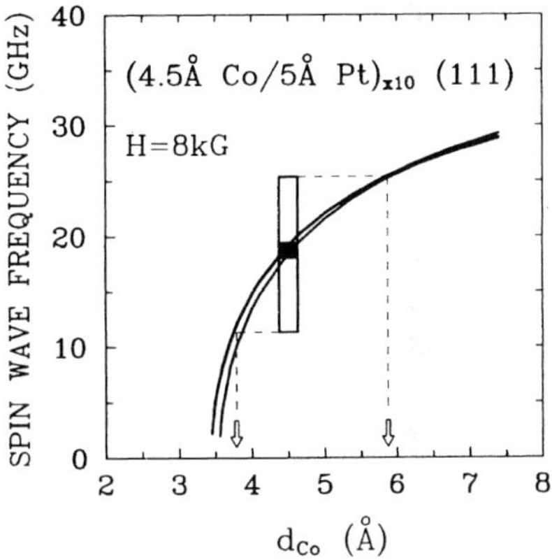

Fig. 3.33. Calculated spin wave frequencies of a Co/Pt multilayer structure with 10 bilayers of varying Co thickness and of 5 Å Pt thickness in an external field of 8 kG as a function of the Co thickness. The experimental, broad mode is shown as a bar with the intensity maximum marked with a black square. From the frequency spread of the mode (range of the bar) the corresponding change in Co layer thickness is estimated as indicated by the dashed lines

experimentally observed linewidth of the peak is indicated as a bar in Fig. 3.33 with the center of the peak as a black square. From the length of the bar the range of spatial variations of d_{Co} of 3.8–5.8 Å is deduced, as illustrated in the figure. This would translate into variations in the interface anisotropy constant of $K_u^s = (0.21–0.32)\,\mathrm{erg/cm^2}$, assuming flat interfaces.

3.3.5.2 Spin Waves in Superlattices with Layer-to-Layer Variations of Internal Fields

We will now discuss the case of multilayer structures, when the individual layer thicknesses vary from layer to layer. This might happen, e.g., due to changing deposition rates in the sample fabrication process. Even for samples with nominal identical thicknesses of all magnetic layers, the "local" thickness, i.e. the thickness on a length scale of the wavelength of the spin waves, may vary from layer to layer in the same manner as the lateral thicknesses vary as discussed above.

We will assume a system with large interface anisotropy values, which therefore exhibits a large dependence of the spin wave frequencies on the layer thickness. Without an external field the uniaxial anisotropy of each layer is assumed to be large enough to point the direction of magnetization perpendicular to the layer planes. Figure 3.34a shows the calculated spin wave frequencies of a "perfect" multilayer structure of eight bilayers with the same thickness of magnetic and nonmagnetic layers of 10 Å. For the magnetic layers the bulk parameters of Co are assumed. With increasing in-plane applied magnetic field the spin wave frequencies first decrease while the direction of magnetization is

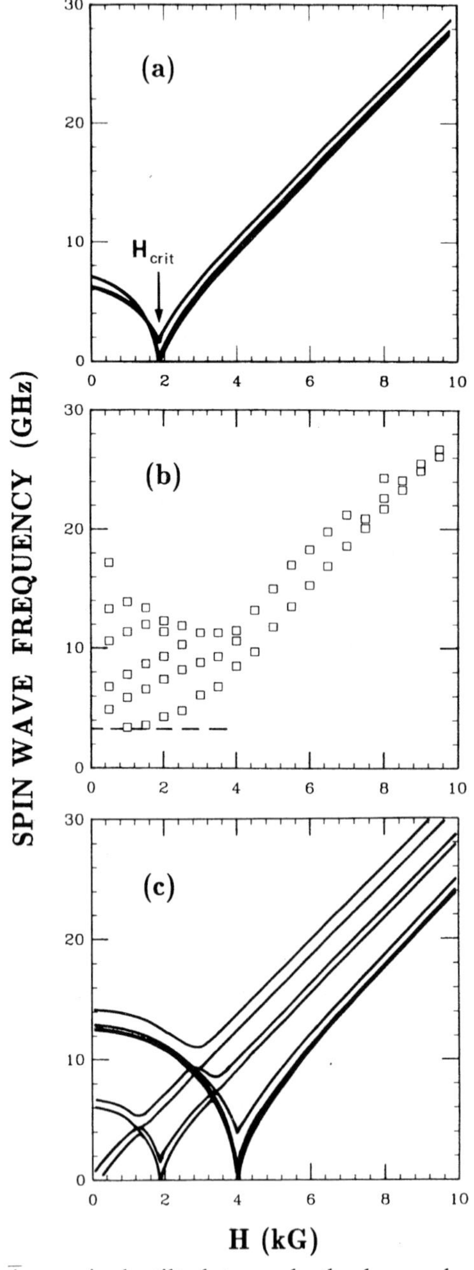

Fig. 3.34. Calculated (**a**, **c**) and experimental (**b**) spin wave frequencies as a function of the in-plane applied magnetic field. Part (**a**) shows the calculated frequencies for a "perfect" multilayer consisting of eight bilayers of some parameters. The thickness of magnetic and nonmagnetic layers is 10 Å. In (**b**) experimental data of Co/Au multilayers with 70 periods and with nominal thicknesses of $d_{Co} = 8.8$ Å and $d_{Au} = 7.5$ Å are shown. The dashed line marks the threshold, below which spin waves observation is inhibited in the Brillouin light scattering experiment due to elastically scattered light. In (**c**) the layer thicknesses of the eight magnetic layers of the multilayer are assumed to be 10, 9, 9, 11, 10, 9, 11 and 9 Å

increasingly tilted towards the layer planes, lying in the layer planes at and above a critical field strength, H_{crit} (Chap. 3.3.3.2). For $H > H_{crit}$ the spin wave frequencies increase approximately linearly with further increasing external field. Near H_{crit} the calculated spin wave frequencies show a sharp minimum with some modes going soft.

Figure 3.34b shows experimental data of a Co/Au multilayer sample consisting of 70 bilayers of 8.8 Å Co and 7.5 Å Au. Above 4 kG the spin wave frequencies increase about linearly with increasing applied field indicative of the saturation magnetization lying in plane. Near $H_{crit} \approx 3.5\,kG$ the spin wave frequencies show a broad minimum, and they are rather widely spread below H_{crit}. The observed behavior is in only rough qualitative agreement with the calculated field dependence of the spin wave modes shown in Fig. 3.34a. We will now show that, by allowing the individual layer thicknesses to have a distribution about the mean thickness value, the calculated spin wave properties much better resemble the experimental data.

Figure 3.34c shows calculated spin wave frequencies for the case where the nonmagnetic layer thickness is fixed at 10 Å, but the magnetic layer thicknesses are 10, 9, 9, 11, 10, 9, 11 and 9 Å, respectively. For this calculation A_{12} has been set to zero. For each layer, first the critical field, H_{crit}, is calculated as well as the direction of the magnetization as a function of the applied in-plane field. H_{crit} varies from layer to layer due to different layer thicknesses. Then the spin wave frequencies of the multilayer stack are calculated using an effective medium approach, using the static orientation of the layer magnetizations as input data. The spin wave modes show zero frequencies at $H_{crit} = 0$, 1.9 and 4 kG. The values of H_{crit} correspond to the chosen thickness values of the magnetic layers of 11, 10, and 9 Å, respectively. The obtained spin wave mode distribution more closely resembles the experimentally observed mode spectrum (Fig. 3.34b) than does the calculation assuming the same parameters for each layer, as shown in Fig. 3.34a. Please note that spin wave modes with frequencies smaller than about 3 GHz (dashed line in Fig. 3.34b) are not accessible in the Brillouin light scattering experiment due to the overlap with elastically scattered laser light.

For a "real" multilayer structure, both thickness variations from layer to layer as well as thicknesses varying laterally due to, e.g., a mosaic spread, contribute to the effect. The Co/Au sample, of which the spin wave data are shown in Fig. 3.34b, was prepared by postannealing the sample in order to gain atomically sharp interfaces for maximizing interface anisotropies [3.165]. On the other hand, evidence has been found that the postannealing process introduces interface corrugations, which might be responsible for local, layer-to-layer thickness variations [3.166]. The pronounced difference in the spin wave properties between a "perfect" structure (Fig. 3.34a) and a "realistic" structure as described above (Fig. 3.34c) is already obtained for a corrugation of ± 1 Å of each layer.

3.3.6 Conclusion and Outlook

We want to conclude by adding some general remarks on the phenomenon of Brillouin light scattering from spin waves in superlattices. Although the penetration depth of light for typical metallic superlattices may be as small as 100 Å, the information depth is given by the perpendicular coherence length of spin waves, which is typically at least a few thousand Ångstroms. This is because the collective spin wave excitations for typical superlattice structures are coherent throughout all magnetic layers. Therefore by probing them in the first few layers, the complete spin wave information on the total stack can be obtained.

The light scattering cross section is proportional to the net fluctuating part of the dipolar moment of the precessing spins within the light scattering interaction volume. Thus, pure exchange-type spin wave modes contribute to the cross section only very weakly. In order to study the exchange interaction, in particular the interlayer exchange interaction, a fair amount of dipolar coupling is necessary. This is the case if modes are studied which in frequency are not separated too much from the dipolar surface mode, or if the net fluctuating part of the dipolar moment averaged over the light penetrated region is sufficiently large.

The broad linewidths observed in many experiments seem to be uncorrelated with intrinsic damping mechanisms of spin waves [3.167]. The line broadening is caused by spatial inhomogeneities on a length scale comparable to and larger than the spin wave wavelength (≈ 3000 Å) and by sampling over many areas with different local properties within the laser spot, which is typically 100 μm in diameter. Loss mechanisms due to direct scattering of spin waves at, e.g., inhomogeneities, are very weak since there are no scattering channels to scatter into.

Many areas could not be covered in this review. We have not reported on the determination of magnetic anisotropies in superlattices, which easily can be performed with Brillouin light scattering [3.109, 110, 127, 128, 139, 140, 164, 166, 168]. Contrary to magnetometry, spin wave frequencies in layered systems composed of different magnetic materials are primarily sensitive to the stiffest magnetic material. Thus, by comparing Brillouin light scattering results with magnetometric investigations, access is gained to the characterization of atomic interface layers with reduced or increased magnetic moments, like magnetically dead layers, or on the contrary, on magnetically polarized spacer layers [3.151, 168]. The same applies to superlattice structures composed of two magnetic materials, which then are strongly exchange coupled [3.109, 110, 169]. Here a new type of collective exchange modes exists. The collective modes are composed of exchange modes of each magnetic layer of one kind of material and they are exchange coupled through the intervening magnetic layers of the other kind.

From the calculated spin wave dispersion the temperature dependence of the saturation magnetization can be derived. The first results for multilayers showing the transition from two- or three-dimensional behavior with increasing interlayer exchange coupling are reported in [3.170]. Here the equation of state (or total energy) is summed over the spin wave modes calculated for the actual system.

The field of magnetic superlattices is advancing very fast. Due to its potential the Brillouin light scattering technique certainly will be of central importance in understanding some of the scientific surprises which forthcoming superlattice magnetism may reveal.

Acknowledgements. We would like to thank P. Baumgart, J. Faßbender, V. Harzer, P. Krams, F. Lauks and R. Lorenz for discussions and experimental work, and F. Nörtemann and R.L. Stamps for discussions and their help with the theory. Support from the Deutsche Forschungsgemeinschaft through the Sonderforschungsbereich 341 is gratefully acknowledged.

3.4 Nuclear Magnetic Resonance in Thin Films and Multilayers

W.J.M. De Jonge, H.A.M. De Gronckel, and K. Kopinga

A nuclear magnetic resonance (NMR) experiment measures the nuclear energy level splitting by inducing transitions via the application of radio frequency electromagnetic (em) radiation. Since the splitting of the energy levels is partially brought about by magnetic and electric interactions with *neighboring* ions, NMR provides, in principle, a local probe of the structural and magnetic properties. This feature of NMR makes this technique suitable to obtain information on layers and interfaces, and also when they are embedded in a multilayered structure. In this chapter we will briefly introduce the basic principles of NMR and focus our attention to the application in thin films and multilayers [3.171, 172].

3.4.1 Basic Principles

The interaction of a nucleus with its surroundings can involve magnetic as well as electrostatic components [3.171]. In this chapter we will only consider the magnetic interaction. The dominant term in this interaction is the hyperfine interaction between the nuclear magnetic moment and the (unpaired) electron moment of the individual atom as well as that of neighboring atoms. The hyperfine interaction is usually expressed by an effective field, the hyperfine field B_{hf}, on the nuclear spin I:

$$B_{hf} = \frac{1}{\gamma \hbar} \cdot A \cdot \langle S \rangle, \qquad (3.102)$$

where $\langle S \rangle$ denotes the averaged electron spin momentum, γ is the gyromagnetic ratio and the tensor A represents the strength of the interaction. For paramagnetic systems $\langle S \rangle \cong 0$ and therefore B_{hf} is small and the splitting of the nuclear levels is mainly brought about by additional external fields. In ferromagnetic materials, however, $\langle S \rangle \neq 0$ and B_{hf} can easily attain values in the order of 10^2 T.

The main contribution to the hyperfine field is contained in the Fermi contact interaction, which is proportional to the spin density *at* the nuclear site, arising from electrons with s symmetry. Core as well as valence s electrons contribute to this term. Neighboring atoms are sensed through the hybridization of the valence s wave functions with the polarized d wave functions of these neighboring atoms [3.173]. This contribution is generally referred to as the transferred hyperfine field and is one of the basic ingredients of the application of NMR as a probe of the local environment, as we shall see later on.

In an NMR experiment, the response of the nuclear magnetization $\gamma\hbar I$ to an applied oscillating field is measured. Since the nuclear levels are split by internal as well as external fields, absorption can be observed at a frequency v given by

$$\Delta E = hv = \hbar\gamma\sum_i \boldsymbol{B}_i = \hbar\gamma|\boldsymbol{B}_{\mathrm{hf}} + \boldsymbol{B}_{\mathrm{appl}} + \boldsymbol{B}_{\mathrm{int}}| \,, \tag{3.103}$$

where $\boldsymbol{B}_{\mathrm{int}}$ represents additional internal fields (dipolar, . . .) which we did not discuss here.

A NMR experiment can be described in two ways. Either one chooses the description based on absorption due to radiation-induced transitions between nuclear levels (as was done in (3.103)) or the description based on the motion of the nuclear moments in a radio frequency (rf) field. Both descriptions are equivalent and can be found in the literature [3.171, 172]. The latter description is preferred when dealing with a pulsed NMR experiment. In such an experiment the rf field (at resonance according to (3.103)) is applied in a sequence of pulses and creates rotations of the nuclear magnetization over angles depending on the duration of the pulses. The transient effects can be monitored by the voltage induced in a receiving coil. A particular sequence of pulses, a $90°$ pulse followed after a time τ by a $180°$ pulse, can bring about an echo intensity at 2τ. This spin-echo intensity is generally monitored in a pulse NMR experiments. This method also offers the possibility to measure the spin–lattice and spin–spin relaxation time of the nuclear system (T_1 and T_2, respectively). Apart from the transition frequency, also these relaxation times can be used to discriminate between the various local nuclear environments. For more details the reader is referred to the literature [3.171, 172].

Within the context of this chapter, an introduction to the applicability of NMR in thin films, we would like to address two experimental aspects related to the feasibility of the observation of NMR. The output of an experiment, in terms of an induction voltage at resonance, V, can be written as:

$$V \approx |1 - \eta|N\gamma^3\,\hbar^2\boldsymbol{B}_{\mathrm{loc}}^2\,I(I+1)/3\,\mathrm{kT} \,, \tag{3.104}$$

where η is an enhancement factor, which we will discuss later on. Intrinsically V is a small number: NMR is basically not as sensitive as ESR, which is mainly due to the small nuclear moment $\gamma\hbar I$. In Table 3.3 some of the relevant NMR parameters for mainly 3d elements are shown [3.174], together with the relative sensitivity according to Eq. 3.104, where Co is taken as reference. For comparison we note that non-magnetic Co requires about 100 monolayers on $1\,\mathrm{cm}^2$ to obtain a signal-to-noise ratio of unity in a field of 5 T. This sensitivity is increased appreciably in ferromagnets, due to the enhancement mediated by the subsystem of electronic moments.

In the case of non-magnetic materials the spin-echo intensity is brought about by the nuclear moments precessing at the resonance radio frequency. In

Table 3.3. Nuclear properties (data taken from [3.174]) and (relative) spin echo NMR sensitivity for some selected elements. A denotes the mass number, I is the nuclear spin, γ is the nuclear gyromagnetic ratio, m is the nuclear magnetic moment (in multiples of the nuclear magneton μ_N). The relative sensitivity refers to the magnitude of the spin-echo signal in a magnetic field of 5 T in the (electronic) non-magnetic state and is normalized to unity for (non-magnetic) Co. The columns subheaded "enriched" and "nat. abund." refer to the sensitivity for a 100% isotopically enriched sample and a sample with natural abundance of the isotopes, respectively. As a reference, at a temperature of 1.4 K about 100 ML (monolayers) on 1 cm^2 are needed to acquire a signal-to-noise ratio of unity for non-magnetic Co in a magnetic field of 5 T; ferromagnetic Co in zero field requires about 2×10^4 times less material, i.e. 0.05 ML on 1 cm^2, due to the large enhancement η (3.104)

Element	A	I	$\gamma/2\pi$ (Mhz/T)	m (μ_N)	Abundance (%)	Relative sensitivity Enriched	Relative sensitivity Nat. abund.
Al	27	5/2	11.094	3.64	100	0.746	0.746
Sc	45	7/2	10.343	4.75	100	1.09	1.09
Ti	47	5/2	2.400	0.66	7.3	0.0076	0.00055
	49	7/2	2.4005	1.18	5.5	0.0136	0.00075
V	50	6	4.2450	3.34	0.24	0.201	0.00048
	51	7/2	11.19	5.14	99.76	1.38	1.38
Cr	53	3/2	2.4065	-0.47	9.55	0.0033	0.00031
Mn	55	5/2	10.501	3.44	100	0.633	0.633
Fe	57	1/2	1.3758	0.09	2.19	0.000122	2.7×10^{-6}
Co	59	7/2	10.054	4.62	100	1	1
Ni	61	3/2	3.8047	-0.75	1.2	0.0129	0.00016
Cu	63	3/2	11.285	2.23	69.1	0.337	0.233
	65	3/2	12.089	2.39	30.9	0.414	0.128
Zn	67	5/2	2.663	0.87	4.1	0.0103	0.00042
Ru	99	5/2	1.9	-0.64	12.7	0.0038	0.00048
	101	5/2	2.1	-0.72	17.1	0.0051	0.00087
Ag	107	1/2	1.723	-0.113	51.8	0.00024	0.00012
	109	1/2	1.981	-0.130	48.2	0.00036	0.00018
Pt	195	1/2	9.153	0.600	33.8	0.0359	0.0121
Au	197	3/2	0.7292	0.143	100	0.000091	0.000091

ferromagnetic materials this rf field also couples to the (relatively large) electronic magnetic moments yielding an enhancement η of the spin-echo intensity [3.175]. The magnitude of this enhancement, which is proportional to the electronic susceptibility, can be of the order of 10. If domains are present in the ferromagnet, an additional enhancement is obtained from the response of the electronic moments located in the domain walls. This enhancement is generally orders of magnitude larger than the (rotational) effect in single domain materials.

As we quoted above, the application of NMR as a local probe is based on the sensitivity of the hyperfine interaction $\boldsymbol{B}_{hf}$ to the immediate surrounding of the nucleus. In the following paragraphs we will discuss some features which are of particular interest in the study of multilayers and thin films. We will then discuss the effect of local structure and symmetry, the introduction of foreign atoms in the nearest neighbor shell (interface topology), and the effect of strain.

The influence of the *local structure and symmetry* on B_{hf} arises from two effects: the sensitivity of the transferred hyperfine interaction to the local environment and the dependence of the orbital and (atomic) dipolar part of B_{hf} on the local symmetry. The former only affects the magnitude of B_{hf}, since it contributes to B_{hf} through the isotropic Fermi contact interaction. The latter mainly determines the anisotropy of B_{hf}, and the specific angular dependence of the resonant frequency will be characteristic of the local symmetry. These features enable one to discriminate uniquely between, for instance, fcc, bcc, and hcp phases, or stacking faults in the pure phase, through the magnitude of B_{hf}, and between fcc, hcp, and trigonal or tetragonal deformed fcc through its anisotropy [3.176, 177].

The presence of a "foreign" atom, or impurity, alters the hyperfine field of its neighboring atoms [3.173]. First, the core polarization and valence s moment contribution to the hyperfine field may change, since substitution of a host atom by a "foreign" atom can alter the magnetic moment of the neighboring atoms. The magnitude and range of this moment disturbance depend very strongly on the (combination of) elements. Secondly, the "foreign" atom alters the hyperfine field of its neighbors because it causes a change in the "transferred" polarization contribution. Although the range of the perturbation can be rather extended, in most experiments only the nearest neighbors of a "foreign" atom can be distinctly observed. If the nearest neighbor shell contains more than one "foreign" atom, the magnitude of B_{hf} changes by an amount roughly proportional to the number of "foreign" atoms, as is evidenced by experiments on alloys (cf. Table 3.4). The sensitivity of B_{hf} to foreign atoms in the nearest neighbor shell enables one to discriminate between atoms at the interfaces in a multilayer and atoms well inside of the constituting layers. Moreover, since the magnitude of B_{hf} depends on the number of "foreign" atoms in the nearest neighbor shell, the hyperfine field distribution, i.e. the intensity distribution over the corresponding resonance lines (satellites), reflects the characteristics of the topology of the interfaces.

Table 3.4. Experimentally observed hyperfine field of Co in some diluted alloys. $B_{hf,i}$ denotes the hyperfine field of a Co atom with i "foreign" element atoms as nearest neighbors. The data refer to liquid helium temperatures

Element	$B_{hf,0}(T)$	$B_{hf,1}(T)$	$B_{hf,2}(T)$	$B_{hf,3}(T)$	Ref.
Ti	-21.6	-17.5	-13.9		3.178
Cr	-21.6	-17.6	-14.0		3.178
Ni	-21.6	-20.9	-20.0		3.178
	-21.62	-20.89	-20.14	-19.4	3.179
Cu	-21.6	-19.7	-17.9		3.180
	-21.62	-19.8	-17.9	-16.2	3.181

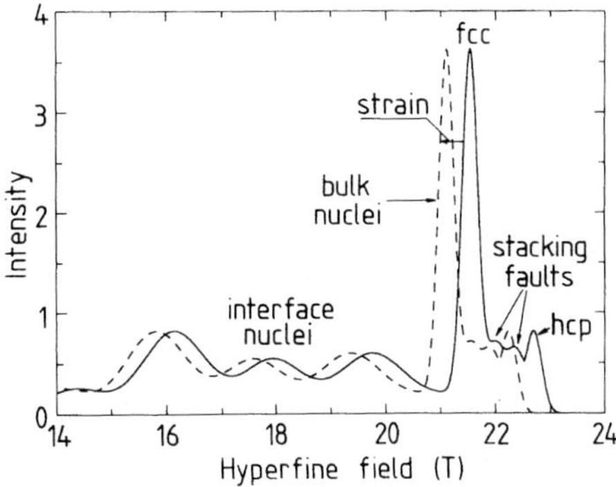

Fig. 3.35. Schematic illustration of the effects of local symmetry, interface or surface, and strain on an NMR spectrum

Since the magnetic hyperfine field is largely the result of a delicate (un)balance between the spatial distribution of the spin-up and spin-down electrons, it will also be sensitive to distortions of this spatial distributions, i.e., (volume) *strain* [3.182]. In the case of a magnetically ordered material under pressure this leads, first of all, to a decrease of the net magnetic moment. This implies a decrease in the exchange polarization and thus the core polarization at the nucleus. The net magnetic moment also determines the valence polarization at the nucleus, albeit in a more complicated way, so one might expect it to decrease, too. However, the 4s electron density at the nucleus increases under pressure due to the (slight) squeezing of the valence electron wave functions, thereby more than compensating the effect of the reduced magnetic moment. As a result, the valence contribution to the hyperfine field increases under pressure and the net change in hyperfine field will be the result of the opposite changes in the core and valence spin densities. In Co and Ni the hyperfine field decreases (becomes more negative) under pressure, whereas in Fe it increases. To summarize, the effects of structure, foreign atoms, and strain on the hyperfine field spectrum are schematically shown in the hypothetical NMR spectrum plotted in Fig. 3.35.

3.4.2 Experimental Results of NMR on Multilayers and Films

In this section we will present some selected examples of the use of NMR in the research on magnetic multilayers and thin films. The examples relate to the *structure* of thin Co films, and to the *interface roughness and topology* and the *strain* in Co multilayers.

3.4.2.1 Structure

The structure of the magnetic layers in a multilayer or thin film has a significant effect on the observed physical properties. However, the actual structural

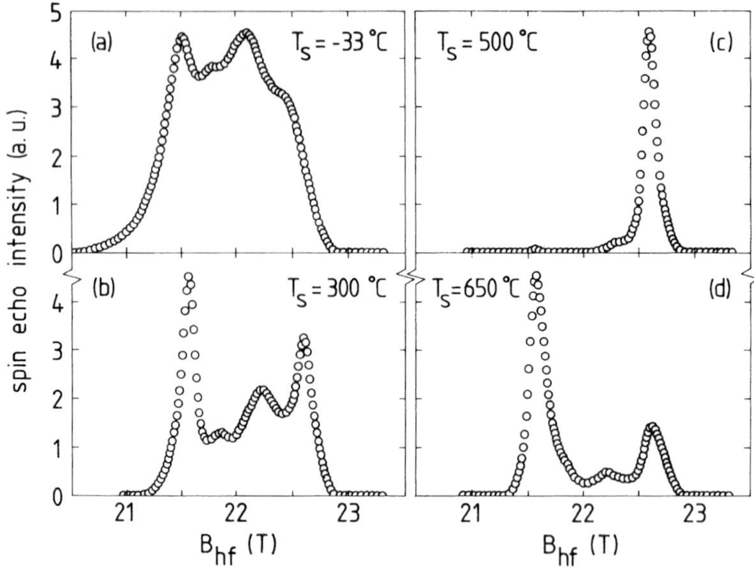

Fig. 3.36. NMR spectra of 1000 Å Co films grown on mica showing the variation in bulk structure as function of the growth temperature. The spectra were recorded at $\nu = 190$ MHz with the magnetic field applied parallel to the film plane and at a temperature of 1.4 K. Data are taken from [3.183]

phase(s) realized in thin films or multilayers may be intrinsically metastable and deviate from the expected bulk phase. Here we will present two examples of NMR studies on the local structure. The first example is a particularly basic one: a Co film [3.183]. In bulk Co, two structures can be obtained [3.184]: hcp Co at $T < 450\,^\circ$C and fcc Co at $T > 450\,^\circ$C. The hyperfine fields for the two phases are known to be 22.6 T for hcp Co and 21.6 T for fcc Co. Figure 3.36 shows the NMR spectra of Co films grown at various substrate temperatures. It is obvious that the actually realized structure depends strongly on the growing conditions. In general a mixture of local structures (hcp, fcc and stacking faults) is obtained at any temperature and structural phases can be stabilized under these conditions outside of their bulk stability region. This type of information can be related to, for instance, the anisotropy of layers, since the volume contributions of fcc Co and hcp Co appreciably differ in magnitude [3.185].

Another interesting example is presented in Fig. 3.37, showing the NMR spectrum of Co/Fe multilayers with variable Co thickness [3.186]. In this case it appears that for small Co thickness even the bcc Co phase, which does not exist in the bulk, can be stabilized. Comparable results have been obtained for Co/Cr multilayers [3.187]. Studies on sputtered Co/Cu multilayers have recently shown that fcc or hcp stacking depends significantly on the Co and Cu layer thickness [3.188].

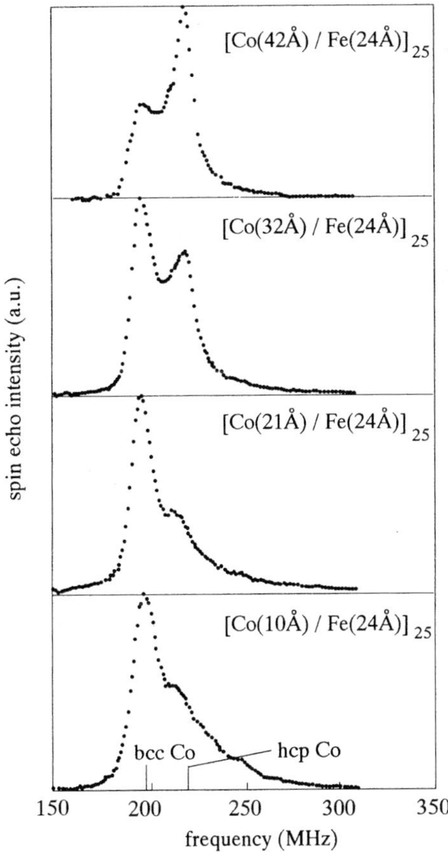

Fig. 3.37. NMR spectra of Co/Fe multilayers for various Co layer thicknesses, showing the stabilization of bcc Co in particular at reduced Co thickness. Data after [3.186]

3.4.2.2 Interface Topology

The interface roughness or topology of the interface has a large influence on the surface anisotropy as well as on the magnetoresistance [3.185]. At present, it is not yet clear whether the overall thickness or the profile (composition) of the interface is the relevant factor, or whether this behavior depends on the specific components of the multilayer. Detailed information on interface roughness can shed light on this relation. Early studies on interface roughness have been reported for Fe/V, Co/Sb, and Fe/Mn multilayers [3.189]. Here we will examplify the use of NMR with some recent results on Co/Cu [3.190] and Co/Ni multilayers [3.191].

Figure 3.38 shows the spin-echo intensity spectrum of a Co/Cu [1 1 1] multilayer, which is schematically depicted in the inset. The main line represents fcc surrounded bulk Co. The Co layers are apparently single phased: no traces of hcp or stacking faults are observed (see also Fig. 3.35). Figure 3.39 shows

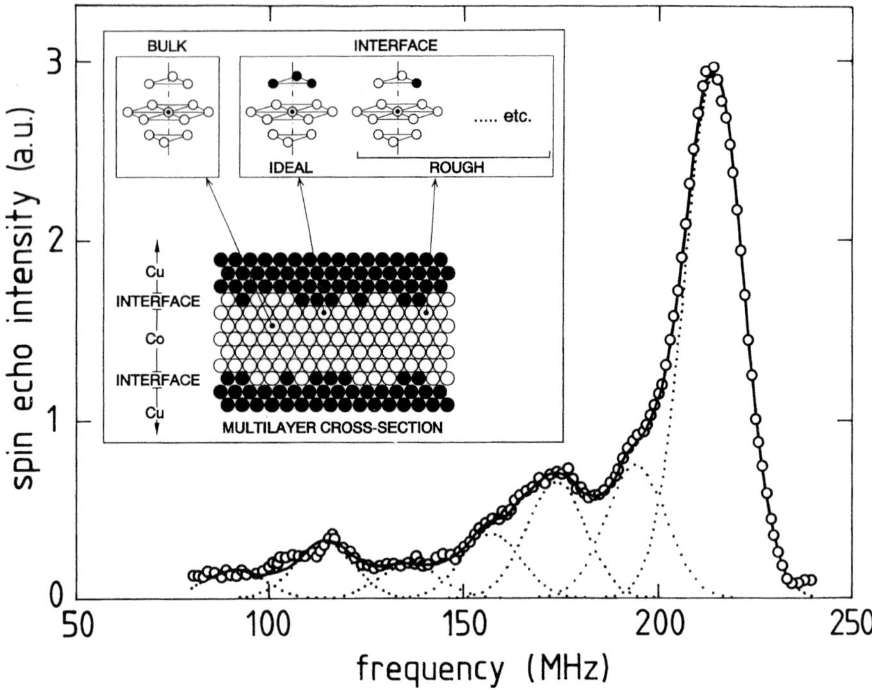

Fig. 3.38. Zero field NMR spectrum of a HV evaporated $40 \times (12\,\text{Å Co} + 42\,\text{Å Cu})$ [1 1 1] multilayer. The solid line represents the result of a fit with seven Gaussian contributions. Each contribution (denoted by a dotted line) corresponds to Co atoms in a specific environment, as illustrated in the inset. Data are taken from [3.190]

the intensity ratio of the main line and the signals at the lower frequencies as a function of t_{Co}. Since this ratio increases systematically with t_{Co}, the low frequency part of the spectrum originates from Co atoms at the interfaces, where one or more nearest neighbor Co atoms are replaced by Cu. This assignment is corroborated by the fact that the spin–spin relaxation time T_2 for these satellites is typically twice the value for the main (bulk) line, evidencing the different origin of the signals. From the inset in Fig. 3.39 it appears that for Co/Cu the intensity ratio varies as $n_{Co} - 2$ (where n is the number of monolayers), unambiguously showing that the mixed interface region is only one monolayer thick. For comparison, Fig. 3.39 also shows the same data for a Co/Ni multilayer, where the intensity varies as $n_{Co} - 4$, indicating that, in contrast to Co/Cu, the mixed interface region is at least two monolayers thick.

To analyze interface spectra as shown in Fig. 3.38 more quantitatively, the various contributions should be identified. Since the spectrum in this particular case originates from Co atoms with one or more Cu atoms in their nearest neighbor shell, it should consist of a number of absorption lines (satellites) shifted with respect to the bulk fcc line by approximately 18 MHz per substituted Cu atom, as deduced from experiments on Co/Cu alloys (cf. Table 3.3). The

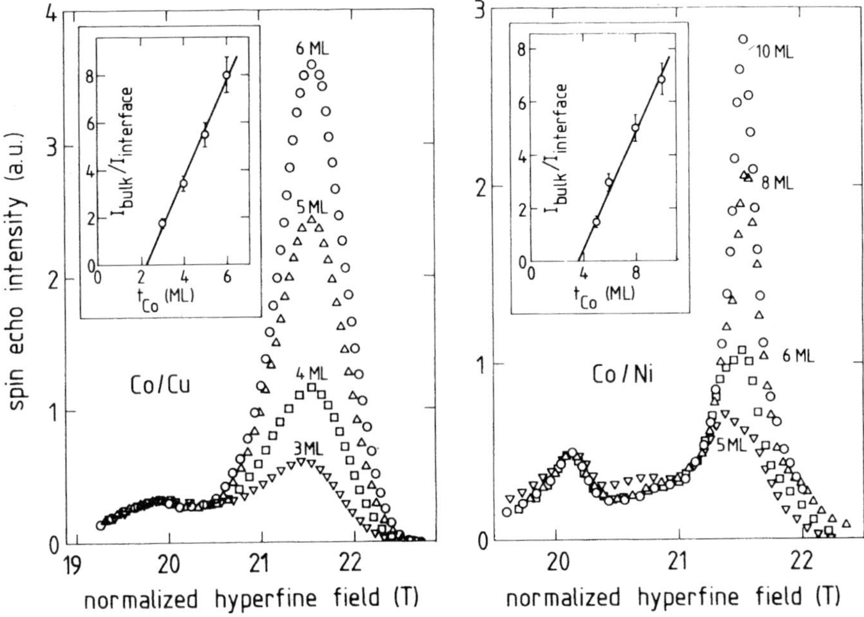

Fig. 3.39. In field spectra of Co_xCu_{21} multilayers (left part of the figure) and Co_xNi_{42} multilayers (right part of the figure) showing the systematic increase in intensity ratio between the main line and the most intense satellite as function of the Co thickness t_{Co} (expressed in monolayers). For easy comparison the spectra have been normalized to the most intense satellite (in intensity as well as in hyperfine field). The insets show the ratio of intensity of bulk- and interface signal as function of Co thickness. Data from [3.190, 191]

solid line in Fig. 3.38 shows that the structure of the spectrum is indeed well-fitted by (seven) approximately equally spaced Gaussians (denoted by dotted lines) with an average spacing of 19 ± 3 MHz.

Although we will not go into detail, the preceding result indicates that the various lines can be assigned to Co nuclei having 12 Co neighbors (" bulk" atoms) and nuclei having 11 to 6 Co neighbors ("interface atoms"). The existence of Co sites with an environment different from 12 or 9 Co neighbors (sites in perfectly flat interface layers have nine Co neighbors for the [1 1 1] oriented growth) is obviously related to interface roughness. Assuming a model for the interface topology, one can now compare the statistical occurrence of various surroundings in the model with the experimental intensity of the satellite spectrum. In the present example of mixing in one monolayer, detailed results about the roughness could be obtained [3.190]. Using a comparable approach, the interface structure of sputtered Co/Cu [1 1 1] multilayers has also been investigated. In this study interface structures up to three monolayers have been modelled and the role of defects has been taken into account [3.188].

NMR studies in which Co/Cu [1 0 0] and [1 1 1] multilayers were compared have also been reported [3.192]. For both cases, clearly different spectra were

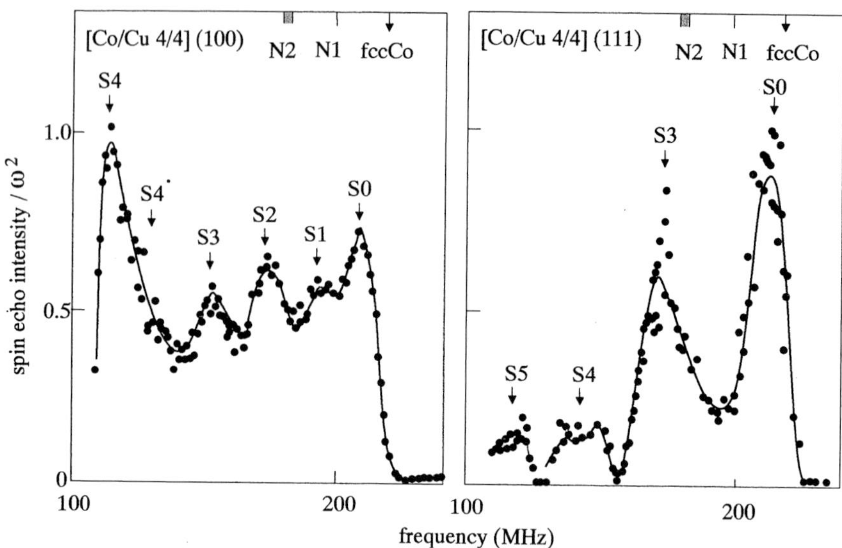

Fig. 3.40. NMR spin-echo spectra of UHV deposited $N \times (4\,\text{Å}\ \text{Co} + 4\,\text{Å Cu})\ [1\,0\,0]$ and $N \times (4\,\text{Å}$ $\text{Co} + 4\,\text{Å Cu})\ [1\,1\,1]$ multilayers. Data from [3.192]

obtained, as shown in Fig. 3.40. One should note that in the ideal case of a perfect flat interface the Co atoms at the interfaces are surrounded by nine Co atoms in the $[1\,1\,1]$ fcc structure and by eight Co atoms in the $[1\,0\,0]$ fcc structure. In Co layers with a thickness of four monolayers, the number of atoms at an (perfect) interface position equals the number with a bulk surrounding. Based on this qualitative argument one might state that the NMR spectra show that the interfaces in Co/Cu $[1\,0\,0]$ are rougher than those in Co/Cu $[1\,1\,1]$ multilayers. More quantitative conclusions would require a systematic comparison involving several layer thicknesses, as shown in Fig. 3.39, and a fit of the whole spectrum to a model of the interface topology.

Recently, a study has been reported relating the magnetoresistance in Co/Cu $[1\,0\,0]$ multilayers with AF coupling to the interface roughness and composition [3.193]. The roughness was manipulated by changing the acceleration voltage in the sputtering process. The interface spectrum was fitted with a topological model in which the width as well as the composition profile of the interface region were variable parameters. The results suggested that no direct relation of the magnetoresistance with the width of the interface (mixed) region could be established, but that the chemical composition of the interface could well be one of the keys to the understanding of the (giant) magnetoresistance.

Strain. To date, the application of NMR to study strain in multilayers has been rather limited. Apart from some preliminary studies on Co/Pd [3.194] and Co/Au [3.195] mulitlayers, extensive studies have only been reported on Co/Cu and Co/Ni [3.190, 191]. To illustrate the effect of strain, in Fig. 3.41 we show the shift of the main absorption line (compare Fig. 3.35) with respect to the bulk

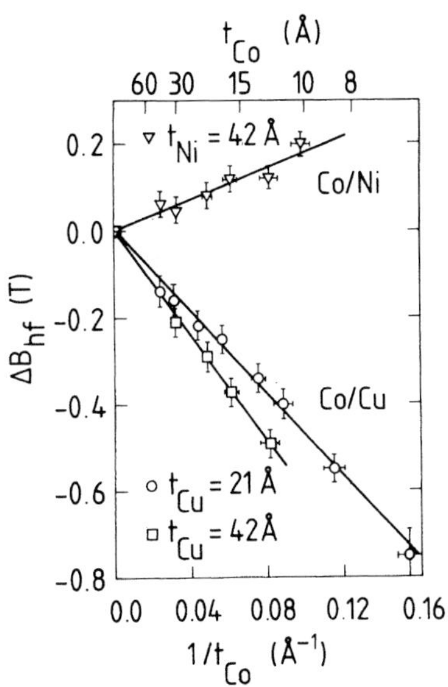

Fig. 3.41. Hyperfine field shift ΔB_{hf} derived from in-field spectra versus inverse Co thickness $1/t_{Co}$ for [1 1 1] Co/Ni and Co/Cu multilayers. The solid lines represent least squares fits of a straight line to the data. Data from [3.191]

value, ΔB_{hf}, versus the thickness of the Co layer. From this figure a systematic shift ΔB_{hf} proportional with $1/t_{Co}$ can be observed. However the sign and magnitude of the shift depend on the composition of the multilayer.

As we quoted above, a reduction of the lattice parameters by isotropic compression yields a shift in $|B_{hf}|$. For Co this shift ΔB_{hf} is positive in case of compression and can be expressed as: $\Delta B_{hf}/B_{hf} = -1.16\,\Delta V/V$ [3.182]. Interpreting the observed shift in Fig. 3.41 as being due to strain, one can conclude that the sign of ΔB_{hf} as well as the relative magnitude of the shift in Co/Ni and Co/Cu multilayers match the lattice mismatch very well (+0.6% for Co/Ni and −2% for Co/Cu). This strongly suggests that the strain is induced by the lattice mismatch. It is worthwhile to note that these shifts can also be observed at the interface sites, thus providing a possibility to probe the variation of strain through the layer.

A quantitative relation between ΔB_{hf} and the change in lattice parameters is less straightforward to establish, since the strain is not isotropic in actual cases. Nevertheless the qualitative behavior, such as the variation of the strain through the layer and its dependence on t_{Co} (or $1/t_{Co}$) and the thickness of the interlayer can be studied in detail, as for instance shown in Fig. 3.41. Such studies can provide relevant information on strain and coherency in multilayers and the validity of various growth models [3.190].

3.4.3 Conclusion

This chapter has focused on the use of NMR as a local probe of structural parameters in a multilayer. The basic principles have been introduced and examples of studies on local structure, interface roughness and topology, and strain have been briefly reviewed. These studies demonstrate that, in spite of the modest sensitivity, the application of NMR indeed can contribute to the determination of these local structural parameters. We would like to mention that, apart from these structural features, in principle NMR also can probe local *magnetic* properties, static as well as dynamic. Therefore, further NMR research on multilayers in the near future might include studies on local magnetization ("dead layers") as well as studies on spin density oscillations in the non-magnetic interlayers.

Acknowledgements. We would like to acknowledge the cooperation with Philips Research in Eindhoven on the research of magnetic multilayers. Thanks are due to P. Panissod, F.J.A. den Broeder, R. Coehoorn, P.J.H. Bloemen and E.A.M. van Alphen for their cooperation and valuable discussions. Part of this work was sponsored by the EEC through SCIENCE project GP^2M^3 and ESPRIT project SM^3S.

References

Section 3.1

3.1   J.R. MacDonald: Proc. Phys. Soc., Sect. A **64**, 968 (1951)
3.2   B. Heinrich, J.F. Cochran, A.S. Arrott, S.T. Purcell, K.B. Urquhart, J.R. Dutcher, W.F. Egelhoff: Appl. Phys. A **49**, 473 (1989)
3.3   B. Heinrich, S.T. Purcell, J.R. Dutcher, J.F. Cochran, A.S. Arrott: Phys. Rev. B **38** 12879 (1988)
3.4   J.J. Krebs, B.T. Jonker, G.A. Prinz: J. Appl. Phys. **61**, 2596 (1987)
3.5   A.G. Gurevitch: Ferrites at Microwave Frequencies (Consultants Bureau, New York, 1963)
3.6   G.T. Rado: Phys. Rev. B **26**, 295 (1982); Phys. Rev. B **32**, 6061 (1985)
3.7   J.F. Cochran, B. Heinrich, A.S. Arrott: Phys. Rev. B **34**, 7788 (1988)
3.8   B. Heinrich, J.F. Cochran, M. Kowalewski, J. Kirschner, Z. Celinski, A.S. Arrott, K. Myrtle: Phys. Rev. B **44**, 9348 (1991)
3.9   M. Benson, D.L. Mills: Phys. Rev. **178**, 839 (1969)
3.10 G. Feher: Bell Syst. Tech. J. **36**, 449 (1957)
3.11 A.T. Starr: *Radio and Radar Technique* (Sir Isaac Pitman & Sons, Limited, London, 1953)
3.12 J.F. Cochran, J.M. Rudd, M. From, B. Heinrich, W. Bennett, W. Schwarzacher, W.F. Egelhoff, Jr.: Phys. Rev. B **45**, 4676 (1992)
3.13 B. Heinrich, A.S. Arrott, J.F. Cochran, K.B. Urquhart, K. Myrtle, Z. Celinski, Q.M. Zhong: Mat. Res. Soc. Symp. Proc. Vol. **151**, 177 (1989)
3.14 B. Heinrich, J.F. Cochran, R. Hasegawa: J. Appl. Phys. **57**, 3690 (1985)
3.15 B. Heinrich, K.B. Urquhart, A.S. Arrott, J.F. Cochran, K. Myrtle, S.T. Purcell: Phys. Rev. Lett. **59**, 1756 (1987)
3.16 G.A. Prinz, B.T. Jonker, J.J. Krebs, J.M. Ferrari, F. Kovanic: Appl. Phys. Lett. **48**, 1756 (1986); J.J. Krebs, F.J. Rachford, P. Lubitz, G.A. Prinz: J. Appl. Phys. **53**, 8058 1982)
3.17 Z. Celinski, B. Heinrich: J. Appl. Phys. **70**, 5935 (1991)
3.18 Yi Li, M. Farle, K. Baberschke: Phys. Rev. B **41**, 9596 (1990)
3.19 J.A.C. Bland, R.D. Bateson, A.D. Johnson, B. Heinrich, Z. Celinski, H.J. Lauter: J. Magn. Magn. Mat. **93**, 331 (1991)
3.20 J.R. Dutcher, J.F. Cochran, B. Heinrich, A.S. Arrott: J. Appl. Phys. **64**, 6095 (1988)

3.21   S.T. Purcell, B. Heinrich, A.S. Arrott: J. Appl. Phys. **64**, 5337 (1988)

3.22   B. Heinrich, Z. Celinski, J.F. Cochran, A.S. Arrott, K. Myrtle: J. Appl. Phys. **70**, 5769 (1991)

3.23   K.B. Urquhart, B. Heinrich, J.F. Cochran, A.S. Arrott, K. Myrtle: J. Appl. Phys. **64**, 5334 (1988)

3.24   B. Heinrich, K.B. Urquhart, J.R. Dutcher, S.T. Purcell, J.F. Cochran, A.S. Arrott, D.A. Steigerwald, W.F. Egelhoff, Jr.: J. Appl. Phys. **63**, 3863 (1988)

3.25   R. Cabanel, P. Etienne, S. Lequien, G. Crewzet, A. Barthelemy, A. Fert: J. Appl. Phys. **67**, 5409 (1990)

3.26   L. Néel: J. Phys. Radium **15**, 227 (1954)

3.27   S. Chikazumi: *Physics of Magnetism* (Robert E. Krieger, Malabar, Florida, 1986)

3.28   C. Chappert, P. Bruno: J. Appl. Phys. **64**, 5736 (1988)

3.29   J.G. Gay, R. Richter: Phys. Rev. Lett. **56**, 2728 (1986)

3.30   B. Heinrich, A.S. Arrott, J.F. Cochran, S.T. Purcell, K.B. Urquhart, K. Myrtle: J. Cryst. Growth **81**, 562 (1987)

3.31   Z.Q. Wang, Y.S. Li, F. Jona, P.M. Marcus: Solid State Commun. **61**, 623 (1987)

3.32   D.T. Jiang, N. Alberding, A.J. Seary, B. Heinrich, E.D. Crozier: Physica B **158**, 662 (1989)

3.33   V.L. Moruzzi, P.M. Marcus: Phys. Rev. B **38**, 1613 (1988)

3.34   J.F. Cochran, J. Rud, W.B. Muir, B. Heinrich, Z. Celinski: Phys. Rev. B **42**, 508 (1990)

3.35   P. Grunberg: J. Appl. Phys. **57**, 3673 (1985)

3.36   Kh.M. Pashaev, D.L. Mills: Phys. Rev. B **43**, 1187 (1991); J. Barnas: J. Magn. Magn. Mat. **102**, 319 (1991)

3.37   B. Heinrich, Z. Celinski, J.F. Cochran, W.B. Muir, J. Rudd, Q.M. Zhong, A.S. Arrott, K. Myrtle, J. Kirschner: Phys. Rev. Lett. **64**, 673 (1990)

3.38   Z. Celinski, B. Heinrich, J. F. Cochran, W.B. Muir, A.S. Arrott, J. Kirschner: Phys. Rev. Lett. **65**, 1156 (1990)

3.39   J.J. Krebs, P. Lubitz, A. Chaiken, G.A. Prinz: Phys. Rev. Lett. **63**, 1645 (1989)

3.40   Z. Celinski, B. Heinrich: J. Magn. Magn. Mat. **99**, L25 (1991)

Section 3.2

3.41   V. Kambersky: Can. J. Phys. **48**, 2906 (1970). Also V. Kambersky, J.F. Cochran, J.M. Rudd: J. Magn. Magn. Mat. 104–107, 2089 (1992)

3.42   B. Heinrich, J.F. Cochran, A.S. Arrott, S.T. Purcell, K.B. Urquhart, J.R. Dutcher, W.F. Egelhoff, Jr.: Appl. Phys. A **49**, 473 (1989)

3.43   S.H. Lu, J. Quinn, D. Tian, F. Jona, P.M. Marcus: Surf. Sci., **209**, 264 (1989)

3.44   J.R. Scandercock: "Trends in Brillouin Scattering: Studies of Opaque Materials, Supported Films, and Central Modes" in *Light Scattering in Solids III*, Topics in Applied physics, Vol. 51. Ed. by M. Cardona, G. Güntherodt (Springer, Berlin Heidelberg, 1982) Chap. 6

3.45   S.M. Lindsay, M.W. Anderson, J.R. Sandercock: Rev. Sci. Instrum. **52**, 1478 (1981)

3.46   J.G. Dil, N.C.J.A. van Hijningen, F. van Dorst, R.M. Aarts: Appl. Opt. **20**, 1374 (1981)

3.47   R. Mock, B. Hillebrands, J.R. Sandercock: J. Phys. E **20**, 656 (1987)

3.48   M. H. Grimsditch: "Brillouin Scattering from Metallic Superlattices" in *Light Scattering in Solids V*, Topics in Applied Physics, Vol 66. ed. by M. Cardona, G. Güntherodt (Springer, Berlin, Heidelberg, 1989) Chap. 7

3.49   F. Nizzoli, J.R. Sandercock: "Surface Brillouin Scattering from Phonons" in *Dynamical Properties of Solids*, Vol. 6. ed. by G.K. Horton, A.A. Maradudin (North-Holland, Amsterdam, 1990) Chap. 5

3.50   This simplification is called the Born approximation.

3.51   H.S. Bennett, E.A. Stern: Phys. Rev. **137**, A 448 (1965)

3.52   R.E. Camley, T.S. Rahman, D.L. Mills: Phys. Rev. B **23**, 1226 (1981)

3.53   J.F. Cochran, J.R. Dutcher: J. Magn. Magn. Mat. **73**, 299 (1988)

3.54   B. Heinrich, S.T. Purcell, J.R. Dutcher, K.B. Urquhart, J.F. Cochran, A.S. Arrott: Phys. Rev. B **38**, 12879 (1988)

3.55 B. Heinrich, J.F. Cochran, A.S. Arrott, S.T. Purcell, K.B. Urquhart, J.R. Dutcher, W.F. Egelhoff, Jr.: Appl. Phys. A **49**, 473 (1989)

3.56 P. Grünberg: "Light Scattering from Spin Waves in Thin Films and Layered Magnetictures" in *Light Scattering in Solids V*, Topics in Applied Physics, Vol. 66. ed. by M. Cardona, G. Güntherodt (Springer, Berlin, Heidelberg, 1989) Chap. 8

3.57 B. Heinrich, Z. Celinski, J.F. Cochran, A.S. Arrott, K. Myrtle: J. Appl. Phys. **70**, 5769 (1991)

3.58 R.P. Erickson, D.L. Mills: Phys. Rev. B **43**, 10715 (1991)

3.59 W. Wettling, M.G. Cottam, J.R. Scandercock: J. Phys. C **8**, 211 (1975)

3.60 P.B. Johnson, R.W. Christy: Phys. Rev. B **9**, 5056 (1974)

3.61 G.S. Krinchik, V.A. Artem ev: zh. Eksp. Teor. Fiz., **53**, 1901 (1967). (English transl: Sov. Phys. JETP, **26**, 1080 1968)

3.62 L. Landau, E. Lifshitz: Phys. Z. Sowjetunion, **8**, 153 (1935)

3.63 William Fuller Brown, Jr: *Micromagnetics* (Robert E. Krieger Publishing Co., Huntington, N.Y. 1978) Chap. 3

3.64 It can be shown using a discrete lattice of interacting spins that the normal derivative of the magnetization density must vanish at a surface in the long wave length limit if there are no surface torques acting on the surface spins [3.65, 66]. It can also be demonstrated using a continuum model that the normal derivatives of the magnetization density must vanish at any surface with which there is associated no surface anisotropy energy term [3.63]

3.65 C. Kittel: Phys. Rev. **110**, 1295 (1958)

3.66 P. Pincus: Phys. Rev. **118**, 658 (1960)

3.67 R.W. Damon, J.R. Eshbach: J. Phys. Chem. Solids **19**, 308 (1961)

3.68 J.D. Jackson: *Classical Electrodynamics*, Second ed. (John Wiley and Sons, N.Y. 1975) Chap. 9

3.69 J.R. Dutcher: "Brillouin Light Scattering Studies of Epitaxial Ferromagnetic Films", PhD Thesis, Simon Fraser University 1988, Pages 254–255

3.70 A.E. Siegman: *An Introduction to Lasers and Masers* (McGraw-Hill, N.Y. 1971) Sect. 8–2

3.71 A.K. Schmidt, J. Kirschner: Ultramicroscopy 42–44, 483 (1992). Andreas Schmidt, Dissertation, Fachbereich Physik, Freie Universität, Berlin, 1991

3.72 (3.64) has been written using the Lorentz guage for which $e = -\,\mathrm{grad}\,V - \dfrac{1}{c}\dfrac{\partial A}{\partial t}$ and div A
$$+ \frac{1}{c}\frac{\partial V}{\partial t} = 0.$$

3.73 F. Hoffmann, A. Stankoff, H. Pascard: J. Appl. Phys. **41**, 1022 (1970)

3.74 J.F. Cochran, J.R. Dutcher: J. Appl. Phys. **64**, 6092 (1988)

3.75 M. Vohl, J. Barnas, P. Grünberg: Phys. Rev. B **39**, 12003 (1989)

3.76 B. Hillebrands: Phys. Rev. B **37**, 9885 (1988); and Phys. Rev. B **41**, 530 (1990)

3.77 B. Dieny, J.P. Gavigan, J.P. Rebouillat: J. Phys. Condens. Matter **2**, 159 (1990)

3.78 B. Dieny, J.P. Gavigan: J. Phys. Condens. Matter **2**, 187 (1990)

3.79 J.F. Cochran, J. Rudd, W.B. Muir, B. Heinrich, Z. Celinski: Phys. Rev. B **42**, 508 (1990)

3.80 D.A. Steigerwald, I. Jacob, W.F. Egelhoff, Jr.: Surf. Sci. **202**, 472 (1988)

3.81 Dr. J.R. Sandercock; Zwillikerstrasse 8, CH-8910 Affoltern a.A., Switzerland

3.82 R.L. Stamps, B. Hillebrands: Phys. Rev. **43**, 3532 (1991)

3.83 J.R. Dutcher, J.F. Cochran, I. Jacob, W.F. Egelhoff, Jr.: Phys. Rev. B **39**, 10430 (1989)

3.84 B. Hillebrands, P. Baumgart, G. Güntherodt: Phys. Rev. B **36**, 2450 (1987)

3.85 J.F. Cochran, B. Heinrich, A.S. Arrott, K.B. Urquhart, J.R. Dutcher, S.T. Purcell: J. Phys. (Paris), Colloque C8, Supplement au no. 12, **49**, C8-1671 (1988)

3.86 B. Heinrich, Z. Celinski, J.F. Cochran, W.B. Muir, J. Rudd, Q.M. Zhong, A.S. Arrott, K. Myrtle: Phys. Rev. Lett. **64**, 673 (1990)

3.87 J.R. Dutcher, B. Heinrich, J.F. Cochran, D.A. Steigerwald, W.F. Egelhoff, Jr.: J. Appl. Phys. **63**, 3464 (1988)

3.88 J.F. Cochran, W.B. Muir, J.M. Rudd, B. Heinrich, Z. Celinski, Tan-Trung Le-Tran, W. Schwarzacher, W. Bennett, W.F. Egelhoff, Jr.: J. Appl. Phys. **69**, 5206 (1991)

3.89 Z. Celinski, B. Heinrich, J.F. Cochran, W.B. Muir, A.S. Arrott, J. Kirschner: Phys. Rev. Lett. **65**, 1156 (1990)

3.90 W.B. Muir, J.F. Cochran, J.M. Rudd, B. Heinrich, Z. Celinski: J. Magn. Magn. Mat. **93**, 229 (1991)

3.91 D. Kerkmann, J.A. Wolf, D. Pescia, Th. Woike, P. Grünberg: Solid State Commun. **72**, 963 (1989)

3.92 B. Hillebrands, P. Baumgart, G. Güntherodt: Appl. Phys. A **49**, 589 (1989)

3.93 P. Baumgart, B. Hillebrands, G. Güntherodt: J. Magn. Magn. Mat. **93**, 225 (1991)

Section 3.3

3.94 J.R. Sandercock: Springer Ser. Topics Appl. Phys. **51**, 173 (1982)

3.95 R. Mock, B. Hillebrands, J.R. Sandercock: J. Phys. E **20**, 656 (1987)

3.96 R.E. Camley, T.S. Rahman, D.L. Mills: Phys. Rev. B **27**, 261 (1983)

3.97 P. Grünberg, K. Mika: Phys. Rev. B **27**, 2955 (1983)

3.98 P.R. Emtage, M.R. Daniel: Phys. Rev. B **29**, 212 (1984)

3.99 G. Rupp, W. Wettling, W. Jantz: Appl. Phys. A **42**, 45 (1987)

3.100 R.P. van Stapele, F.J.A.M. Greidanus, J.W. Smits: J. Appl. Phys. **57**, 1282 (1985)

3.101 L. Dobrzynski, B. Djafari-Rouhani, H. Puszkarski: Phys. Rev. B **33**, 3251 (1986)

3.102 E.L. Albuquerque, P. Fulco, E.F. Sarmento, D.R. Tilley: Solid State Commun. **58**, 41 (1986)

3.103 L.L. Hinchey, D.L. Mills: Phys. Rev. B **33**, 3329 (1986)

3.104 K. Vayhinger, H. Kronmüller: J. Magn. Magn. Mat. **62**, 159 (1986)

3.105 K. Vayhinger, H. Kronmüller: J. Magn. Magn. Mat. **72**, 307 (1986)

3.106 J. Barnaś: J. Phys. C **21**, 1021 (1988)

3.107 J. Barnaś: J. Phys. C **21**, 4097 (1988)

3.108 J. Barnaś: Phys. Rev. B **45**, 10427 (1992)

3.109 B. Hillebrands: Phys. Rev. B **37**, 9885 (1988)

3.110 B. Hillebrands: Phys. Rev. B **41**, 530 (1990)

3.111 R.L. Stamps, B. Hillebrands: J. Appl. Phys. **69**, 5718 (1991)

3.112 R.L. Stamps, B. Hillebrands: Phys. Rev. B **44**, 5095 (1991)

3.113 R.L. Stamps, B. Hillebrands: J. Magn. Magn. Mat. **93**, 616 (1991)

3.114 G.T. Rado, R.J. Hicken: J. Appl. Phys. **63**, 3885 (1988)

3.115 J.F. Cochran, J.R. Dutcher: J. Appl. Phys. **63**, 3814 (1988)

3.116 R.L. Stamps, B. Hillebrands: Phys. Rev. B **44**, 12417 (1991)

3.117 F. Hoffmann, A. Stankoff, H. Pascard: J. Appl. Phys. **41**, 1022 (1970)

3.118 F. Hoffmann: Phys. Status Solidi **41**, 807 (1970)

3.119 G.T. Rado, J.R. Weertman: J. Phys. Chem. Solids **11**, 315 (1959)

3.120 We have corrected the Hoffman boundary condition by adding the term (a $\partial M_{n'}/\partial n_{n'}$) to the square bracket in the second term, in order to obtain a consistent form for the strong interlayer-coupling case

3.121 This requirement is not fulfilled, if the effective internal field, i.e., the sum of the external field, the demagnetizing field and the anisotropy fields, is close to zero

3.122 In [3.110] the right hand sides of (20, 21) are accidentally permuted

3.123 N.S. Almeida, D.L. Mills: Phys. Rev. B **38**, 6698 (1988)

3.124 F.C. Nörtemann, R.L. Stamps, R.E. Camley, B. Hillebrands, G. Güntherodt: Phys. Rev. B **47**, 3225 (1993)

3.125 B. Hillebrands, A. Boufelfel, C.M. Falco, P. Baumgart, G. Güntherodt, E. Zirngiebl, J.D. Thompson: J. Appl. Phys. **63**, 3880 (1988)

3.126 C.M. Falco: J. Phys. (Paris) Colloq. **45**, C5-499 (1984)

3.127 B. Hillebrands, P. Baumgart, R. Mock, G. Güntherodt, A. Boufelfel. C.M. Falco: Phys. Rev. B **34**, 9000 (1986)

3.128 B. Hillebrands, P. Baumgart, G. Güntherodt: Appl. Phys. A **49**, 589 (1989)

3.129 This holds for zero anisotropies. A general rule for the existence of a distinct stack surface mode is given in [3.112]

3.130 W.B. Zeper, F.J.A.M. Greidanus, P.F. Carcia, C.R. Fincher: J. Appl. Phys. **65**, 4971 (1989)

3.131 The light scattering cross section is proportional to the net part of the fluctuating magnetization, which is zero for exchange-type modes. Only dipolar contributions contribute to the cross section

3.132 C. Demangeat, D.L. Mills: Phys. Rev. B **16**, 2321 (1977)

3.133 J.G. LePage, R.E. Camley: Phys. Rev. Lett. **65**, 1152 (1990)

3.134 P. Grünberg, R. Schreiber, Y. Pang, M.B. Brodsky, H. Sowers: Phys. Rev. Lett. **57**, 2442 (1986)

3.135 M. Vohl, J. Barnaś, P. Grünberg: Phys. Rev. B **39**, 12003 (1989)

3.136 J. Barnaś, P. Grünberg: J. Magn. Magn. Mat. **82**, 186 (1989)

3.137 B. Heinrich, S.T. Purcell, J.R. Dutcher, K.B. Urquhart, J.F. Cochran, A.S. Arrott: Phys. Rev. B **38**, 12879 (1988)

3.138 J.F. Cochran, J.R. Dutcher: J. Appl. Phys. **64**, 6092 (1988)

3.139 J.V. Harzer, B. Hillebrands, R.L. Stamps, G. Güntherodt, C.D. England, C.M. Falco: J. Appl. Phys. **69**, 2448 (1991)

3.140 A possible enhancement of the Co moment for ultrathin Co layers is not included, since here large interface anisotropies dominate over the saturation magnetization anyway.

3.141 B. Hillebrands, J.V. Harzer, G. Güntherodt, C.D. England, C.M. Falco: Phys. Rev. B **42**, 6839 (1990)

3.142 B. Hillebrands, J.V. Harzer, R.L. Stamps, G. Güntherodt, C.D. England, C.M. Falco: J. Magn. Magn. Mat. **93**, 211 (1991)

3.143 B. Hillebrands, J.V. Harzer, G. Güntherodt, D. Weller, B.N. Engel, J. Magn. Soc. Jpn. 17 Sup. Sl, 17 (1993)

3.144 S.S.P. Parkin, N. More, K.P. Roche: Phys. Rev. Lett. **64**, 2304 (1990)

3.145 G. Binasch, P. Grünberg, F. Saurenbach, W. Zinn: Phys. Rev. B **39**, 4828 (1989)

3.146 B. Heinrich, Z. Celinski, J.F. Cochran, W.B. Muir, J. Rudd, Q.M. Zhong, A.S. Arrott, K. Myrtle: Phys. Rev. Lett. **64**, 673 (1990)

3.147 J.F. Cochran, J. Rudd, W.B. Muir, B. Heinrich, Z. Celinski: Phys. Rev. B **42**, 508 (1990)

3.148 D.H. Mosca, F. Petroff, A. Fert, P.A. Schroeder, W.P. Pratt Jr., R. Laloee: J. Magn. Magn. Mat. **94**, L1 (1991)

3.149 S.S.P. Parkin, Phys. Rev. Lett. **67**, 3598 (1991)

3.150 S.S.P. Parkin, R. Bhadra, K.P. Roche: Phys. Rev. Lett. **66**, 2152 (1991)

3.151 F. Petroff, A. Barthelemy. D.H. Mosca, D.K. Lottis, A. Fert, P.A. Schroeder, W.P. Pratt Jr., R. Loloee, S. Lequien: Phys. Rev. B **44**, 5355 (1991)

3.152 S.S.P. Parkin, A. Mansour, G.P. Felcher: Appl. Phys. Lett. **58**, 1473 (1991)

3.153 S. Demokritov, J.A. Wolf, P. Grünberg: Europhys. Lett. **15**, 881 (1991)

3.154 J. Unguris, R.J. Celotta, D.T. Pierce: Phys. Rev. Lett. **67**, 140 (1991)

3.155 S.T. Purcell, W. Folkerts, M.T. Johnson, N.W.E. Mc Gee, K. Jager, J. ann de Stegge, W.B. Zeper, W. Hoving, P. Grünberg: Phys. Rev. Lett. **67**, 903 (1991)

3.156 P. Grünberg, S. Demokritov, A. Fuß, R. Schreiber, J.A. Wolf, S.T. Purcell: J. Magn. Magn. Mat. **106**, 1734 (1992)

3.157 Z.Q. Qiu, J. Pearson, A. Berger, S.D. Bader: Phys. Rev. Lett. **68**, 1398 (1992)

3.158 A. Fuß, S. Demokritov, P. Grünberg, W. Zinn: J. Magn. Magn. Mat. **103**, L221 (1992)

3.159 M.N. Baibich, J.M. Broto, A. Fert, F. Nguyen Van Dau, F. Petroff, P. Eitenne, G. Creuzet, A. Friederich, J. Chazelas: Phys. Rev. Lett. **61**, 2472 (1988)

3.160 S.S.P. Parkin, Z.G. Li, D.J. Smith: Appl. Phys. Lett. **58**, 2710 (1991)

3.161 B. Dieny, V.S. Speriosu, S. Metin, S.S.P. Parkin, B.A. Gurney, P. Baumgart, D.R. Wilhoit: J. Appl. Phys. **69**, 4774 (1991)

3.162 S.S.P. Parkin, D. Mauri: Phys. Rev. B **44**, 7131 (1991)

3.163 J. Faßender, F. Nörtemann, R.L. Stamps, R.E. Canley, B. Hillebrands, G. Güntherodt, S.S.P. Parkin: Phys. Rev. B **46**, 5810 (1992)

3.164 J.V. Harzer, B. Hillebrands, R.L. Stamps, G. Güntherodt, D. Weller, Ch. Lee, R.F.C. Farrow, E.E. Marinero: J. Magn. Magn. Mat. **104–107**, 1863 (1992)

3.165 F.J.A. den Broeder, D. Kuiper, A.P. van de Mosselaer, W. Hoving: Phys. Rev. Lett. **60**, 2769 (1988)
3.166 P. Krams, B. Hillebrands, G. Güntherodt, K. Spörl, D. Weller: J. Appl. Phys. **69**, 5307 (1991)
3.167 R.L. Stamps, R.E. Camley, B. Hillebrands, G. Güntherodt: Phys. Rev. B **47**, 5072 (1993)
3.168 R. Van Leeuwen, C.D. England, J.R. Dutcher, C.M. Falco, W.R. Bennett, B. Hillebrands: J. Appl. Phys. **67**, 4910 (1990)
3.169 H. Litschke, M. Schilberg, Th. Kleinefeld, B. Hillebrands: J. Magn. Magn. Mat., **104–107**, 1807 (1992)
3.170 R.L. Stamps, B. Hillebrands: J. Magn. Magn. Mat. **104–107**, 1868 (1992)

Section 3.4

3.171 An extensive mathematical and quantum mechanical treatment of NMR can be found in: A. Abragam: *Principles of Nuclear Magnetism* (Oxford University Press, London, 1961); C.P. Slichter: *Principles of Magnetic Resonance*, in Springer Ser. Solid-State Sci. Vol. **1** (Springer, Berlin, Heidelberg, 1980)
3.172 General reviews including solid state applications of NMR can be found in: M.A.H. McGausland, I.S. Mackenzie: Adv. Phys. **28**, 305 (1979): P.C. Riedi Hyp. Int. **49**, 335 (1989), P. Panissod: Chapter 12 in *Microscopic Methods in Metals*, ed. by U. Gonser (Springer, Berlin, Heidelberg 1986)
3.173 H. Akai, M. Akai, S. Blügel, B. Drittler, H. Ebert, K. Terakura, R. Zeller, P.H. Dederichs: Prog. Theor. Phys. Suppl. **101**, 11 (1990)
3.174 K. Lee, W. Anderson: in *CRC Handbook of Chemistry and Physics*, ed. by R.C. Weast, M.J. Astle, W.H. Beyer (CRC Press, Boca Raton, 1988) p. E80–E85
3.175 E.A. Turov, M.P. Petrov: *Nuclear Magnetic Resonance in Ferro- and Antiferromagnets* (Wiley, New York, 1972)
3.176 A.J. Freeman, C. Li, R.Q. Wu; "Electronic Structure and Magnetism of Metal Surfaces, Overlayers and Interfaces" in *Science and Technology of Nanostructured Magnetic Materials*, ed. by G.C. Hadjipanayis, G.A. Prinz (Plenum Press, New York, 1991) p 1 ff.
3.177 H. Brömer, H.L. Huber: J. Magn. Magn. Mat. **8**, 61 (1978)
3.178 T.M. Shavishvili, I.G. Kiliptari: Phys. Status Solidi B **92**, 39 (1979)
3.179 P.C. Riedi, R.G. Scurlock: J. Appl. Phys. **39**, 1241 (1968)
3.180 S. Nasu, H. Yasuoka, Y. Nakamura, Y. Murakami: Acta Metall. **22**, 1057 (1974)
3.181 K. Le Dang, P. Veillet, Hui He, F.J. Lamelas, C.H. Lee, R. Clarke: Phys. Rev. B **41**, 12902 (1990)
3.182 J.F. Janak: Phys. Rev. B **20**, 2206 (1979)
3.183 E.A.M. van Alphen, H.A.M. de Gronckel, P.J.H. Bloemen, A.S. van Steenbergen, W.J.M. de Jonge: Proceedings of the E-MRS Symposium on Ultra Thin Films, Multilayers and Surfaces (Lyon 1992), published in J. Magn. Magn. Mat. **121**, 77 (1993)
3.184 C.R. Houska, B.L. Averbach, M. Cohen: Acta Metall. **8**, 81 (1960)
3.185 For instance, the contribution in this book and references therein by W.J.M. de Jonge, P.J.H. Bloemen, F.J.A. den Broeder
3.186 J. Dekoster, E. Jedryka, G. Meny, C. Langouche: Proceedings of the E-MRS Symposium on Ultra Thin Films, Multilayers and Surfaces (Lyon 1992), published in J. Magn. Magn. Mat. **121**, 69 (1993)
3.187 Ph. Houdy, P. Boher, F. Giron, F. Pierre, C. Chappert, P. Beauvillain, K. Le Dang, P. Veillet, E. Velu: J. Appl. Phys. **69**, 5667 (1991)
3.188 C. Meny, P. Panissod, R. Loloee: Phys. Rev. B **45**, 12269 (1992)
3.189 These results are reviewed in: H. Yasuoka: Chapter 5 in *Metallic Superlattices*, ed. by T. Shinjo, T. Takada (Elsevier, Amsterdam, 1987)
3.190 H.A.M. de Gronckel, K. Kopinga, W.J.M. de Jonge, P. Panissod, J.P. Schillé, F.J.A. den Broeder: Phys. Rev. B **44**, 9100 (1991)

3.191 H.A.M. de Gronckel, B.M. Mertens, P.J.H. Bloemen, K. Kopinga, W.J.M. de Jonge: J. Magn. Magn. Mat. **104–107**, 1809 (1992)
3.192 Y. Suzuki, T. Katayama, H. Yasuoka, J. Magn. Magn. Mat. **104–107**, 1843 (1992)
3.193 K. Inomata, Y. Saito, S. Hashimoto: in Proceedings of the E-MRS Symposium on Ultra Thin Films, Multilayers and Surfaces (Lyon 1992), published in J. Magn. Magn. Mat. **121**, 350 (1993)
3.194 H.A.M. de Gronckel, C.H.W. Swüste, K. Kopinga, W.J.M. de Jonge: Appl. Phys. A **49**, 467 (1989)
3.195 C. Cesari, J.P. Faure, G. Nihoul, K. le Dang, P. Veillet, D. Renard: J. Magn. Magn. Mat. **78**, 296 (1989)

4. Magneto-Optical Effects in Ultrathin Magnetic Structures

S.D. Bader and J.L. Erskine

The magneto-optic Kerr effect has provided an important new means of probing a broad range of thin film magnetic properties. This chapter covers recent developments and new applications of magneto-optical techniques with an emphasis on phenomena encountered in thin film structures. No attempt has been made to include all relevant topics nor to reference all important work in the field. The objective has been to provide a balanced summary of magneto-optic Kerr effect applications that complement topics covered in other chapters of this two volume set on ultrathin magnetic structures.

4.1 Microscopic Basis

Magneto-optical effects in ferromagnetic materials are produced by a combination of the net spin polarization that exists in the ferromagnetic state and the spin–orbit coupling [4.1]. The spin–orbit interaction couples the spin components of the electron wavefunctions to the spatial components which govern the electric dipole matrix elements and optical selection rules.

Manifestations of ferromagnetic behavior observed as changes of polarization and/or intensity when light is reflected from a magnetic material are called magneto-optic Kerr effects. The general property that distinguishes magneto-optic Kerr effects from other magneto-optical effects in solids is that all manifestations of the Kerr effect are proportional to the magnetization $M(T)$ and vanish at temperatures above the Curie temperature T_C.

The magneto-optical response of a magnetic material can be calculated (in principle) in the same manner that the optical response of a non-magnetic material is calculated, the difference being that calculations of the magneto-optical response require carrying out the evaluation of dipole matrix elements to first order in the spin–orbit terms (for both the wavefunctions and the momentum operator) [4.2]. Various regimes [4.1–8] have been used as a basis for calculating the magneto-optical response. In the long-wavelength limit, a Drude-like treatment [4.6] based on the Boltzmann equation has been used. In the visible region [4.5, 6] the optical and magneto-optical response is governed by the electronic structure, and realistic models must be based on conduction band wavefunctions. In the vacuum-UV/X-ray region, where core

B. Heinrich and J.A.C. Bland (Eds.)
Ultrathin Magnetic Structures II
© Springer-Verlag Berlin Heidelberg 1994

level excitations occur, simple models [4.7] based on atomic wavefunctions and band density of states can be used to obtain qualitative results.

Formal approaches for calculating magneto-optical effects are generally based on calculating the difference in absorption of left (LCP) and right circularly (RCP) polarized light [4.2, 6, 7]. In a magnetized material, time-reversal symmetry is broken, and separate wave vectors are required to describe the propagation of polarized light with left or right helicity. This introduces off-diagonal elements in the conductivity tensor. Taking the $\bar{z}$ direction along the magnetization, M, the conductivity tensor becomes

$$\tilde{\sigma}(\omega) = \begin{pmatrix} \sigma_{xx} & \sigma_{xy} & 0 \\ -\sigma_{yx} & \sigma_{xx} & 0 \\ 0 & 0 & \sigma_{zz} \end{pmatrix}, \tag{4.1}$$

where the diagonal terms are even functions of M (independent of M to first order), and the off-diagonal terms are odd functions of M (linear dependence on M to first order). All elements of $\sigma(\omega)$ are complex quantities. The absorptive part of the diagonal terms is proportional to the total optical absorption (sum of the absorption for LCP and RCP light). The absorptive part of the off-diagonal terms is proportional to the difference in absorption of LCP and RCP light. Under certain circumstances (i.e., at low photon energies where specific bands near E_F can be identified as the initial and final states), the sign of σ_{xy} can be used to determine the spin polarization. This property was used to argue [4.8] that early spin-polarized photoemission experiments had been incorrectly interpreted.

Formal expressions for calculating magneto-optical absorption have been derived by various authors. The wavelength dependence of σ_{xy} has been derived based on a specific band model of ferromagnetic iron [4.5] over a limited energy range (states restricted to those near E_F). Numerical evaluations [4.9, 10] of σ_{xy} for transition metals also have been performed over the entire visible optical range (covering the d-bands of ferromagnetic transition metals). The treatments based on specific band models are not particularly transparent in terms of elucidating how spin–orbit effects and a net spin polarization produces the magneto-optical effects. However, the formal expressions [4.2, 6] and the first calculations predicting magneto-optical effects (MCD) in the UV/X-ray region [4.7] based on a simple atomic model provide clear examples.

While it is possible, in principle, to obtain information about the band structure of magnetic metals based on their optical and magneto-optical response, angle-resolved photoemission, including experiments that detect spin polarization, provides a more incisive tool for detailed studies of the spin resolved electronic structure. Therefore, most of the applications of magneto-optical spectroscopy to ferromagnetic metals have not required detailed calculation of the magnetic contribution to the optical response.

4.2 Macroscopic Formulas

Macroscopic descriptions [4.11] of the optical and magneto-optical response relate measurable parameters such as reflectance, polarization changes, and optical phase shifts, to general parameters that describe the media response, i.e., the conductivity tensor, dielectric tensor, or index of refraction. The descriptions for magnetic materials are much more complicated than for nonmagnetic materials because the direction of the magnetization with respect to the plane of incidence and the angles of refraction enter the formulas. For example, in Fig. 1, three distinct magneto-optical configurations are shown. In the polar and longitudinal configurations, the magneto-optic effect consists of an M-dependent change in elliptical polarization of the reflected beam. In the transverse configuration, the "p" component (the component confined to the plane of incidence) of the reflected beam exhibits an M-dependent change in intensity. In all cases, the changes are proportional to M. The goal of macroscopic formalisms is to obtain formulas that describe these observable effects in terms of the magnetization directions, incident angles and properties of the media characterized by $\sigma(\omega)$. In thin film systems, the film thickness and film structure (i.e., a multilayer) must also be treated within the macroscopic formalism, which introduces additional parameters and complexity.

Macroscopic formulas in optics are based on Snell's law and utilize Fresnel transmission and reflection coefficients. In this manner, both the boundary conditions for the components of the electromagnetic fields at interfaces and the phase relations of the light propagating from one medium to another can be satisfied. In magnetically active materials, the same approach is taken, but the off-diagonal couplings of the dielectric tensor add algebraic complexity to the problem. The index of refraction takes on different values for the left and right circularly polarized components of the light. If the light impinges on the film at an oblique angle, rather than at normal incidence, two distinct beams enter the film at slightly different angles. Many formalisms have been developed to describe magneto-optical effects in magnetic materials. It has been suggested

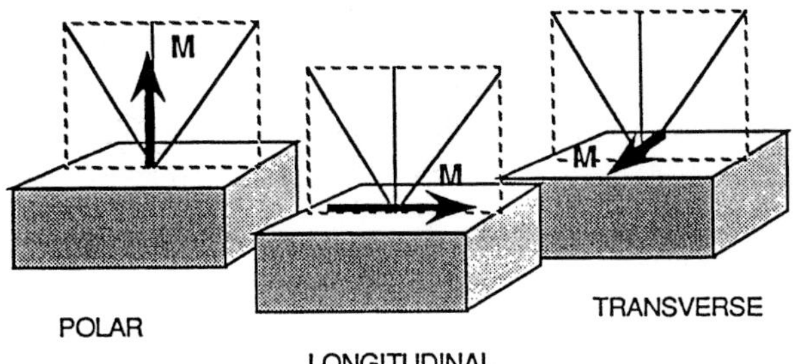

Fig. 4.1. Three high symmetry configurations used in magneto-optic Kerr effect measurements

that it is sometimes easier to derive such expressions oneself than to follow the derivations of others. In that vein *Zak* and coworkers [4.12, 13] recently set out to provide useful guidelines appropriate to the ultrathin film regime and more generally to multilayer film configurations. The result is a very general formulation of the multiple reflection problem that can describe magneto-optic effects (to first order in the magnetic field) in layered magnetic materials having magnetization vectors in each layer in arbitrary directions.

Zak et al. start by considering that a beam of light traveling from medium 1 to medium 2 conserves the tangential components of its electric E_x, E_y and magnetic H_x, H_y fields, where the xy plane is the boundary between the two media. Expressed in terms of the electric fields of the incident (i) and reflected (r) waves, we have

$$F = \begin{pmatrix} E_x \\ E_y \\ H_x \\ H_y \end{pmatrix} = AP = A \begin{pmatrix} E_s^{(i)} \\ E_p^{(i)} \\ E_s^{(r)} \\ E_p^{(r)} \end{pmatrix}, \tag{4.2}$$

where the 4×4 matrix A that connects column vectors F and P is referred to traditionally as the *medium boundary matrix*, analogous to the refraction matrix D in polarized neutron reflection (PNR). See the Chap. 6 by *Bland* in Volume 1. The matrix elements of A are constructed from the geometric angles of the problem and from the N and Q-values of the medium, where N is the refractive index in the absence of a net magnetization, and Q is the magneto-optic Voigt constant that describes the off-diagonal couplings. For a two-medium, one-boundary problem, the boundary matching condition becomes

$$A_1 P_1 = A_2 P_2. \tag{4.3}$$

If there is more than one boundary, the wave propagation inside of the medium at depth z from the interface is described using the *medium propagation matrix* $\bar{L}$ where

$$P_2(z = 0) = \bar{L}_2(z) P_2(z). \tag{4.4}$$

The analogous propagation matrix also appears in PNR, as discussed in Chap. 6, Volume I. For a multilayer system, the light originates in the initial medium i, goes through the multilayer stack, and ends up in the substrate or final medium f. The information of interest for l layers in the stack is contained in the expression

$$A_i P_i = \prod_{m=1}^{l} (A_m \bar{L}_m A_m^{-1}) A_f P_f. \tag{4.5}$$

If this expression is put in the form $P_i = TP_f$, where

$$T = A_i^{-1} \prod_m A_m \bar{L}_m A_m^{-1} A_f \equiv \begin{pmatrix} G & H \\ I & J \end{pmatrix}, \tag{4.6}$$

then the 2×2 matrices G and I can be used to obtain the Fresnel transmission t and reflection coefficients, since

$$G^{-1} = \begin{pmatrix} t_{ss} & t_{sp} \\ t_{ps} & t_{pp} \end{pmatrix} \quad \text{and} \quad IG^{-1} = \begin{pmatrix} r_{ss} & r_{sp} \\ r_{ps} & r_{pp} \end{pmatrix}. \tag{4.7}$$

The Kerr rotation ϕ' and ellipticity ϕ'' for s- and p-polarized light are then expressed as

$$\phi_s = \phi'_s + i\phi''_s = \frac{r_{ps}}{r_{ss}} \quad \text{and} \quad \phi_p = -\phi'_p + i\phi''_p = \frac{r_{sp}}{r_{pp}}. \tag{4.8}$$

Prescriptions for constructing the A and $\bar{L}$ matrices appear in [4.13] and in the references cited therein.

The above expressions simplify and provide useful insights in the ultrathin limit, which is defined by

$$\frac{2\pi}{\lambda}|N|d \ll 1, \tag{4.9}$$

where λ is the wavelength of the light and d is the thickness of the layer [4.14]. Equation (4.9) defines the condition generally understood to distinguish what has been known as the surface magneto-optic Kerr effect (SMOKE) from the traditional effect associated with reflection from a magnetized bulk sample – the magneto-optic Kerr effect (MOKE). In the ultrathin limit for a magnetic overlayer, characterized by N, Q and t, on a non-magnetic substrate with refractive index N_{sub}, the polar (POL) and longitudinal (LON) Kerr effects become

$$\phi^{POL} = -\left(\frac{4\pi}{\lambda}\right)\left(\frac{N^2}{1-N_{sub}^2}\right)Qt \quad \text{and} \quad \phi^{LON} = \left(\frac{4\pi}{\lambda}\right)\left(\frac{N_{sub}}{1-N_{sub}^2}\right)\theta Qt, \tag{4.10}$$

where θ, the angle of incidence measured from the surface normal, is assumed small. Equation (4.10) illustrates that $\phi^{POL} > \phi^{LON}$ because of the extra N-factor and the lack of a θ-factor in ϕ^{POL}. Note that ϕ^{LON} is independent of the refractive index of the magnetic layer. This implies that ϕ^{LON} can be enhanced by choosing a substrate with an appropriate value of N_{sub}. Equation (4.10) also demonstrates that the SMOKE signal is proportional to t, as in the Faraday effect, and unlike the bulk Kerr effect, which is independent of t. In this important respect SMOKE is distinct from MOKE.

For a multilayer stack in the ultrathin limit there is an additivity law whereby the total ϕ^{POL} and ϕ^{LON} are represented by the appropriate expressions in (4.10) summed over the magnetic layers:

$$\phi^{POL} \propto \sum_m N_m^2 Q_m t_m \quad \text{and} \quad \phi^{LON} \propto \sum_m Q_m t_m. \tag{4.11}$$

However, for (4.11) to be valid, the condition specified by (4.9) becomes

$$\frac{2\pi}{\lambda}\left[\sum_n |N_n|\,d_n + \sum_m |N_m|\,t_m\right] \ll 1, \tag{4.12}$$

where the index n denotes the additional sum over any non-magnetic interlayers and d_n refers to the non-magnetic layer thickness. Thus, the total N-weighted thickness of the film has to be small with respect to λ for the additivity law to be valid. This is demonstrated in Fig. 4.2, where longitudinal ellipticity data for Fe/Ag superlattices are plotted as the number of bilayers in the films are increased [4.15]. Data are shown for four superlattices with different thicknesses of the individual Fe and Ag layers that make up the bilayer repeat unit. The results are plotted in Fig. 4.2a as a function of the total thickness of the superlattice, while in Fig. 4.2b the same results are plotted versus total Fe-layer thickness only. The data in Fig. 4.2b initially superimpose to form the same line.

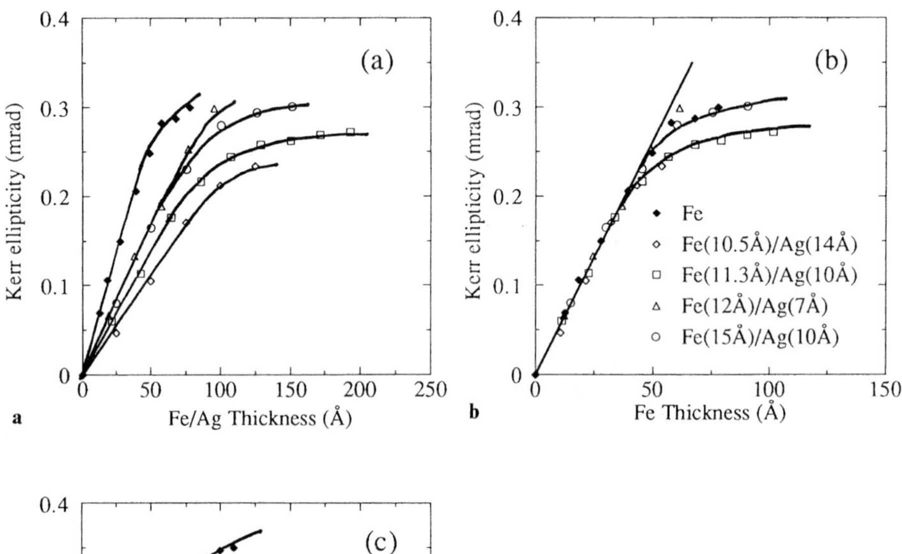

Fig. 4.2. Kerr ellipticities for Fe/Ag superlattices. The experimental data are for Fe(1 1 1)/Ag(1 1 1) superlattices versus total superlattice thickness (**a**), and Fe-only thickness (**b**). The curves are guides to the eye, but the initial linearity demonstrates the additivity law in the ultrathin regime. Simulations are shown in (**c**) utilizing the full matrix formalism described in the text

This is in accord with the expectation due to the additivity law. Having established the validity of the additivity law, it is interesting to focus on further details that appear in the data. Note that the breakaway from linearity occurs at a different Fe thickness for each of the four superlattices. This is due to the limitation that the additivity law is only applicable in the ultrathin regime, and that regime is defined with respect to the total film thickness (cf. (4.12)). Note also that the curves in Fig. 4.2c are simulations based on the full matrix formalism presented above. The input for the simulations are the literature values tabulated for N and Q taken at the He–Ne-laser wavelength [4.14]. The agreement between experiment and simulation is quite good. This is especially interesting because there is no reason *a priori* to expect that bulk optical constants would describe such thin films. Indeed, *Suzuki* et al. [4.16] recently have observed new magneto-optical transitions for ultrathin Fe films, but their effect is most pronounced at shorter wavelengths than studied in Fig. 4.2.

The superlattice periodicity introduces new symmetries that can simplify the magneto-optic expressions further. For a superlattice (SL) of magnetic and non-magnetic layers of thicknesses d_1 and d_2, respectively, that satisfy the ultrathin criterion, it can be shown that if $N_1 \approx N_2$, there is a remarkably simple relationship to the Kerr effect of bulk films [4.17]:

$$\frac{\phi_{\mathrm{SL}}}{\phi_{\mathrm{bulk}}} = \frac{d_1}{d_1 + d_2}. \tag{4.13}$$

Thus, the non-magnetic layer acts as a dilution factor, and the signal scales as the ratio of magnetic layer thickness to the total thickness. Expressions such as the above provide guidelines for understanding the systematics of real materials. Of course, the most interesting experiments often are those that challenge simple expectation in a way that advances our state of knowledge. Such results will be discussed with respect to Co/Pt superlattices in Sect. 4.4.4 on magneto-optical media.

4.3 Instrumentation, Techniques, and Sensitivity

Two general classes of magneto-optical studies can be carried out on magnetic materials. One class of studies can be characterized as spectroscopic investigations in which the principal objective is to determine $\sigma_{xy}(\omega)$, the frequency dependent magneto-optical response. Such studies are relevant to seeking optimized magneto-optical materials and characterizing electronic excitations. Magnetic circular-dichroism experiments fall into this general class. A second general class of magneto-optical studies can be characterized as polarimetry in which the optical wavelength remains fixed, and the magneto-optic Kerr effect is used to study the magnetic response of a system as other parameters are varied (such as the temperature, film thickness, direction of applied field, etc.). This class of experiments explores critical phenomena, magnetic anisotropy, and probes for ferromagnetism at high sensitivity.

Successful application of magneto-optical techniques requires knowledge of factors that affect the performance of instrumentation, which is the subject of this subsection. Figure 4.3 illustrates the principal elements of a visible light magneto-optic Kerr effect polarimeter/spectrometer. Magneto-optic Kerr effect spectrometers require a variable wavelength source and detector, whereas polarimeters typically employ a fixed wavelength laser source. Optical viewports, which are required for *in situ* measurements, are not shown, but careful attention to birefringence/depolarization effects introduced by the viewports is important when considering quantitative measurements involving the ellipsometric parameters (rotation and ellipticity). Suitable strain/birefringence-free viewports have been developed and are commercially available [4.18].

The optimum settings and ultimate sensitivity of the Kerr effect instrument is easy to determine [4.19] based on a few simplifying assumptions. Using the parameters defined in Fig. 4.3, the detected intensity is given by

$$I = I_{0r} \sin^2(\gamma_a + \phi_k) + I_{rr} \tag{4.14}$$

where I_{0r} is the intensity of reflected light and I_{rr} is the residual intensity transmitted through the analyzing polarizer when set for minimum transmission. When depolarization effects from the sample, viewports, and phase shifter (assumed set to compensate the ellipticity) are neglected, $I_{0r}/I_{rr} = \varepsilon$, the extinction ratio of the polarizers.

In order to measure a hysteresis loop using the magneto-optic Kerr effect, one requires that the detected intensity, I, be proportional to the magnetization, M. Therefore, it is necessary to set $\gamma_a > \phi_k$ where γ_a is the analyzing prism offset angle. Under this condition, a Taylor series expansion of $I(\gamma_a)$ around γ_0 has a first order (linear) term in rotation angle ϕ_k. In practice, both ϕ_k and γ_a are small and $\sin^2(\gamma_a + \theta_k) \sim (\gamma_a + \phi_k)^2$. The figure of merit for the system is the contrast

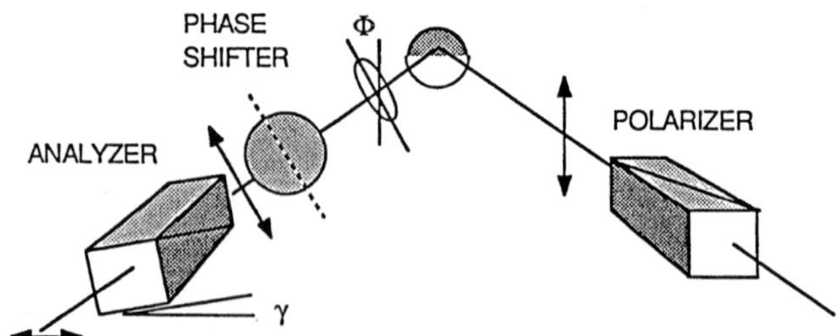

Fig. 4.3. Schematic representation of a magneto-optic Kerr effect polarimeter/spectrometer showing parameters that govern its operation and sensitivity: γ is the analyzer off-set angle, and ϕ is the magneto-optic Kerr effect rotation

C defined by

$$C \equiv \frac{\Delta I}{I} = \frac{2\gamma_a \phi_k + \phi_k^2}{\gamma_a^2 + \varepsilon}.$$ (4.15)

The maximum contrast is obtained by setting

$$\frac{\partial C}{\partial \gamma_a} = 0,$$ (4.16)

which yields

$$\gamma_m = \frac{\phi_k}{2}\left[\left(1 + \frac{4\varepsilon}{\phi_k^2}\right)\right].$$ (4.17)

In cases of practical interest, $4\varepsilon \gg \phi_k^2$ and γ_m is simply equal to $\sqrt{\varepsilon}$. In this case, the contrast becomes

$$C/C_m = \frac{2(\gamma_a/\gamma_m)}{(\gamma_a/\gamma_m)^2 + 1}$$ (4.18)

where C_m is the maximum contrast that occurs at $\gamma_a = \gamma_m$. This analysis shows that the extinction ratio governs the performance of the optical system.

Polarimeters that have been carefully set up (with attention to vibrational stability, magnetic shielding of the detector, and the selection of optics including high extinction ratios) yield sensitivities that approach the statistical limit imposed by the photon flux from the laser. The resulting sensitivity is sufficient to permit measurements of critical exponents of monolayer films.

In order to take full advantage of Kerr effect polarimetry, it is desirable to be able to apply high magnetic fields in any of the three distinct configurations shown in Fig. 4.1. In practice, a single magnet can achieve any two of the configurations by a simple rotation. Rotation between polar and transverse or polar and longitudinal configurations permit studies of the novel effects associated with perpendicular anisotropy [4.20, 21] in ultrathin magnetic films. Corresponding rotations of the magnet between the longitudinal and transverse configurations permit analysis of in-plane, thin film magnetic anisotropy [4.22]. Successful applications of both in-vacuum magnets [4.21] (using special alloys and Kapton wire) and external coil magnets [4.19, 20] with pole caps inside of the vacuum chamber have been reported. Acceptably high fields ~ 3 kOe from electromagnets having a $1/2''$ gap can be obtained with a magnet having a volume of about 30 cubic inches. Because of the large gap and relatively low fields, the magnetic field produced by these electromagnets is generally a linear function of current. However, accurate thin film hysteresis loops require field calibrations or an *in situ* probe of H at the sample. In cases where a thin film must be magnetically saturated along the surface normal direction (and no perpendicular easy axis exists), a superconducting magnet may be required.

4.4 Thin Film Phenomena and Applications

4.4.1 Monolayer Magnetism

Three related topics pertinent to single layer magnetic behavior are discussed in this subsection: (1) enhanced magnetic moments and magnetic dead layers, (2) substrate hybridization effects, and (3) thin film, epitaxy-induced ferromagnetism. These topics are also treated theoretically by *Gay* and *Richter* (Sect. 2.1, Volume 1), and by *Bland* (Chap. 6, Volume 1). Issues related to thin film anisotropy and critical phenomena in monolayer and ultrathin magnetic films are discussed in separate subsections, which follow. Corresponding theoretical treatments are found in *Mills*, Chap. 3, Volume 1.

Free-standing monolayer magnetic films exist only as a theoretical novelty; they cannot be realized in practice. However, *ab initio*, all-electron calculations that explore the magnetism of both free-standing films and epitaxial layers have shown that substrate effects on magnetic behavior can vary from extremely weak (where conditions approaching free-standing films can occur) to very strong (where magnetic behavior can be quenched). Magnetic-optic Kerr effect spectroscopy can, in principle, probe details of substrate–overlayer electronic hybridization (or the absence of these effects) through measurements and analysis of $\sigma_{xy}(\omega)$. This has not yet been attempted. Spin-polarized photo-emission (refer to *Siegmann* and *Kay*, Sect. 4.2, Volume 1 and *Hopster*, Sect. 4.1, Volume 1) provides a more detailed and incisive probe of exchange of exchange-split conduction bands in ferromagnetic materials, and, therefore, is preferred over magneto-optical methods. However, magnetic circular-dichroism studies of core-level states in ferromagnetic materials offer important opportunities which are discussed in a following subsection. Magneto-optic Kerr effect polarimetry has been effectively used to explore the existence or absence of magnetism in novel thin film systems, as well as to explore specific predictions related to strong or weak overlayer/substrate electronic hybridization.

The Stoner criterion for the existence of ferromagnetism provides some insight into the differences in magnetic behavior that are expected to exist in thin films. This criteria is

$$JD(E_{\mathrm{F}}) > 1, \tag{4.19}$$

where $D(E_{\mathrm{F}})$ characterizes the density of states at the Fermi energy, and J is the exchange integral. The Stoner model correctly predicts that ferromagnetism should exist only for Fe, Co, and Ni in the 3d and 4d series of the periodic table.

Thin film materials, and particularly epitaxial monolayer films, may exhibit novel magnetic behavior. The lower atomic coordination associated with atoms at a surface or in a monolayer film can lead to reduced overlap of d electron wavefunctions and a related reduction in bandwidth and increase of $D(E_{\mathrm{F}})$. Corresponding effects can result from stretching the bulk lattice constant by growing (strained layer) epitaxial films on a suitable substrate. The reduced overlap and narrower bands can lead to enhanced magnetic moments, and in favorable cases could lead to ferromagnetism in monolayer films of metals that fail to order magnetically in their bulk form. Several examples of this behavior have been theoretically predicted, and have been tested using magneto-optic Kerr effect polarimetry.

Specific examples of predicted ferromagnetic behavior in monolayer epitaxial films of normally non-magnetic elements include V on Ag [4.23], Pd on Ag [4.24], Rh on Au [4.25], and Ru on Au [4.26]. Electron-capture spectroscopy, inverse photoemission experiments and impedance measurements of thin V layers on various substrates suggested the existence of a ferromagnetic state, but subsequent calculations, and careful attempts [4.27] to observe ferromagnetism in V films based on Kerr effect and polarized-electron measurements, have shown V on Ag(1 0 0) to be non-ferromagnetic, and probably antiferromagnetic. Monolayer Rh films on Au(1 0 0) [4.28], Ag(1 0 0) [4.29] have also been carefully studied by Kerr effect polarimetry with null results, and, to date, the theoretical predictions have not been verified (or changed). Corresponding calculations [4.30] for Rh sandwiched between Ag suggest that an overlayer of Ag should quench the ferromagnetic state; and, there is some experimental evidence [4.31] that Ag atoms do migrate from the substrate to surface sites on a Rh film. Therefore, at present, it is not clear if all of the requirements for ferromagnetism in Rh films on Ag(1 0 0) have been successfully met in experiments conducted to date.

The search for enhanced or suppressed magnetic moments at surfaces and in ultrathin epitaxial films has played an important role in the field of two-dimensional magnetism. Early experiments [4.32] reported magnetic dead layers at magnetic surfaces and at the interface between magnetic and non-magnetic materials, as well as transitions between magnetic and non-magnetic behavior as film thickness and oxygen exposure were varied. Early theoretical [4.23, 31] work attempted to explain the disparate experimental results based on calculations of surface, interface, and thin film electronic properties with emphasis on exploring the role of the substrate in determining magnetic behavior. Specific predictions that guided subsequent experimental studies included (1) a prediction that a Ni monolayer is magnetic on Cu(1 0 0) but paramagnetic on Cu(1 1 1) [4.31], (2) predictions of enhanced magnetic moments in thin film Fe layers on noble metal substrates [4.23], (3) the prediction that substrate hybridization effects were weak for the Fe on Ag(1 0 0) [4.23], and (4) the prediction that hybridization effects were strong enough in the Fe on W(0 0 1) system [4.32] to quench magnetism of a one-monolayer film.

Magneto-optic Kerr effect measurements have addressed several of these theoretical predictions. In many cases, the thin film structure and growth behavior are understood well enough to argue that experimental results can be meaningfully compared with the corresponding theoretical predictions. Epitaxial growth of magnetic layers, film stability, and atomic level structural characterization are topics discussed separately in other sections of this volume, and will not be treated in detail in this section. Fortunately, most of the recent magneto-optic Kerr effect studies have been carried out on thin film structures that are reasonably well-characterized by standard surface science techniques. A few of the more recent magneto-optic Kerr effect experiments related to magnetic moment enhancement/reduction and the substrate dependencies are described below.

Substrate and layer dependencies in the magnetic behavior of Ni on Ag(1 1 1) and Ag(1 0 0) were carried out using magneto-optical techniques. These experiments discovered [4.33] that a p(1 × 1) monolayer of Ni on Ag(1 1 1) was non-magnetic at 30 K whereas a monolayer of Ni film (which did not exhibit good epitaxy) was ferromagnetic at 110 K. The preferred spin orientation (easy axis of magnetization) of thin (two to five layer films) on both Ag(1 1 1) and Ag(1 0 0) is in the film plane. While magneto-optic Kerr effect measurements found that a monolayer of p(1 × 1) Ni on Cu(1 1 1) is ferromagnetic (in disagreement with the predicted paramagnetism of this system), the trend in magnetic behavior observed for Ni on Ag(1 1 1) and Ag(1 0 0) may still be understood as a consequence of sp-d induced quenching.

A second system in which both experiments [4.33, 34] and *ab initio* calculations [4.32] indicate strong substrate–film hybridization effects is the p(1 × 1) Fe on W(0 0 1) system. The high surface energy of W(1 0 0) ($\gamma_W = 2.9$ J/m^2, γ_{Fe} $= 2.0$ J/m^2) strongly favors monolayer nucleation rather than other possible nucleation modes, and extensive studies of film growth and structure of 1 and 2 ML films have established the high quality of epitaxial Fe films on W(1 0 0) and their structural parameters. Spin-polarized, angle-resolved photoemission studies [4.33] of p(1 × 1) Fe on W(0 0 1) failed to detect remanent magnetization in films less than 1 ML thick, but detected spin polarization for films greater than 1 ML thick. *Ab initio* calculations [4.32] explored the effects of both lattice strain ($a_w = 3.165$ Å, $a_{Fe} = 2.866$ Å yielding a 9.4% strain in the Fe film) and d-band hybridization on magnetism of p(1 × 1) Fe on W(0 0 1). The giant ferromagnetic moment ($3.1\mu_B$) predicted for a free-standing Fe monolayer film is effectively quenched (to $< 0.1\mu_B$) by film–substrate hybridization effects. This quenching has been verified by magneto-optic Kerr effect studies [4.34] of p(1 × 1) epitaxial Fe films on W(0 0 1).

While magneto-optical effects can probe for magnetism or the absence of magnetism at high sensitivity ($0.02\mu_B$/Å), it is not particularly well-suited for probing enhanced moments directly. Layer dependencies of the magnetic moment per atom in ultrathin films can, in principle, be obtained from careful layer-dependent measurements of the magneto-optic Kerr effect signals in cases where one can effectively argue that there are no changes in magnetic anisotropy versus film thickness, and where it is clear that the film is magnetically saturated. Because of the rapid variation of the Curie temperature with film thickness, the measurements must also be carried out at temperatures significantly below the Curie temperature of the thinnest film. A procedure has been proposed [4.35] for absolute moment measurements that combines magneto-optic Kerr effect measurements of ultrathin films prior to and after capping with a suitable protective overlayer followed by SQUID magnetometer studies. The procedure appears to have yielded evidence of the predicted enhanced moments for Fe on Ag(1 0 0 1). A related discussion involving spin-polarized neutrons and ferromagnetic resonance techniques may be found in Chap. 6 by *Bland* in Volume 1.

4.4.2 Thin Film Anisotropy

The magneto-optic Kerr effect provides an ideal method for studying thin film magnetic anisotropy. Several other techniques also offer the sensitivity and versatility to probe magnetic anisotropy of ultrathin films; for example, refer to *Heinrich* (Sect. 3.1) (FMR), and *Pierce, Unguris* and *Celotta* (Sect. 2.3) (SEMPA). Magnetic hysteresis loops can be obtained using any of the three distinct magneto-optic Kerr effect configurations illustrated in Fig. 4.1. These configurations differ by the orientation of M with respect to the plane of incidence. Magneto-optic Kerr effect studies that examine hysteresis loops using both polar and longitudinal configurations can discriminate between either a perpendicular or in-plane easy axis of magnetization. Corresponding studies in which the direction of the applied field is changed between the transverse and longitudinal configuration (or in which the sample is rotated around the normal direction using either of the two in-plane Kerr effect configurations) make possible the determination of the in-plane anisotropy.

Magneto-optical effects have been used for many years in studies of domain structure at surfaces. (The technique is known as Kerr effect microscopy and is capable of spatial resolution governed by the diffraction limit of light). Macroscopic formulas [4.22, 36] required to interpret the reflected intensity of polarized light in terms of the magnetization direction (domain structure) have been developed for Kerr effect microscopy applications. The present subsection outlines several new areas of application-related magnetic anisotropy of ultrathin magnetic films.

One of the more interesting recent discoveries in the field of thin film magnetism was the experimental manifestation of perpendicular anisotropy in ultrathin films [4.38, 39], and subsequent first-principles calculations [4.40] showing that the phenomena is energetically possible in certain cases. (*Mills*, Chap. 3, Volume 1 for a nice theoretical discussion of anisotropy in ultrathin ferromagnetic films.) Magneto-optic experiments have played an important role in characterizing this new phenomena, as illustrated by the material presented in this subsection. While perpendicular magnetic anisotropy in ultrathin films has been studied by Kerr effect measurements in several thin film systems [4.21, 37], the Fe on Ag(1 0 0) system provides a good example for historical reasons as well as because it has been the most extensively studied.

. The absence of spin-polarization in spin-polarized photoemission studies of ultrathin Fe on Ag(1 0 0) [4.38] led to the conclusion that the film had remanent magnetism perpendicular to the surface. This novel behavior was subsequently verified by a variety of experimental techniques [4.39], including magneto-optic Kerr effect measurements [4.37], and placed on a reasonably sound theoretical

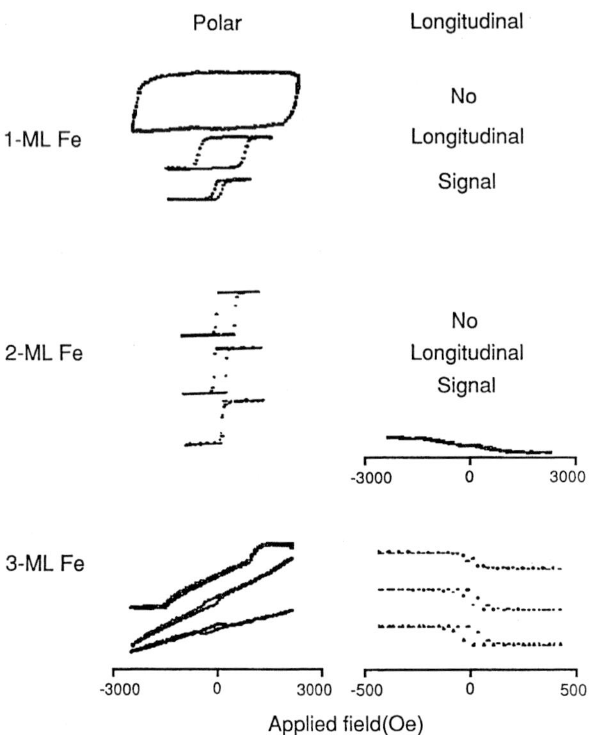

Fig. 4.4. Hysteresis loop behavior for p(1 × 1) Fe thin films on Ag(1 0 0) showing the change in magnetic anisotropy from perpendicular (1 and 2 ML) to in-plane at 3 ML

basis by *ab initio* calculations [4.40]. Figure 4.4 displays hysteresis loops for several ultrathin epitaxial Fe films on Ag(1 0 0) obtained from magneto-optic Kerr effect experiments. The variation of hysteresis loop characteristics as a function of film thickness obtained from polar and longitudinal configuration measurements of the same sample represents a clear manifestation of perpendicular magnetic anisotropy in the 1 and 2 ML films. The behavior exhibits a rather sharp transition from perpendicular to in-plane behavior at a thickness of 3 ML.

Interesting discrepancies were noted [4.37] between experimental observations of perpendicular anisotropy in ultrathin p(1 × 1) Fe on Ag(1 0 0) as determined by magneto-optic Kerr effect studies and spin-polarized photoemission experiments. For example, the polarized-electron experiments suggested that perpendicular magnetization exists only for film thickness between 3.5 and 5 ML at 30 K, and that above 100 K, no perpendicular remanence exists for any film thickness. These conclusions are clearly inconsistent with data displayed in Fig. 4.4.

These discrepancies have now been explained. Additional magneto-optic Kerr effect studies [4.41] have explored the effects of low doses of oxygen on the

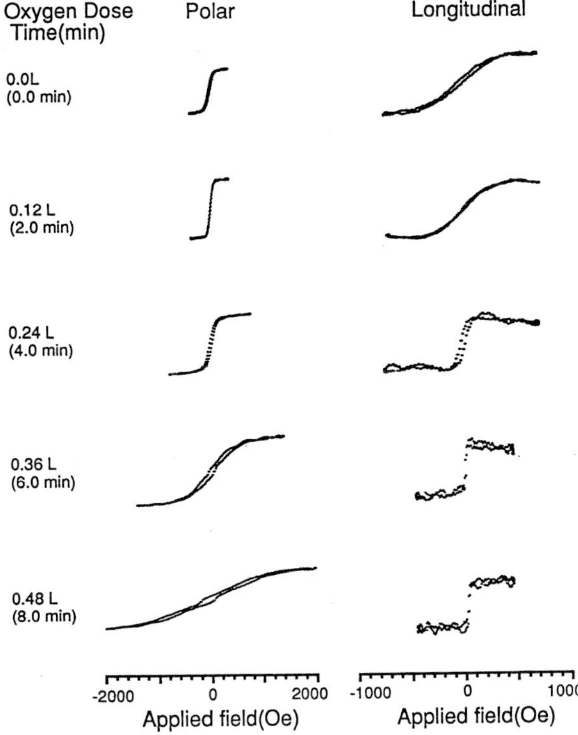

Fig. 4.5. Effects of oxygen adsorption on magnetic properties of p(1 × 1) Fe on Ag(1 0 0). The oxygen dose is measured in Langmuirs (1 L = 1 × 10^{-6} Torr s)

magnetic behavior of ultrathin p(1 × 1) Fe films on Ag(1 0 0). Figure 4.5 displays the principal result. The easy direction of magnetization for 2 ML films clearly change from perpendicular to in-plane after exposure to 0.25 Langmuir of oxygen. Many of the apparent disagreements between the polarized-electron and Kerr effect experiments can be reconciled if the polarized-electron experiments were carried out on oxygen-contaminated samples.

While the precise growth behavior and structure of thin film p(1 × 1) Fe on Ag(1 0 0) remains a topic of debate [4.41, 42] (refer to *Heinrich*, Sect. 3.1), the thickness and temperature dependent magnetic anisotropy of this system appears to be stable over a reasonably broad range of growth conditions. This implies that the film system is structurally stable, as well. In contrast to this behavior, the p(1 × 1) Fe on Cu(1 0 0) system has been shown by Kerr effect studies [4.28] to exhibit novel magnetic behavior that depends on both thickness and growth temperature. Figure 4.6 illustrates the broad range of magnetic response as determined by hysteresis loops measured by the Kerr effect showing a dependence on film thickness and growth temperature. The precise relationships between atomic level structures and the various magnetic signatures are being presently established by structural analysis techniques [4.43], but it

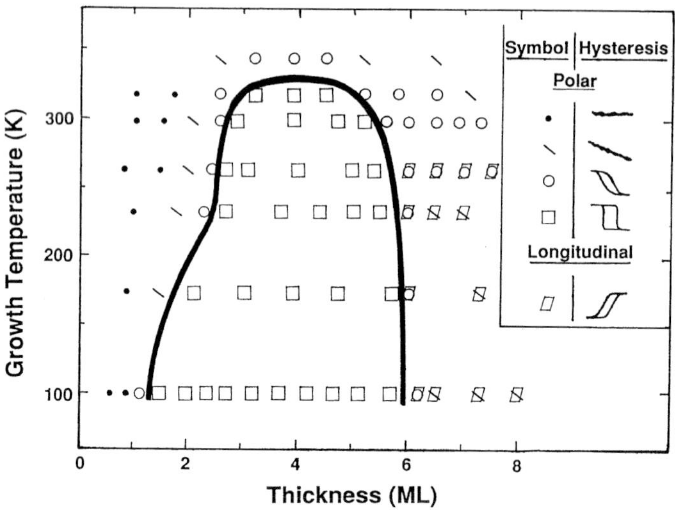

Fig. 4.6. Hysteresis loop phase diagram for $p(1 \times 1)$ Fe on Cu(1 0 0) showing magnetic manifestations of different material phases produced under different growth conditions

already is clear from Low Energy Electron Diffraction (LEED), Auger, and forward scattering experiments that surface roughness, interface interdiffusion, and adlayer agglomeration all play a role in affecting the magnetic behavior [4.44].

Demagnetizing factors (shape anisotropy) generally force M to lie in the plane for all but the very thinnest films (which can manifest perpendicular anisotropy). In these cases, magneto-optical techniques continue to offer important opportunities for exploring in-plane magnetic anisotropies. Several options are available for probing in-plane anisotropy. Rotating the crystal in a polarimeter operating in a particular magneto-optic Kerr effect configuration is one option; utilizing both the longitudinal and transverse magneto-optical Kerr effects (by rotating the magnet around an axis parallel to the crystal normal direction) is a second option. Either approach can yield direct information about in-plane magnetic anisotropy. The latter approach has recently been applied to the study of magnetization reversal in single-crystal-Fe films grown on GaAs(1 0 0) [4.22].

An important factor driving the interest in thin film magnetism is the capability to deliberately modify a structural property of a magnetic material (by using various thin film technologies) which subsequently affects the magnetic behavior. As a final example, illustrating the application of magneto-optical techniques to thin film magnetic anisotropy, we consider the effects associated with regular atomic level steps on a surface. Figure 4.7 displays hysteresis loops obtained [4.45] from an ultrathin epitaxial Fe film grown on a W(0 0 1) crystal having two different surfaces accessible on the same face: an atomically flat surface (half the crystal face) and a stepped surface (step every 25 Å). Various

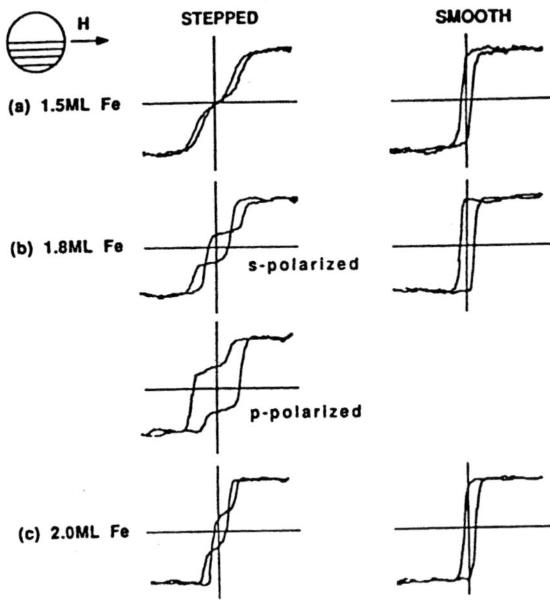

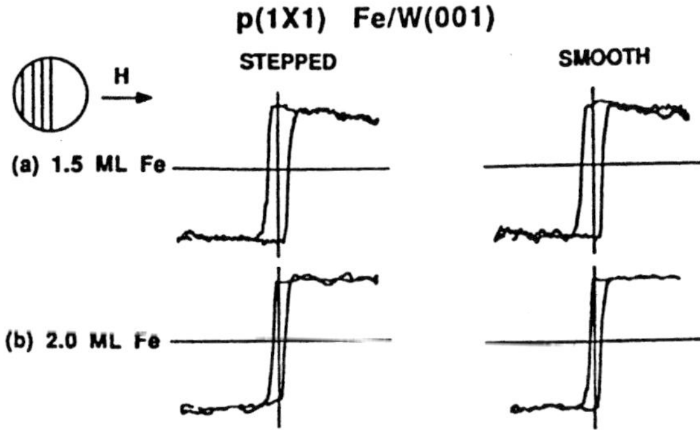

Fig. 4.7. Hysteresis loops for p(1 × 1) Fe on a stepped W(1 0 0) surface showing step-induced anisotropy. The loops indicate an easy axis perpendicular to the steps

factors (referred to previously) favor excellent epitaxy, and good control of step uniformity and film perfection of thin Fe films grown on W(0 0 1). The dramatically different hysteresis loops obtained from the smooth and stepped surfaces can be explained based on a strong uniaxial anisotropy ($M \perp$ to the step edges) in the film plane induced by the steps (refer to *Heinrich*, Chap. 3.1). This direction of easy axis ($M \perp$ step edges) is opposite to what would be predicted based on simple demagnetizing factor arguments. Therefore, the induced anisotropy is considered a novel result suggesting an interesting new field of micromagnetics will emerge from deliberately modified structures on a sub-micron scale.

4.4.3 Critical Phenomena

Surface magneto-optic Kerr effect polarimetry has been utilized in two types of thin film studies associated with the magnetic phase transition in the vicinity of the Curie temperature T_C. The first application measures the variation of T_C as a function of film thickness; the second application attempts to quantify the value of the magnetization exponent. The focus is to realize model two-dimensional (2D) systems. The goal is to make contact with the body of theoretical work that treats the universal aspects of phase transition phenomena which are governed by the spatial and order parameter dimensionality, without reference to the microscopic details of the interactions.

Studies of the thickness dependence of T_C include investigations of epitaxial Ni/Cu(1 1 1) [4.19, 46], Fe/Pd(1 0 0) [4.47], Fe(1 1 0)/Ag(1 1 1) [4.48], and Co/Cu(1 0 0) [4.49]. The data for Ni were fitted to the bulk relation to describe finite thicknesses:

$$\frac{T_C(\text{bulk}) - T_C(n)}{T_C(n)} \propto n^{-\lambda}, \tag{4.20}$$

where n is the number of monolayers in the film, and λ is referred to as the shift exponent. It is interesting that the expression can be used to describe films as thin as 1 and 2 ML. For the Fe and Co films there is a question about the existence of monolayer magnetism. It was argued that the lack of ferromagnetic hysteresis for monolayer-range Co films was due to the disappearance of long-range magnetic order. In two dimensions, isotropic Heisenberg systems disorder at finite temperature. This is known as the Mermin–Wagner theorem. However, four-fold in-plane anisotropy should be present for Co/Cu(1 0 0) which violates the conditions of the theorem. For Fe(1 1 0)/Ag(1 1 1) the lack of ferromagnetic hysteresis at the 1 ML level was attributed instead to superparamagnetism, associated with a 2D island structure. Superparamagnetism is traditionally characteristic of samples with isolated grains; such samples possess long-range ferromagnetic order, but thermal excitations can randomize the magnetization direction between each island or grain. Support for the application of this model to Fe/Ag(1 1 1) comes from the analysis of its Mössbauer spectra [4.50].

Interest in Fe/Pd(1 0 0) stems from the observation of long-range order even into the submonolayer Fe regime. This is a particularly fascinating result. It suggests that a 2D island model of the submonolayer morphology may be inappropriate. The T_C value of Fe/Pd(1 0 0) continues to drop below a mono-layer of Fe coverage, as shown in Fig. 4.8. A random-site-vacancy model has been used to describe these results. In this picture submonolayer coverages are visualized by starting with a fully occupied monolayer, and then removing Fe atoms at random to create a vacancy structure. It is also anticipated from band-

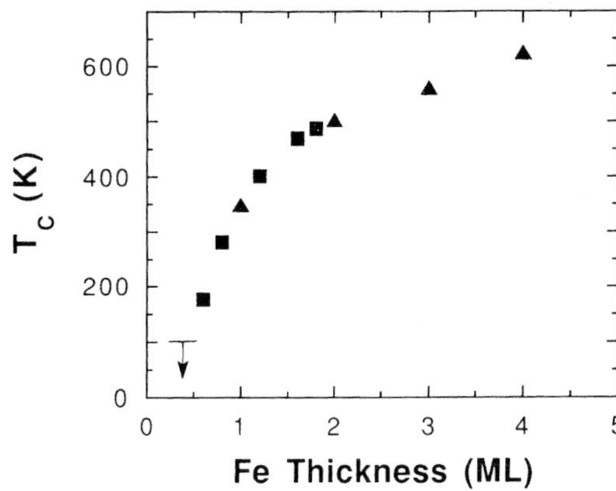

Fig. 4.8. T_C variation for Fe/Pd(1 0 0) as a function of Fe thickness

structure calculations, and from the high susceptibility of bulk Pd, that the interfacial Pd will be polarized and possess small moments. This polarization should have the effect of raising T_C in the monolayer range. This is because the interfacial Pd spins increase the coordination number of nearest neighbor spins around the Fe sites. Thus, Fe/Pd(1 0 0) is a good candidate for follow-up studies. Studies to date involve both Monte Carlo simulations [4.51] to estimate the enhancement of the ordering temperature due to the coupling to the substrate spins, as well as photoemission investigations [4.52] of Fe/Pd(1 0 0) utilizing physisorbed Xe, in order to examine the applicability of the random-site-vacancy model.

The same four systems discussed above have been studied to extract a magnetization exponent. For a second order phase transition, the order parameter is expected to approach zero on warming as $T \rightarrow T_C$ according to the power law expression:

$$M \propto (1 - T/T_C)^{\beta_c}, \tag{4.21}$$

where β_c is the critical exponent. The order parameter for a ferromagnet is the magnetization M. Since M is proportional to the Kerr rotation and ellipticity, the SMOKE technique can be used to obtain β_{eff}, which is the effective exponent obtained experimentally. The goal is to relate β_{eff} with the theoretical quantity β_c. However, the quantity β_{eff} is subject to a number of uncertainties, including random experimental errors, as well as to systematic errors associated with the available temperature range for applying the fitting procedure, and with the method used to fix a precise value of T_C.

The theoretically defined quantity β_c in 2D takes on the values 1/8, 1/9, and 1/12 for 2-, 3-, and 4-state Potts models, respectively. The different models are associated with the number of degrees of freedom of the order parameter. The 2-state Potts model corresponds to a spin system that can take on two orientations $+1$ and -1 (e.g., up–down or left–right), and is better known as the famous Ising model [4.53]. Other models that have been studied extensively include 4-, 5-, and 6-state clock models, in which the availability of the states is limited to follow a clockwise progression. These clock models also exhibit Ising transitions. The XY model permits all spin orientations within the 2D XY plane. In its isotropic form the 2D XY model has a Kosterlitz–Thouless transition, but in the presence of cubic anisotropy it possesses a non-universal exponent that depends on the anisotropy strength. It is interesting to note that the 2D β_c exponents (1/8, 1/9, and 1/12) are all much smaller in magnitude than those corresponding to 3D systems (e.g., ~ 0.3–0.4), or those associated with the surface of a semi-infinite solid (e.g., ~ 0.7–0.8).

The first ultrathin overlayer system to be studied with the goal of characterizing β_{eff} by the SMOKE technique was Ni/Cu(1 1 1) [4.53]. However, the small magneto-optic response of Ni made it difficult to approach T_C closely. *Li and Baberschke* [4.54] have recently shown, via magnetic resonance measurements, that Ni(1 1 1)/W(1 1 0) can be described by the 2D Ising model for Ni thickness ≤ 4 ML. They found that thicker films show crossover to 3D behavior. Fe/Pd(1 0 0) and Fe(1 1 0)/Ag(1 1 1) also display β_{eff} values consistent with a 2D Ising description of the phase transition. For Fe/Pd(1 0 0) the understanding is quite straightforward in that the samples possess vertical easy axes of magnetization. Thus, the spin orientations are limited to up–down. For Fe/Ag(1 1 1), the easy axes are in-plane, and the in-plane surface anisotropy again stabilizes only two-state switching. For Co/Cu(1 0 0) the easy axes also are in-plane, but there is no surface anisotropy present. The bulk anisotropy imposes four-fold symmetry on the problem. The β_{eff} values extracted from experiment should reflect this symmetry constraint.

There are many interesting questions to pursue in this area. Those already alluded to involve constraints on the data reduction methodology to simultaneously extract β_{eff} and T_C from the measurements. For example, the transition can be rounded because the critical correlation length ξ_c cannot diverge as $T \rightarrow T_C$:

$$\xi_c = \xi_0(1 - T/T_C)^{\nu}, \tag{4.22}$$

where the exponent $\nu = 1$ for the 2D Ising model, and ξ_0 is of the order of an interatomic spacing. The critical fluctuations can be limited in spatial extent due to surface defects (i.e. steps) or to finite size effects. Another question involves the temperature range ΔT over which critical fluctuations dominate the temperature dependence of the order parameter. For example, it is well known that spin wave models are appropriate to describe $M(T)$ as $T \rightarrow 0$. Thus, ΔT certainly is not a large fraction of T_C. In 3D phase transitions ΔT often tends to be prohibitively small. This makes it very difficult to carry out meaningful measurements in all but the most demanding experiments. However, studies of the Ising model show that a relatively broad T range should be accessible in 2D ($\Delta T \sim T_C/10$). Also, because β_c tends to be small in 2D systems, the transition is rather abrupt, and M takes on a relatively large magnitude over much of the temperature range below T_C. This is fortunate for experimentalists because it makes critical-exponent studies generally more feasible in 2D rather than in 3D systems.

The last question of interest to be discussed here involves the exploration of coupled 2D layers, such as are represented by Fe/Pd(1 0 0), where a 2D sheet of interfacial Pd atoms is magnetic. The correspondence of β_{eff} with the 2D Ising value of β_c suggests that very few Pd layers are magnetically active. Otherwise the transition would be characterized by a different universality class. The expectation that only a few Pd layers are ferromagnetic is supported also by band structure calculations [4.24]. However, the bands are calculated at $T = 0$. Thus, the magnetization of the interfacial Pd remains a challenge to characterize at finite temperatures, especially near T_C. We will see in subsequent sections of this chapter that magnetic circular-dichroism measurements hold great promise for providing such a characterization.

4.4.4 Coupled Layers

The physics of coupled magnetic layers has attracted attention recently because of the discovery of two intriguing phenomena in ferromagnetic transition metal films spaced apart by metallic layers. (Topics related to this subject are also treated by *Hathaway* (Chap. 2.1), and by *Fert* and *Bruno* (Chap. 2.2).) The first effect is that the ferromagnetic layers can couple antiparallel to each other to form an antiferromagnetic (AF) structure [4.55]. The second is that the AF-coupled system can exhibit a giant, negative magnetoresistance (MR) [4.56]. (Ferromagnetic films ordinarily exhibit a small, positive MR, associated with the conduction electrons following spiral trajectories along the magnetic field lines.) The magnetic coupling in these new systems oscillates between ferromagnetic and AF as a function of spacer layer thickness [4.57]. This fact immediately suggests the involvement of the RKKY (Rudermann–Kittel–Yosida–Kasuya) interaction in any explanation of the underlying physics [4.58]. The RKKY interaction is long-ranged and oscillatory, and has been successfully invoked to describe the indirect coupling of both electronic and nuclear magnetic moments by conduction electrons. The RKKY interaction is well known from dilute alloy theory. It gives rise to static spin density oscillations in free-electron host materials to enable dissolved magnetic impuries to "see" each other. It is worth noting that the two phenomena – AF coupling and giant MR – are not necessarily prerequisites for each other. Giant MR effects can be present without intrinsic AF coupling; and, all AF-coupled films do not have a giant MR.

In the systems of interest the magnetic and non-magnetic spacers are limited in thickness to the nanometer scale. Thus, an explanation of the coupling in terms of simple magnetic domain formation that can mimic AF structures can be ruled out. This is because domain walls themselves tend to be many times thicker than the trilayer films of interest. Fe/Cr/Fe is the prototype system. It was the first system to be explored, and, except for Co/Cu/Co [4.59], it can sustain the largest MR anomaly reported to date. However, Cr as a spacer layer provides two levels of added complexity. First, it is a transition metal with a complex Fermi surface. Secondly, Cr possesses its own magnetic moments. Cu

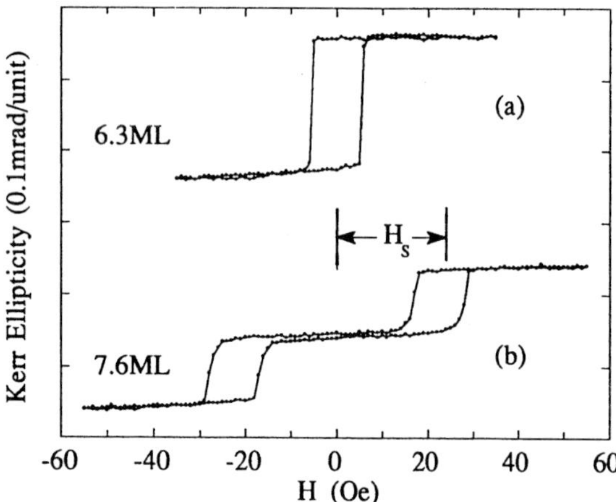

Fig. 4.9. Hysteresis loops for epitaxial Fe(14 ML)/Mo/Fe(14 ML) trilayers grown on Mo(1 0 0). The average Mo spacer layer thicknesses are shown. The loops demonstrate ferromagnetic and antiferromagnetic coupling examples along wedge-shaped Mo spacer layers. The switching field H_s is defined for the antiferromagnetic case as the offset from zero of the loop centroid

serves as a simpler example of the free-electron medium of simple RKKY theory. For a general background to this exciting area of current research, the reader is referred to other chapters in the volume and its companion volume. In the present context we will highlight the manner in which magneto-optic Kerr effects have been used to explore coupled magnetic layers.

The SMOKE technique provides a straightforward approach to search for AF-coupled systems. The nanometer-scale total thickness of a trilayer puts such films in the ultrathin limit where the magneto-optic signals from each magnetic layer are additive. Thus, AF-coupled films are expected to have zero net Kerr signal in remanence if the trilayer exhibits a prescribed magnetic symmetry (i.e. if the two magnetic layers are of equal thickness and interface effects are negligible). Hysteresis loops also can be used to measure the switching field H_s associated with lining up the two magnetic layers along the applied field direction. The idea is that the interlayer magnetic coupling is weak enough that a laboratory field can be used to overcome the intrinsic AF-coupling interaction. This is a reason for the interest of the device community; such structures can be *atomically engineered* to switch from antiparallel to parallel alignment in fields that could make them useful as magnetic sensors and pick-up heads.

Many research groups have used magneto-optical methods to study trilayers [4.60, 61] and superlattices [4.62]. Figure 4.9 shows two SMOKE loops for Fe/Mo/Fe grown epitaxially on Mo(1 0 0) [4.63]. The difference between ferromagnetic and AF coupling is quite apparent. These trilayers are grown in a wedged configuration to provide a continuous variation in Mo spacer thickness. This is achieved by translating the sample behind a mask during growth. Typical thickness gradients in these experiments are only ~ 1 Å/mm, but with a laser spot size of ~ 0.1 mm and a crystal length of ~ 1 cm, of the order 10^2 distinct

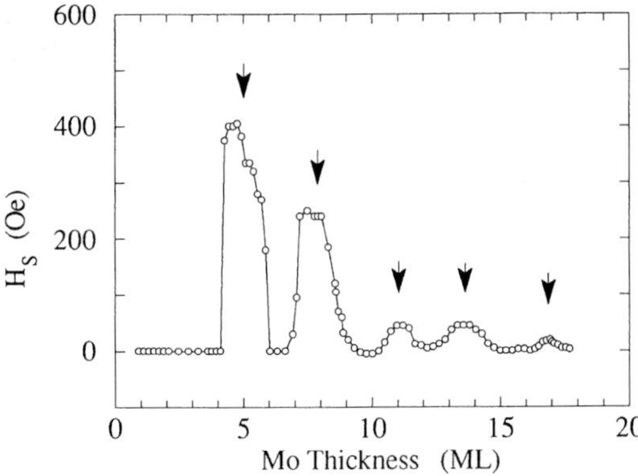

Fig. 4.10. The H_S values for Fe/Mo/Fe wedges, as in Fig. 4.9. The average Mo thickness, shown to the right, represents a linear vernier along a wedge. The arrows locate the five antiferromagnetic oscillations with a short period of ~ 3 ML. H_S values of zero are characteristic of ferromagnetic or non-coupled regions

regions can be probed along each wedge. Figure 4.10 summarizes the H_S data for this system obtained from experiments on a series of such wedges. Ferromagnetic (or non-coupled) films are characterized by $H_S = 0$. The oscillation periodicity in this study can be seen to be ~ 3 ML Mo. This is in the range expected for RKKY oscillations. Most studies on films with rougher interfaces show longer-period oscillations, the understanding of which is an added challenge. In particular, related work on sputtered superlattices of Fe/Mo have ~ 11 Å periodicity [4.64]. Thus, it is reassuring to observe that the simple RKKY picture appears as an appropriate starting point to understand the physics of the magnetic coupling. Also, note that the growth of wedged sandwiches is becoming a popular way to examine thickness-dependent magnetic properties [4.65].

4.4.5 Magneto-Optical Media

In most of this chapter magneto-optical effects have been used as a tool to explore thin film magnetism. However, from our theoretical treatments it is clear that new *magneto-optical* (MO) properties should be realizable in the ultrathin regime. In this section we discuss MO materials opportunities more thoroughly. MO materials are useful in magnetic data storage applications [4.66]. There are two advantages to MO-based data storage concepts. First, head crashes can be avoided because the MO reading process utilizes a laser, rather than a pick-up head. Secondly, the polar Kerr effect gives rise to a strong signal that is easy to detect. To make use of the polar geometry, the medium needs to have vertical easy axes. This is very desirable, also, from another point of view. It is believed that vertical easy axes are needed to get higher bit-packing densities than are ultimately possible with conventional longitudinal magnetic media. Since higher density also would require closer contact between the head and the medium, the advantage of a non-contact laser-reading strategy becomes even more important. We now know that it is possible to stabilize vertical easy axes via surface and interface anisotropies. Hence, interest has increased in the search for candidate MO materials based on magnetic superlattices. The magnetic layers need to be thin in order to preserve the vertical easy axes. The most promising superlattice system [4.67] presently is Co/Pt, where, typically, $d_{Co} \sim$ 3–4 Å and $d_{Pt} \sim$ 13–17 Å. Advantages of this particular system are its corrosion resistance, and that its polar signal is comparable to that of pure Co and is large at short wavelength. The short-wavelength behavior is important for digital information storage, since the wavelength of the light eventually becomes a limiting factor in improving bit-packing densities. Note, however, that for Co/Pt to be a viable medium in the marketplace, the economic situation has to be examined. There is competition due both to the continued improvement of traditional Winchester technology, and to the success of the present generation MO media. Presently, amorphous TbFeCo alloy films are in commercial use [4.68].

The scientifically intriguing aspect of Co/Pt superlattices is that the MO signal cannot be understood based on the known properties of bulk Co. Equation (4.13) indicates that the superlattice signal should be reduced from its bulk value by the dilution factor $\sim d_{Co}/(d_{Co} + d_{Pt})$. This expression is valid when $N_{Co} \sim N_{Pt}$, which *is* the case in the vicinity of the He–Ne wavelength. But the signal level reported at 4 eV photon energy can be $\sim$4–5 times as large as this factor would suggest. The main reason for this is that the interfacial Pt itself becomes magnetic due to its proximity to the Co [4.69], similar to the situation discussed earlier with respect to Fe/Pd. Even though the Pt moment is relatively small ($\sim 0.3\mu_B$ in the interfacial monolayer) the high atomic number Z of Pt endows it with a strong spin–orbit interaction. The MO response is proportional to both the magnetization *and* the wavelength-dependent spin–orbit weighted optical matrix element. It is interesting to note that the Co/Pd system is similar magnetically, but the Pt system outperforms it magneto-optically because of the higher Z of Pt than Pd. Co/Cu superlattices, on the other hand, show the trend with thickness anticipated from the $d_{Co}/(d_{Co} + d_{Cu})$ dilution factor [4.70]. Cu is not polarizable in the same manner as Pt and Pd. While Cu can sustain RKKY static spin density oscillations, Pt and Pd are so highly ferromagnetically polarizable that they are sometimes referred to as *incipient ferromagnets*. However, it would be very instructive to have a direct proof of the ferromagnetism of the interfacial Pt or Pd. Magnetic cirular-dichroism, as discussed in the next section, offers such a possibility.

4.4.6 Magnetic Circular-Dichroism

The availability of synchrotron radiation having high intensity and well defined polarization properties has recently made feasible magneto-optical measurements using vacuum ultraviolet, soft X-ray and hard X-ray photon energies. Magneto-optical effects associated with core excitations in magnetic materials offer an element and site specific probe of magnetism, a feature that can be effectively exploited in studies of magnetic alloys and compounds.

A simple model calculation [4.7] applied to the M_{23} absorption edge in ferromagnetic Ni was the first indication that strong magneto-optical effects could be observed in core level excitations of ferromagnetic solids. These numerical estimates of magnetic circular-dichroism at the M_{23} edge of Ni and consideration of the flux available from existing synchrotron radiation sources indicated that the calculated signals were detectable in principle. Several applications of core level, magneto-optical studies were anticipated by the model calculation, including the determination of the spin polarization at E_F, element specific probes of the sublattice magnetization, and temperature dependent studies that could separately measure the contribution of the two mechanisms that decrease the magnetism as $T \to T_C$: the variation of the exchange splitting and the excitation of spin waves.

Several attempts to observe magneto-optical effects associated with core level excitations followed the model calculations. Evidence of 4f excitations in Gd was reported [4.71, 72] but the first attempt [4.73] to observe magnetic circular-dichroism (MCD) associated with the M_{23} edge in Ni (as predicted by the model calculation) failed. Current generation synchrotron facilities achieve performance levels orders of magnitude above those of Tantalus I (used in the first experiments). The higher performance achieved in the light sources and monochromators have now revolutionized MCD measurements, and have vastly expanded the anticipated range of applications. Methods for calculating MCD effects also have been significantly advanced from the early model calculation and offer greater opportunities for extracting information from the measurements. A few of the key advancements are outlined below.

The rapidly expanding applications of MCD effects is a direct consequence of improvements in electron storage ring beamline instrumentation that provides well-characterized linearly or circularly polarized synchrotron radiation. All of the work to date has been carried out using out-of-plane circularly polarized or in-plane linearly polarized synchrotron radiation from a bending magnet. Higher intensity will soon be available from magnetic insertion devices (crossed or helical undulators) and asymmetric wigglers, and from combinations consisting of undulators and novel optical devices such as multilayer reflector/quarter wave plate devices. A reasonably extensive literature describing creation and characterization of circularly polarized synchrotron radiation is cited in a recent instrumentation paper describing a high performance soft X-ray monochromator [4.74]. Recent experiments based on multiple reflection polarimeters [4.75, 76] have begun to accurately characterize the available polarization characteristics of bending magnets and insertion devices permitting more accurate quantitative determination of MCD effects.

The first successful measurements of core-level-based magneto-optical effects were carried out in the hard X-ray region [4.77, 78]. These studies observed relative differences in absorption for left and right circularly polarized X-rays of order 5×10^{-4} above the K edge of Fe. Stronger differential absorption (10^{-2}) was reported for the L_3 and L_2 edges of Gd. These observations verified the order of magnitude for magneto-optical effects resulting from core level excitations in ferromagnets predicted by the original calculation [4.6]. These experiments were the first to explore spin densities of unoccupied bands and local magnetic structure based on deep core level MCD techniques.

The hard X-ray region experiments were soon followed by corresponding experimental measurements in the soft X-ray region [4.79–81]. The measurements [4.80] of absorption and MCD at the L_{23} edge of Ni revealed that the L_3/L_2 MCD intensity ratio is not $-1:1$ as predicted in the original simple model calculation, but $-1.60:1$. In addition, a secondary absorption feature was observed a few eV above the main L_3 and L_2 absorption thresholds in the MCD spectra.

These features of the L_{23} absorption and MCD spectra were examined in a series of simulations [4.80] based on a Slater–Koster approach including extensions incorporating spin–orbit splitting and taking account of relativistic effects on dipole selection rules. The simulations were able to model measured absorption and MCD spectra only when effective values for ξ, the spin–orbit parameter relevant to conduction bands, were assumed much larger than the ground state value, and values of Δ_{ex}, the exchange splitting, was assumed smaller. These discrepancies are consistent with expectations based on many-body, dynamical effects and were interpreted to represent the failure of one-electron band theory in predicting core level spectra.

An important additional factor supports the conclusion that correlation effects are apparent in MCD spectra. The simulation apparently accounted for features above the L_3 and L_2 absorption edges attributed to density of states effects but yielded no hint of the corresponding features (which appear at different energies) above the MCD absorption edges that were attributed to correlation satellites. This important result suggests that MCD spectroscopy offers a new means of exploring correlation phenomena in magnetic materials.

4.5 Outlook

The future for applications of magneto-optical techniques appears to be very good. Experimental techniques and macroscopic formalisms are well developed, and recent advances in first-principles approaches now permit inclusion of spin–orbit effects in a meaningful way. Thin film anisotropy and magneto-optical excitations can now be addressed within a microscopic framework. More importantly, new magnetic phenomena are being discovered as novel materials are synthesized and explored. These studies are a source of an expanding class of thin film materials and magnetic phenomena that can be probed by magneto-optical techniques.

Acknowledgements. Work at Argonne supported by US DoE Basic Energy Sciences – Materials Sciences under Contract No. W-31-109-ENG-38. Work at UT Austin supported by NSF DMR-8922359.

References

4.1　P.N. Argyres: Phys. Rev. **97**, 334 (1955)

4.2　H.S. Bennett, E.A. Stern: Phys. Rev. **137**, A448 (1965)

4.3　Y.R. Shen: Phys. Rev. **133**, A51 (1964)

4.4　J. Halpern, B. Lax, Y. Nishina: Phys. Rev. B **4**, A140 (1964)

4.5　B.R. Cooper: Phys. Rev. **139**, A1505 (1965)

4.6　J.L. Erskine, E.A. Stern: Phys. Rev. B **8**, 1239 (1973)

4.7　J.L. Erskine, E.A. Stern: Phys. Rev. B **12**, 5016 (1975)

4.8　J.L. Erskine, E.A. Stern: Phys. Rev. Lett. **30**, 1329 (1973)

4.9　N.V. Smith, S. Chiang: Phys. Rev. B **19**, 5013 (1979)

4.10　J.M. Sexton, D.W. Lynch, R.L. Benhow, N.V. Smith: Phys. Rev. B **37**, 2879 (1988)

4.11　A.V. Sokolov: *Optical Properties of Metals* (American Elsevier Publishing Co., Inc., NY, 1967)

4.12　J. Zak, E.R. Moog, C. Liu, S.D. Bader: J. Magn. Magn. Mat. **89**, 107 (1990)

4.13　J. Zak, E.R. Moog, C. Liu, S.D. Bader: Phys. Rev. B **43**, 6423 (1991)

4.14　J. Zak, E.R. Moog, C. Liu, S. D. Bader: J. Magn. Magn. Mat. **88**, L261 (1990)

4.15　Z.Q. Qiu, J. Pearson, S.D. Bader: Phys. Rev. B **45**, 7211 (1992)

4.16　Y. Suzuki, T. Katayama, S. Yoshida, K. Tanaka, K. Sato: Phys. Rev. Lett. **68**, 3355 (1992)

4.17　J. Zak, E.R. Moog, C. Liu, S.D. Bader: Appl. Phys. Lett. **58**, 1214 (1991)

4.18　A.A. Studna, D.E. Aspnes, L.T. Florez, B.J. Wilkens, J.P. Harbison, R.E. Tyan: J. Vac. Sci. Technol. A **7**, 3291 (1989)

4.19　C.A. Ballentine, R.L. Fink, J. Araya-Pochet, J.L. Erskine: Appl. Phys. A **49**, 459 (1989)

4.20　J. Araya-Pochet, C.A. Ballentine, J.L. Erskine: Phys. Rev. B **38**, 7846 (1988)

4.21　C. Liu, E.R. Moog, S.D. Bader: Phys. Rev. Lett. **60**, 2422 (1988); S.D. Bader: J. Magn. Magn. Mat. **100**, 440 (1991)

4.22　J.M. Florczak, E.D. Dahlberg: J. Appl. Phys. **67**, 75201 (1990); Phys. Rev. B **44**, 9338 (1991)

4.23　S. Ohnishi, C.L. Fu, A.J. Freeman: J. Magn. Magn. Mat. **50**, 161 (1985); A.J. Freeman, C.L. Fu: J. Appl. Phys. **61**, 3356 (1987)

4.24　S. Blügel, M. Weinert, P.H. Dederichs: Phys. Rev. Lett. **60**, 1077 (1988); S. Blügel, B. Drittler, R. Zeller, P.H. Dederichs: Appl. Phys. A **49**, 547 (1989)

4.25　M.J. Zhu, D.M. Bylander, L. Kleinman: Phys. Rev. B **43**, 4007 (1991); O. Eriksson, R.C. Albers, M.A. Boring: Phys. Rev. Lett. **66**, 1350 (1991)

4.26　S. Blügel: Phys. Rev. Lett. **68**, 851 (1992)

4.27　R.L. Fink, C.A. Ballentine, J.L. Erskine, J.A. Araya-Pochet. Phys. Rev. B **41**, 10175 (1990)

4.28　C. Liu, S.D. Bader: Phys. Rev. B **44**, 12062 (1991)

4.29　G.A. Mulhollan, R.L. Fink, J.L. Erskine: Phys. Rev. B **44**, 2393 (1991)

4.30　R. Wu, A.J. Freeman: Phys. Rev. B **45**, 7222 (1992)

4.31　P.J. Schmitz, W.-Y. Lenng, G.W. Graham, P.A. Thiel: Phys. Rev. B **40**, 11477 (1989); T.J. Raeker, D.E. Sanders, A.E. DePristo: J. Vac. Sci. Technol. **8**, 3531 (1990); J. Tersoff, L.M. Falicov: Phys. Rev. B **26**, 6186 (1982)

4.32　L. Liebermann, J. Clinton, D.M. Edwards, J. Mathon: Phys. Rev. Lett. **25**, 232 (1970); R. Wu, A.J. Freeman: Phys. Rev. B **45**, 7532 (1992)

4.33　J. Araya-Pochet, C.A. Ballentine, J.L. Erskine: "Dead Layers in Thin-Film Magnetism: p(1 × 1)Ni on Ag(1 0 0) and Ag(1 1 1)" in *Springer Proc. Phys. Vol. 50 Magnetic Properties of Low Dimensional Systems II*, ed. by L.M. Falicov, F. Mejía-Lira, J.L. Morán-López (Springer, Berline Heidelberg 1990); R.L. Fink, G.A. Mulhollan, A.B. Andrews, G.K. Walters, J. Appl. Phys. **69**, 4986 (1991); Phys. Rev. B **43**, 13645 (1991)

4.34 J. Chen, J.L. Erskine: Phys. Rev. Lett. **68**, 1212 (1982)

4.35 J.T. Markert, C.L. Wooten, J. Chen, G.A. Mulhollan, J.L. Erskine: Bull. Am. Phys. Soc. **37**, 360 (1992)

4.36 W. Rave, R. Schäfer, A. Huber: J. Magn. Magn. Mat. **65**, 7 (1987)

4.37 J. Araya-Pochet, C.A. Ballentine, J.L. Erskine: Phys. Rev. B **38**, 7846 (1988)

4.38 B.T. Jonker, K.H. Walker, E. Kisker, G.A. Prinz, C. Carbone: Phys. Rev. Lett. **57**, 142 (1986)

4.39 M. Stampanoni, A. Vaterlaus, M. Aeschlimann, F. Meir: Phys. Rev. Lett. **59**, 2483 (1987); N.C. Koon, B.T. Jonker, F.A. Volkening, J.J. Krebs, G.A. Prinz: Phys. Rev. Lett. **59**, 2463 (1987); R. Cabanet, P. Etienne, S. Lequien, G. Creuzet, A. Barthelemy, A. Fert: J. Appl. Phys. **67**, 5409 (1990); B. Heinrich, K.B. Urquhart, A.S. Arrott, J.F. Cochran, K. Myrtle, S.T. Purcell: Phys. Rev. Lett. **59**, 1756 (1987); J.R. Dutcher, J.F. Cochran, B. Heinrich, A.S. Arrott: J. Appl. Phys. **64**, 6095 (1988); D.L. Mills: Phys. Rev. B **39** 12306 (1989); R.C. O'Handley, J.P. Wood: Phys. Rev. B **42**, 6568 (1990); C.L. Fu, A.J. Freeman, T. Oguchi: Phys. Rev. Lett. **54**, 2700 (1985); F.J.A. den Broeder, W. Hoving, J.P. Bloemen: J. Magn. Magn. Mat. **93**, 562 (1991)

4.40 J.G. Gay, Roy Richter: Phys. Rev. Lett. **56**, 2728 (1986); J. Appl. Phys. **61**, 3362 (1987)

4.41 J. Chen, M. Drakaki, J.L. Erskine: Phys. Rev. B **45**, 3636 (1992); B. Heinrich, A.S. Arrott, J.F. Cochran, K.B. Urquhart, K. Myrtle, Z. Celinski, Q.M. Zhong: Mat. Res. Symp. Proc. **151**, 177 (1989); B. Heinrich, A.S. Arrott, J.F. Cochran, Z. Celinski, K. Myrtle: *Science and Technology of Nanostructured Magnetic Materials*, ed. by G.C. Hadjipanayis, G.A. Prinz (Plenum, New York, 1991)

4.42 H. Li, Y.S. Li, J. Quinn, D. Tian, S. Sokolov, F. Jona, P.M. Marcus: Phys. Rev. B **42**, 9195 (1990)

4.43 P. Xhonneux, E. Courtens: Phys. Rev. B **46**, 556 (1992)

4.44 D.A. Steigerwald, W.F. Egelhoff, Jr.: Phys. Rev. Lett. **60**, 2558 (1988)

4.45 J. Chen, J.L. Erskine: Phys. Rev. Lett. B **68**, 1212 (1992)

4.46 C.A. Ballentine, R.L. Fink, J. Araya-Pochet, J.L. Erskine: Phys. Rev. B **41**, 2631 (1990)

4.47 C. Liu, S.D. Bader: J. Appl. Phys. **67**, 5758 (1990)

4.48 Z.Q. Qiu, J. Pearson, S.D. Bader: Phys. Rev. Lett. **67**, 1646 (1991)

4.49 G.J. Mankey, M.T. Kief, R.F. Willis: J. Vac. Sci. Technol. A **9**, 1595 (1991); M.T. Kief, G.J. Mankey, R.F. Willis: J. Appl. Phys. **69**, 5000 (1991)

4.50 Z.Q. Qiu, S.H. Mayer, C.J. Gutierrez, H. Tang, J.C. Walker: Phys. Rev. Lett. **63**, 1649 (1989)

4.51 K.J. Strandburg, D.W. Hall, C. Liu, S.D. Bader: Phys. Rev. B **46**, 10818 (1992)

4.52 C. Liu, S.D. Bader: Phys. Rev. B **44**, 2205 (1991)

4.53 B.M. McCoy, T.S. Wu: *The Two-Dimensional Ising Model* (Harvard, Cambridge, 1973)

4.54 Y. Li, K. Baberschke: Phys. Rev. Lett. **68**, 1208 (1992)

4.55 P. Grünberg, R. Schreiber, Y. Pang, M.B. Brodsky, C.H. Sowers: Phys. Rev. Lett. **57**, 2442 (1986)

4.56 M.N. Baibich, J.M. Broto, A. Fert, F. Nguyen Van Dau, F. Petroff, P. Etienne, G. Creuzet, A. Friederich, J. Chazelas: Phys. Rev. Lett. **61**, 2472 (1988)

4.57 S.S.P. Parkin, N. More, K.P. Roche: Phys. Rev. Lett. **64**, 2304 (1990)

4.58 Y. Wang, P.M. Levy, J.L. Fry: Phys. Rev. Lett. **65**, 2732 (1990); D.M. Edwards, J. Mathon, R.B. Muniz, M.S. Phan: Phys. Rev. Lett. **67**, 493 (1991); P. Bruno, C. Chappert: Phys. Rev. Lett. **67**, 1602 (1991); R. Coehoorn: Phys. Rev. B **44**, 9331 (1991); D.M. Deaven, D.S. Rokhsar, M. Johnson: Phys. Rev. B **44**, 5977 (1991)

4.59 S.S.P. Parkin, R. Bhadra, K.P. Roche: Phys. Rev. Lett. **66**, 2152 (1991); S.S.P. Parkin, Z.G. Li, D.J. Smith: Appl. Phys. Lett. **58**, 2710 (1991)

4.60 M. Rührig, R. Schäfer, A. Hubert, R. Mosler, J.A. Wolf, S. Demokritov, P. Grünberg: Phys. Status Solidi (a) **125**, 635 (1991); S.T. Purcell, W. Folkerts, M.T. Johnson, N.W.E. McGee, K. Jager, J. ann de Stegge, W.B. Zeper, W. Hoving, P. Grünberg: Phys. Rev. Lett. **67**, 903 (1991); S.T. Purcell, M.T. Johnson, N.W.E. McGee, R. Coehoorn, W. Hoving: Phys. Rev. B **45**, 13064 (1992); M.T. Johnson, S.T. Purcell, N.W.E. McGee, R. Coehoorn, J. ann de Stegge, W. Hoving: Phys. Rev. Lett. **68**, 2688 (1992); A. Fuss, S. Demokritov, P. Grünberg, W. Zinn, J. Magn. Magn. Mat. **103**, L221 (1992)

4.61 W.R. Bennett, W. Schwarzacher, W.F. Egelhoff, Jr.: Phys. Rev. Lett. **65**, 3169 (1990); B. Heinrich, Z. Celinski, J.F. Cochran, W.B. Muir, J. Rudd, Q.M. Zhong, A.S. Arrott, K. Myrtle, J. Kirschner: Phys. Rev. Lett. **64**, 673 (1990)

4.62 J.E. Mattson, C.H. Sowers, A. Berger, S.D. Bader: Phys. Rev. Lett. **68**, 3252 (1992); B. Heinrich, J.F. Cochran, M. Kowalewski, Z. Celinski, A.S. Arrott, K. Myrtle: Phys. Rev. B **44**, 9348 (1991)

4.63 Z.Q. Qiu, J. Pearson, A. Berger, S.D. Bader: Phys. Rev. Lett. **68**, 1398 (1992)

4.64 M.E. Brubaker, J.E. Mattson, C.H. Sowers, S.D. Bader: Appl. Phys. Lett. **58**, 2306 (1991)

4.65 J. Ungaris, R.J. Celotta, D.T. Pierce: Phys. Rev. Lett. **67**, 140 (1991)

4.66 D.S. Bloomberg, G.A.N. Connell: "Magnetooptical Recording" in *Magnetic Recording Handbook: Technology and Applications*, ed. by C.D. Mee, E.D. Daniel (McGraw-Hill, New York, 1989) pp. 530–634

4.67 P.F. Carcia: J. Appl. Phys. **63**, 5066 (1988); S. Hashimoto, Y. Ochiai. J. Magn. Magn. Mat. **88**, 211 (1990)

4.68 B.S. Krusor, G.A.N. Connell: "Thin Film Rare Earth-Transition Metal Alloys for Magnetooptical Recording" in *Physics of Thin Films*, Vol. 15, ed. by M.H. Franscombe, J.L. Vossen (Academic, Boston, 1991) pp. 143–217

4.69 E.R. Moog, J. Zak, S.D. Bader: J. Appl. Phys. **69**, 880 (1991)

4.70 E.R. Moog, J. Zak, S.D. Bader: J. Appl. Phys. **69**, 4559 (1991)

4.71 J.L. Erskine: AIP Conf. Proc. #24 (1975) p. 190

4.72 J.L. Erskine: Phys. Rev. Lett. **37**, 157 (1976)

4.73 J.L. Erskine, F.C. Brown (unpublished). This attempt was based on the work of C. Gähwiller and F.C. Brown, Phys. Rev. B **2**, 1918 (1970) and used the Tantalus I storage ring

4.74 C.T. Chen: Rev. Sci. Instrum. **63**, 1229 (1992)

4.75 T. Koide, T. Shidara, M. Yuri, N. Kandaka, K. Yamaguchi, H. Fukutani: Nucl. Inst. Methods A **308**, 635 (1991)

4.76 E. Gluskin, J.E. Mattson, S.D. Bader, P.J. Viccaro, T.W. Barbee, Jr., N.B. Brookes, A. Pitas, R. Watts: SPIE Conf. Proc., Vol. 1548 (1991) p. 56

4.77 G. Schütz, W. Wagner, W. Wilhelm, P. Kienle, R. Zeller, R. Frahm, G. Materlik: Phys. Rev. Lett. **58**, 737 (1987)

4.78 G. Schütz, R. Wienke, W. Wilhelm, W. Wagner, P. Kienle, R. Zeller, R. Frahm: Z. Phys. **75**, 495 (1989)

4.79 C.T. Chen, F. Sette, Y. Ma, S. Modesti: Phys. Rev. B **42**, 7262 (1990)

4.80 C.T. Chen, N.V. Smith, F. Sette: Phys. Rev. B **43**, 6785 (1991)

4.81 J.G. Tobin, G.D. Waddill, D.P. Pappas: Phys. Rev. Lett. **68**, 3642 (1992)

5. Mössbauer Spectroscopy as a Means of Characterizing Surfaces, Thin Films, and Superlattices

J.C. WALKER

In the study of thin films and surfaces as well as the examination of superlattices, the careful characterization of these systems including their structural, magnetic, transport, and other properties has been absolutely crucial to the advancement of the field. As the means of sample preparation have progressed, techniques for evaluating the flatness, continuity, crystallinity, etc. of thin films and surfaces have become ever more necessary to understand the resulting magnetic and electronic properties. Iron is often a constituent of *magnetic* thin films and the isotope ^{57}Fe shows a strong Mössbauer effect over a wide temperature range, enabling the technique of Mössbauer spectroscopy to offer much to the study of surfaces, thin films, and superlattices. Indeed, the techniques outlined below have contributed outstandingly to our understanding of these systems.

The Mössbauer effect was first seen by Rudolph Mössbauer in Germany in 1958. It involves the recoilless resonant emission and absorption of gamma rays by nuclei bound in solids. The Nobel Prize was awarded to Mössbauer in 1961 in recognition of his discovery.

Besides ^{57}Fe Mössbauer spectroscopy for thin films there is considerable interest in several rare earth isotopes (^{161}Dg, ^{156}Gd, etc.) which also show significant Mössbauer effects. In the discussion of the method given here, ^{57}Fe will be used as the primary example as it has seen most of the applications in thin film science and is a particularly easy isotope to use.

Mössbauer spectroscopy depends on the achievement in a very special way of nuclear resonance florescence. Atomic resonance florescence is a common occurrence, as an atom emitting a photon recoils very little because atomic transitions are in the range of a few electron volts in energy and little momentum and therefore recoil energy is imparted to the atom. The emitted photon overlaps in energy (Fig. 5.1) with the energy necessary for resonance absorption by a ground state atom. In the case of nuclear transitions, photon energies are typically in the range from tens of thousands of electron volts to millions. Typically, recoil energies in both the emission and absorption processes are greater than the linewidths and very little resonance florescence takes place. Mössbauer was studying this unlikely phenomenon when he made his discovery: when nuclear photon energies are not *too* high (~ 10–100 keV) and the atom containing the excited nucleus is firmly bound in a stiff solid at moderate-to-low temperatures, there is a significant probability that the only lattice-vibrational

B. Heinrich and J.A.C. Bland (Eds.)
Ultrathin Magnetic Structures II
© Springer-Verlag Berlin Heidelberg 1994

normal mode (phonon) excited as a result of the photon emission is the zero-frequency ($\omega = 0$) mode. This implies that the recoil momentum is taken up by the entire lattice, not just the recoiling nucleus. In this case there is negligible recoil energy loss. The same process can also occur in resonance absorption. In both cases momentum conservation is accomplished by the whole crystal lattice with a negligible recoil energy shift. An additional bonus – very important for Mössbauer spectroscopy – is the fact that the first order Doppler line broadening caused by thermal vibration of the emitting nucleus disappears in the Mössbauer effect.

For ^{57}Fe, with a 14.4 keV nuclear transition energy to the ground state, a recoil-free fraction of gamma emissions in Fe metal is 0.81 at room temperature. This high fraction can be attributed to the rather low nuclear transition energy and to the significant lattice stiffness associated with Fe metal. Using a model by Debye and Waller in which the lattice stiffness is characterized by a Debye temperature θ, we have for the recoilless fraction

$$f = \exp\left[\frac{-E^2}{2MC^2k\theta}\left\{\frac{3}{2} + \frac{\pi^2 T^2}{\theta^2}\right\}\right], \quad T \ll \theta.$$

Here, E is the gamma energy and M the mass of an atom of the lattice. This theory was originally developed to explain the recoilless scattering of X-rays by

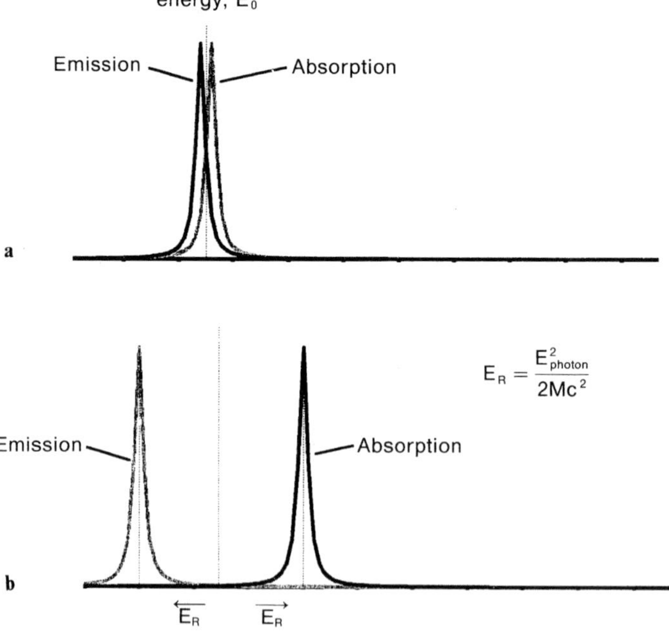

Fig. 5.1. (a) Represents atomic photon emission with negligible recoil. **(b)** Shows nuclear emission with recoil greater than the γ linewidth

crystals which produces Bragg diffraction. The physics is quite similar and the reader interested in more detail at this point is advised to consult references on the Debye–Waller factor in X-ray scattering [5.1].

As mentioned above, the Doppler broadening of the emitted or absorbed gamma linewidth cancels out to first order in v/c, where v is the speed of the emitting or absorbing nucleus associated with thermal vibrations. This eliminates the major source of gamma line broadening, and for a high-quality crystalline source and absorber, the gamma linewidth will be close to the value given by the uncertainty principle. Using the 100 ns lifetime of the 14.4 keV level of ^{57}Fe as the measurement time uncertainty: $\Delta E \sim 10^{-8}$ eV. This implies a resolution $E/\Delta E$ of about 10^{12} which offers an explanation for the efficacy of Mössbauer spectroscopy for looking at small energy shifts associated with solid state systems.

5.1 Elements of Mössbauer Spectroscopy

Mössbauer spectroscopy makes use of gamma ray emission from an excited nucleus bound in a solid. As the gamma transitions are usually very short-lived (^{57}Fe is long at 100 ns), the excited nuclear state which normally decays to the nuclear ground state is usually fed by a conveniently long-lived beta emitting state which decays to the relevant gamma emitting state. In the case of ^{57}Fe the beta-emitting parent, ^{57}Co, decays primarily to the 136 keV state of ^{57}Fe (Fig. 5.2). This second excited state decays very rapidly primarily to the 14.4 keV first excited state with the emission of a 122 keV gamma ray. The subsequent decay of the 14.4 keV state provides the gamma rays used for Mössbauer spectroscopy in ^{57}Fe. It should be noted that only 2% of natural Fe is ^{57}Fe. The

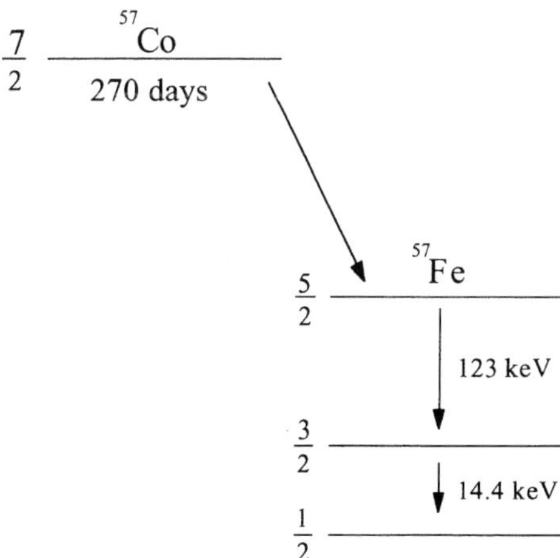

Fig. 5.2. Decay scheme for ^{57}Co to ^{57}Fe

98%, isotope, ^{56}Fe, shows *no* Mössbauer effect. This can be exploited in a very interesting way as we show below.

In the case of ^{57}Fe the 14.4 keV transition is from a $J = \frac{3}{2}$ nuclear level to a $J = \frac{1}{2}$ ground state. The nuclear moments, magnetic and electric quadrupole, associated with these levels interact with magnetic and electric fields at the nucleus to produce a hyperfine splitting of the nuclear levels. The excited state can split into four levels labelled with quantum number m_J: $+\frac{3}{2}, +\frac{1}{2}, -\frac{1}{2}, -\frac{3}{2}$. The ground state can split into two levels: $+\frac{1}{2}, -\frac{1}{2}$. For ^{57}Fe, the two nuclear moments associated with the splitting are the nuclear magnetic dipole moment, and the nuclear electric quadrupole moment. These moments are independently known, therefore a determination of the hyperfine splittings for a particular sample determine the magnetic field and electric field gradient at the nucleus. Both of these quantities provide very useful information when studying thin films, surface, or superlattices. The magnetic hyperfine interaction produces a nuclear Zeeman effect with equal spacing of the magnetic sublevels of the $J = \frac{3}{2}$ excited state (Fig. 5.3). The spacing of the two sublevels of the $J = \frac{1}{2}$ ground state is different because the nuclear magnetic moment of the ground state differs from that of the excited state. The electric quadrupole interaction splits the $\pm\frac{3}{2}$

(a)

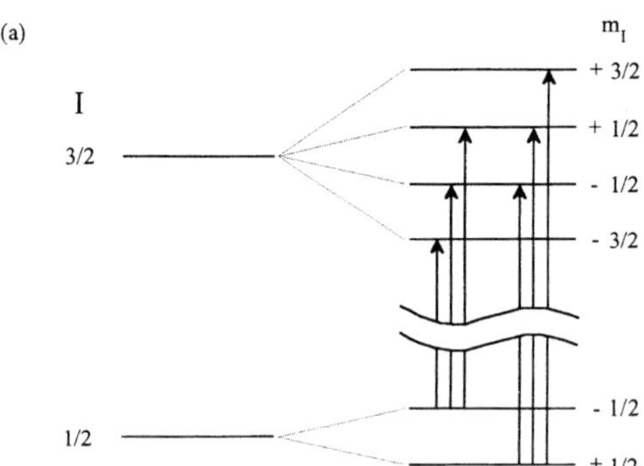

(b)

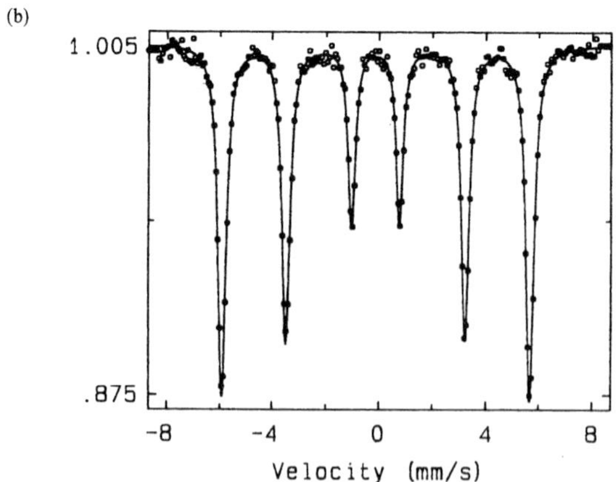

Fig. 5.3. (a) Magnetic hyperfine splittings of the ground state and first excited state of ^{57}Fe. The six allowed M1 transitions are indicated. A typical absorption spectrum from transmission Mössbauer spectroscopy is shown in (b)

levels from the $\pm\frac{1}{2}$ levels in the excited state and leaves the $\pm\frac{1}{2}$ ground state levels unaffected.

5.2 Mössbauer Spectrometers

In order to observe the nuclear hyperfine splittings by Mössbauer spectroscopy an apparatus is used which enables an "unsplit" Mössbauer source of recoilless gamma rays to be swept in energy across the range in which recoilless nuclear resonance absorption can take place (Fig. 5.4). The easiest method is to record

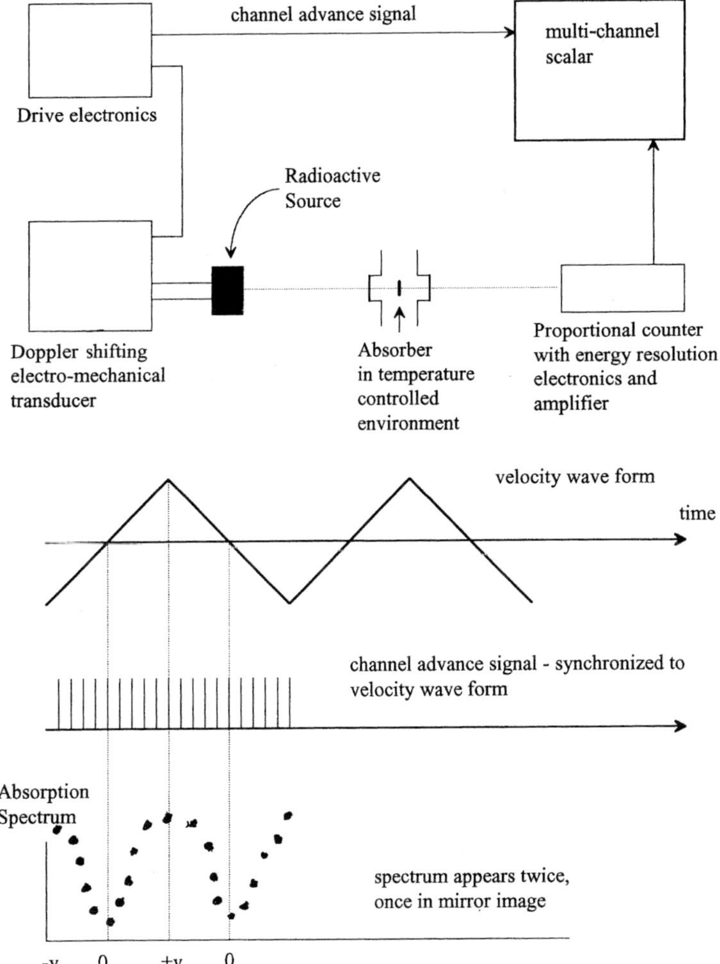

Fig. 5.4. Block diagram and signal details for a transmission Mössbauer spectrometer

the number of 14.4 keV gamma rays detected by a counter (typically a gas proportional counter) which can be made to "see" only the 14.4 keV gamma rays and not other, perhaps more intense, photons (6.4 keV Fe X-rays and the 122 keV gamma rays). The monochromatic gamma source (Co^{57} in a host which produces neither magnetic nor electric quadrupole splitting of the sublevels of the $\frac{3}{2}$–$\frac{1}{2}$ transition) is Doppler shifted by a mechanical velocity transducer to cover the energy range appropriate to the hyperfine splitting: for ^{57}Fe in Fe metal $\Delta E_{gamma} \approx \pm 5 \times 10^{-7}$ eV which corresponds to a Doppler velocity of ± 1.0 cm/s since ΔE_{gamma} $v/c \times E_{gamma}$.

The typical Mössbauer spectrometer includes a velocity transducer with associated electronics that follows a *triangular* wave form for velocity (a sinusoidal wave form would be easier for the transducer to follow, but would not spend equal times at each velocity). A multichannel scaler whose address channels are synchronized to a particular velocity accepts the counts from the gamma ray detector. A typical absorption spectrum is also shown in Fig. 5.3. A resonance absorption condition between source and absorber results in a smaller number of gammas reaching the detector and being recorded in that channel. By running the spectrometer until the baseline of the multichannel scaler shows $>10^6$ counts/channel, absorptions of 0.1% can be seen. The multichannel scaler is *constantly* read out visually onto a CRT screen enabling spectrum to be observed as it develops. In some cases resonant scattering rather than resonant absorption is detected. In this case the spectrum will appear upside down. The resonances are peaks rather than valleys, but the information contained is the same. We will later discuss different spectrometer configurations relevant to different experimental problems.

The problem of driving the transducer according to a precise wave form and synchronizing the multichannel scalar with the transducer so that counts for a particular Doppler velocity are counted in a specific channel has been solved in a number of ways. Commerical Mössbauer spectrometers which function well are available in the United States and in Europe. Many investigators have built their own systems using widely available published circuits and modern electronic components. Obsolete desktop computers are often used as multichannel analyzers and multiscalars making use of easily available electronic cards which can plug into the computers.

More complex than the Doppler shifting spectrometers are the arrangements for temperature control of the relevant Mössbauer effect absorbers. To obtain maximum information about a magnetic system using Mössbauer spectroscopy requires the ability to take spectra with the sample kept accurately at temperatures ranging from 4.2 (normal liquid He) to 600–700 K (temperatures above which irreversible changes in thin film or superlattice systems can occur). Various techniques have been devised to deal with this matter: some groups are satisfied by obtaining just three temperature values — 4.2 K (liquid He), 77 K (liquid N_2) and 295 K (room temperature). Flow cryostats with resistance heaters, thermal sensors, and electronic temperature controllers can hold temperatures to 0.1 K indefinitely at significant cost for liquid coolants (particularly at the lower temperatures).

We have found the most satisfactory solution in using a helium gas refrigerator system [5.2] that requires no liquid coolants and includes temperature control as good or better than flow cryostats. Refrigerator systems can have vibration problems due to vigorous motion of the displacer piston in the cooling head. This can cause serious line broadening in the Mössbauer spectra. Modern commercial units have overcome this completely by isolating the cooling station from the mechanical part of the system and conducting heat away from the station using cold helium gas. No mechanical connection means no line broadening in the system and the operator can easily stabilize any temperatures between 10 and 400 K. For higher temperatures a specialized temperature controlled oven is used.

5.3 Information Obtainable From Mössbauer Spectra

A typical Mössbauer spectrum shown in Fig. 5.3 contains a great deal of information. The nuclear transition from the $j = \frac{3}{2}$ 14.4 keV excited state to the $j = \frac{1}{2}$ ground state is an M1 (magnetic dipole) transition with selection rules $\Delta m = 0, \pm 1$. In the case in which a magnetic splitting of the $\frac{3}{2}$ level into $\pm \frac{3}{2}$ and $\pm \frac{1}{2}$ levels and the $\frac{1}{2}$ level into $\pm \frac{1}{2}$ levels occurs, six of the possible eight transitions are allowed. This shows in the spectrum as six lines. The transitions included by magnetic dipole selection rules $m = +\frac{3}{2} \rightarrow m = -\frac{1}{2}$ and $m = -\frac{3}{2} \rightarrow m + \frac{1}{2}$ do not appear. The relative intensities of the transitions are also very interesting. The transition probabilities are the *squares* of Clebsch–Gordan coefficients appropriate to these angular momentum states. They depend not only on the j and m values but also on the angle θ between the direction of the detected gamma ray and the direction of the magnetic field producing the Zeeman splitting of the nuclear levels. For an *unmagnetized* Fe absorber in which the direction of the magnetic field at the nucleus is random (due to domains) the six lines have intensity ratios $3:2:1:1:2:3$. Of course the *unmagnetized* Fe is still magnetically ordered within a domain so that the nucleus still "sees" a net magnetic hyperfine field produced by the surrounding electrons. Bulk *Fe* shows this magnetic order below the Curie temperature of 1040 K.

Thin films and superlattices are usually magnetized films of Fe thinner than 100 layers, often have single domains, and in the absence of significant surface anisotropy are magnetized in-plane. In this case the ratios of Mössbauer spectral lines are $3:4:1:1:4:3$. When large surface anisotropy effects overcome the dipolar shape effects and lead to a magnetization perpendicular to the film surface (and therefore parallel to the gamma ray direction in a transmission Mössbauer measurement) the transition probabilities for the $+\frac{1}{2} \rightarrow +\frac{1}{2}$ and $-\frac{1}{2} \rightarrow -\frac{1}{2} \Delta m = 0$ transitions become zero. In this interesting case the line intensities become $3:0:1:1:0:3$.

In general the line intensity relation is given by

$$3:x:1:1:x:3, \quad x = 4\sin^2\theta/(1 + \cos^2\theta),$$

where θ is the angle between the gamma ray and the direction of the magnetization M. In pure bulk Fe the magnetic hyperfine field is 341 kOe at 4.2 K and 333 kOe at 295 K. Because the magnetic moments of the nuclear $\frac{3}{2}$ and $\frac{1}{2}$ levels are independently known, the Mössbauer spectrum shown in Fig. 5.3 yields values for the hyperfine fields from an appropriate least-squares fit to the spectrum.

The magnetic hyperfine field at the nucleus depends in a complicated way on the magnetic properties of host lattice. In a paramagnetic metal or alloy, the hyperfine field at the nucleus rapidly relaxes in a spatial direction due to the spin–spin relaxation associated with the thermal lattice energy kT. This results in no *net* magnetic field at the nucleus over the time comparable to the Larmor precession period of the nuclear moment in the (non-static) magnetic field produced by the electrons around the nucleus. In making a ^{57}CO radioactive source for obtaining the initial resonant gamma rays for Mössbauer spectroscopy this is used to produce a magnetically "unsplit" single line gamma source. Typically ^{57}CO is diffused into rhodium. The resulting paramagnetic alloy also has cubic symmetry so that the source line is "unsplit" either by magnetic or electric quadrupole hyperfine interactions.

5.4 Isomer Shift

A careful look at a Mössbauer spectrum (Fig. 5.3) shows that the centroid of the spectral lines often occurs at other than *zero* Doppler velocity. This so-called isomer shift (or chemical shift) arises both from a different finite nuclear size in the nuclear excited and ground states *and* differences in the electron density at the nucleus between the source and absorber. Without both of these differences simultaneously no shift is seen. The expression for the isomer shift is

$$\Delta E_{1.5} = \frac{4\pi}{5} Ze^2 R^2 \left(\frac{\delta R}{R}\right) \left\{ |\psi(0)|^2_{\text{absorber}} - |\psi(0)|^2_{\text{source}} \right\}.$$

The quantity δR represents the difference in mean nuclear radius, R, between nuclear excited and ground states, and $|\psi(0)|^2$ represents the total electron density at the nucleus. Only s electrons contribute significantly to that density. In the case of ^{57}Fe changes in the 4s electron density can be estimated by observing isomer shifts.

5.5 Conversion Electron Mössbauer Spectroscopy (CEMS)

In this variant of conventional transmission Mössbauer spectroscopy, advantage is taken of the fact that nature provides a means of de-excitation of the 14.4 keV nuclear excited state of ^{57}Fe other than photon emission. An excited nucleus may de-excite by transferring its energy to one of the inner electrons which have some appreciable density at the nucleus (typically s electrons). In the case of ^{57}Fe this "internal conversion" occurs *nine* times more often than emission of a 14.4 keV photon.

This forms the basis for Mössbauer spectroscopy in cases in which the thin film or superlattice system is not transparent to 14.4 keV gammas. This occurs, for example with very dense or thick single crystal substrates such as tungsten or GaAs. A scattering geometry is appropriate with 14.4 keV gamma rays from a Co^{57} source incident on the Mössbauer scatterer which is often much smaller than a transmission geometry absorber. When the source is Doppler shifted into resonance with the scatterer, a recoilless Mössbauer absorption takes place, temporarily leaving some ^{57}Fe nuclei in the scatterer in an excited state. These nuclei de-excite primarily by "internal conversion" which results in the ejection of a κ-electron of about 7 keV energy. The resulting excited atom with a κ-shell hole will de-excite by emitting a 6.4 keV X-ray or (about 50% of the time) by emitting a number of outer shell electrons by the Auger process. The spectrum of electrons from these processes is not sharply defined in energy, but tends to have lower energy than "noise" electrons from photoelectric or Compton processes.

More sophisticated Conversion Electron Mössbauer Spectrometers (CEMS) pass these electrons through a rather broadband electron spectrometer before they are detected by a channeltron or other suitable electron detector. The signal-to-noise ratio for such a sophisticated system can be very high, although dense substrates such as tungsten can produce significant noise problems. The high sensitivity and large signal-to-noise ratios have permitted Mössbauer spectroscopy of Fe films of monolayer thickness, while the thinnest films for which transmission spectra are reported are about two monolayers thick. Because electrons in this energy range are not very penetrating, the entire spectrometer must be evacuated, while transmission experiments don't usually require vacuum except as part of the absorber cooling system. The need for a vacuum for CEMS has been turned to advantage by some groups by having the film production inside the same vacuum system as the Mössbauer spectrometer. In principle, this might mean that no protective covering layer would have to be put over the Fe before Mössbauer measurements. In practice, a great deal of time (~ 15 h) must elapse before enough counts are recorded to provide a Mössbauer spectrum with good counting statistics. It is not possible to keep a bare Fe surface free of contamination during this period of time.

Because one is counting electrons which cannot escape from very deep in the sample, CEMS is more surface sensitive than are transmission experiments. However, in practice, surface sensitivity is achieved by making Fe films from isotopically pure ^{56}Fe which shows no Mössbauer effect and then depositing one or two atomic layers of isotopically pure ^{57}Fe at or near the Fe film surface [5.3]. In this way genuine surface sensitivity is achieved by both Mössbauer spectroscopy geometries.

5.6 Magnetic Relaxation in Thin Films and Superlattices

A phenomenon quite familiar to those who examine ultrathin films by Möss-bauer spectroscopy is the presence of an unsplit component in a Mössbauer spectrum of what should be a ferromagnetically ordered sample (Fig. 5.5). Such features were once associated with non-magnetic oxides, but the similarity of the spectra to those of small magnetic particles provided strong clues to their real origin. It was learned early that small ferromagnetic particles can have such reduced magnetocrystalline anisotropies that the direction of the magnetization of the entire single-domain grain can thermally "relax" or change in a time short compared with the measuring time of the magnetization. This so-called "super paramagnetism" results from the spins of the ferromagnetic atoms being strongly exchange coupled to each other, but their mutual direction is not fixed in space. Each grain acts like a giant paramagnetic moment because the thermal energy kT exceeds the magnetic anisotropy energy which locks the magnetism to a particular crystalline ("easy") direction. This phenomenon of *super para-magnetic relaxation* in very small magnetic grains ($r \sim 10$ nm) has been exten-sively studied by Mössbauer spectroscopy. The time of measurement of the magnetization is typically the Larmor precession time of the nuclear moment in the magnetic hyperfine field due to the surrounding electrons. When the

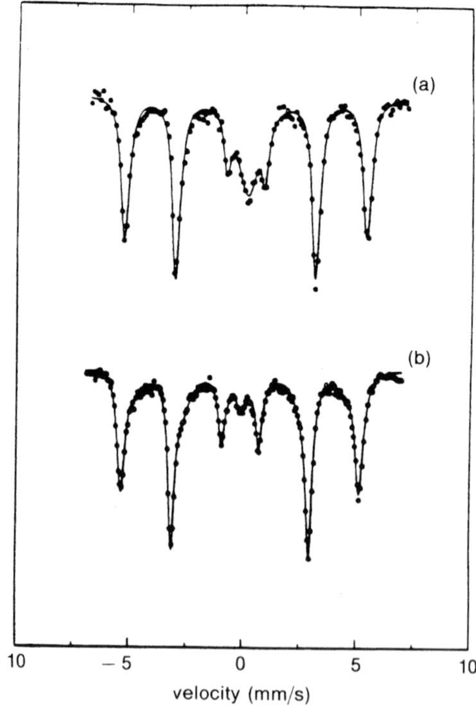

Fig. 5.5. (a) Mössbauer spectrum of Fe(1 1 0) showing magnetic relaxation effects (central spectral feature); (b) reduction of the feature by applica-tion of 5 kG external magnetic field to increase the effective magnetic aniso-tropy

magnetic relaxation of this hyperfine field occurs in a time shorter than this Larmor precession time and nucleus sees only a time-averaged field. For the super paramagnetic grains this time-average will be zero and an unsplit line will result. The Larmor precession time and the super paramagnetic relaxation times are close for conveniently-sized grains, enabling this phenomenon to have been well-studied by Mössbauer spectroscopy.

The relevant equation describing this behaviour is [5.4]: $1/\tau = f_0 \exp(\sim KV/kT)$, where τ is the relaxation time for the process with $f_0 \approx 10^9 \, s^{-1}$, V the volume of the particle and K the volume anisotropy. Relaxation models following from these considerations provide good fits to relaxation Mössbauer spectra taken over a wide temperature range.

Ultrathin ferromagnetic films can also have low volume anisotropies. The very thinnest films (less than three atomic layers for $Fe(1\,1\,0)$ also may have an "island" structure (like flat pancakes) which reduces the dipolar contribution to the anisotropy (so-called "shape" anisotropy). Under these circumstances magnetic relaxation may also occur. In this case the relaxation of magnetic spins is two-dimensional in the plane of the thin film. Here a single line may appear in the Mössbauer spectra taken at higher temperature ≈ 300 K, while at lower temperatures a fully split spectrum appears. Great caution must be taken to separate relaxation effects from the effects of a lower magnetic ordering temperature (T_c) for an ultrathin ferromagnetic system.

One way of identifying magnetic relaxation is to measure the temperature dependence of the magnetic hyperfine field. For a magnetically relaxing thin film system this dependence will be quasi-linear [5.5] with the relationship

$$\frac{M(T)}{M(O)} \approx 1 - \frac{k_B}{2KA_i t(1 + t_0/2t + M_0 H/k)} \times T.$$

In this model t is the film thickness, A_i is the basal area of the "islands", and $K = 4.5 \times 10^5$ erg/cm^3 is the volume anisotropy for Fe. M_0 is the bulk magnetization and H the external magnetic field. This is similar to the signature of two-dimensional ferromagnetism, but with an important difference: For a magnetically relaxing system the slope of the M versus T line will change noticeably with the application of a small external magnetic field in the plane of the thin film. For a true two-dimensional ferromagnet no slope change is seen with application of an external field. Mössbauer spectroscopy is a unique way of sorting out these differences.

5.7 Examples of Mössbauer Spectroscopy Applied to the Study of Magnetic Thin Films, Surfaces, and Superlattices

We conclude our discussion of Mössbauer spectroscopy applied to the study of magnetic thin films, superlattices, and surfaces with a discussion of three experiments in which Mössbauer spectroscopy provided the critical information about these magnetic systems:

(I) J. Korecki, U. Gradman: "*In Situ* Mössbauer Analysis of Hyperfine Interactions near Fe(1 1 0) Surfaces and Interfaces," Phys. Rev. Lett. **55**, 22, 2491 (1985).

In this classic study Fe(1 1 0) was grown on W(1 1 0). This is an interesting system because Fe "wets" tungsten and flat, layered growth occurs. By growing Fe to a thickness greater than ten layers the stretched pseudomorphic Fe(1 1 0) gives way to a normal Fe structure. In this work the body of the Fe film was made of isotopically pure ^{56}Fe which shows no Mössbauer effect. A single probe layer of ^{57}Fe was grown either at the upper surface of the film or at positions interior to the film. The upper film surface was covered with Ag(1 1 1) to prevent oxidation. The use of the dense W(1 1 0) single crystal substrate effectively precluded the use of a transmission geometry for Mössbauer spectroscopy. Instead a sophisticated Conversion Electron Mössbauer Spectrometer (CEMS) was used. Samples were made in one part of the vacuum system and transferred to the spectrometer section for analysis.

Gamma rays from a Doppler shifted source outside of the vacuum envelope impinged on the small absorber/scatterer which achieved resonant Mössbauer absorption when the source was Doppler shifted into resonance. The de-excitation of an absorber/scatterer nucleus followed by the highly probable electron conversion process which resulted in an 7.3 keV electron and several lower-energy Auger electrons. The spectrum of the emitted electrons does not show the distinct peaks of the source gamma ray spectrum. In this experiment the electrons were passed through a cylindrical electron spectrometer roughly tuned to the expected energy range of the emitted electrons and focused on a channeltron detector.

This system had the advantage that samples could be made in one part of the system, moved into the position as absorber/scatterer and the detector could be remote from the absorber/scatterer. In some cases the resultant Fe surfaces could be examined without a covering layer of Ag(1 1 1). However, the reactivity of Fe and the background pressure of $\sim 10^{-9}$ Torr implied that residual gases were absorbed on this Fe surface during the necessary measuring time of about 15 h (using a 100 M currie source).

The interesting result of this experiment was the first observation of an electric quadrupole splitting at the surface of a cubic metal occasioned by the reduced symmetry from the presence of the surface itself. This small effect is essentially missing in the second layer from the surface, so the monolayer resolution of the ^{57}Fe probe in the ^{56}Fe film is crucial to its observation.

The authors also confirmed a reduction of the hyperfine field at the surface first seen in a lower resolution transmission geometry experiment [5.6]. This work offered at least a qualitative experimental comparison with the calculations of *Ohnishi* et al. [5.7]. The results also confirm an earlier [5.3] finding of a positive isomer shift at the surface compared with bulk values. This implies a reduced density of 4s electrons (since δR is negative for ^{57}Fe) at the Fe surface, as expected from the same calculations. These measurements show how Mössbauer spectroscopy offers unique information on epitaxial metal surfaces.

(II) N.C. Koon : "Direct Evidence for Perpendicular Spin Orientations and Enhanced Hyperfine Fields in Ultrathin Fe(1 0 0) Films on Ag(1 0 0)," Phys. Rev. Lett., **59**, 2463 (1987).

In this experiment CEMS was also used, although not *in situ* in the sample growth chamber. The investigators took advantage of the fact that Fe(1 0 0) matches the surface of Ag(1 0 0) with a 45° rotation of the Fe net compare with the Ag(1 0 0). The Ag substrate was grown on a buffer layer of ZnSe which had been grown on a GaAs(1 0 0) substrate. This procedure yielded good surface morphology of the Ag(1 0 0) as indicated by reflection high energy electron diffraction (RHEED) patterns. The Fe(1 0 0) layers were grown with thicknesses from one monolayer (ML) to 5.5 ML in the form of superlattices with intervening Ag(1 0 0) layers of from four to 7 ML. As many as 45 superlattice periods of the Fe and Ag layers were grown. The signal conversion electrons could come from >60 nm depth in the sample, and therefore several Fe layers could contribute to the signal, enhancing the sensitivity. In this experiment enriched ^{57}Fe was used exclusively for the Fe to further enhance the signal.

For such a superlattice structure it is possible to monitor the quality of successive Fe layers by RHEED during the growth to insure that the RHEED patterns for the nth film element are the same as for the first. In this experiment the sample was finished by a thin Al coating to prevent oxidation when the superlattice was removed from the growth vacuum and placed in the CEMS measuring chamber. The Al was thin enough (7.5 nm) to pass most of the signal electrons.

The conversion electron Mössbauer spectrometer had a cylindrical mirror electrostatic analyzer to also focus the signal electrons into the counter. It had an acceptance angle of 30°–60° and an energy resolution of about 4%. The superlattice samples were mounted on a Cu block connected to a closed cycle helium refrigerator by flexible Cu braid so that the sample could be cooled as low as 15 K without significant vibration from the refrigerator.

An interesting example of the *Koon* et al. data is shown in Fig. 5.6. The thin superlattice sample was made up of ^{57}Fe(1 0 0) films of 2.4 ML thickness. The first thing to be seen from the spectrum (where the γ-rays were incident 90° to the film surface) is that the second and fifth lines of the magnetic hyperfine sextet are very weak. As mentioned above the relative intensities are given by $3:x:1:1:x:3$ where $x = 4\sin^2\theta/(1 + \cos^2\theta)$ with θ the angle between the incident γ-ray direction and the direction of the magnetic hyperfine field. For the thick Fe film (≈ 50 nm) the line intensities are $3:4:1:1:4:3$ as expected if the hyperfine field (and magnetization) are in the film plane. For the thin 2.4 ML film the near vanishing of the second and fifth lines means that the hyperfine field is nearly normal (90°) to the film surface. This is particularly interesting because of calculations by *Gay* and *Richter* [5.8] which predict a strong magnetic anisotropy for a free-standing monolayer of Fe(1 0 0). This perpendicular anisotropy

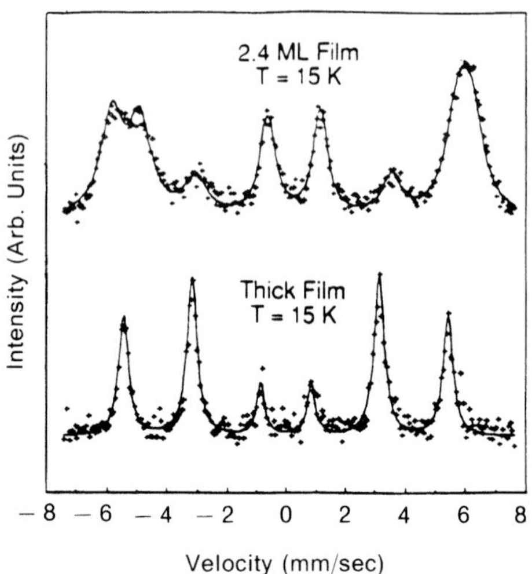

Fig. 5.6. CEMS Mössbauer spectra from *Koon* et al. The supper spectrum indicates that the magnetization is pointing well out of the film plane (lines 2 and 5 are very weak). The lower spectrum shows in-plane alignment with $3:4:1:1:4:3$ intensity ratios

could overcome the dipolar shape anisotropy which normally assures that M lies in the film plane if the film was thick enough for the volume of dipoles to be insufficient to overcome the surface perpendicular anisotropy.

The results of *Koon* et al. seem to support this prediction. The 2.4 ML film shows a magnetization which points nearly normal to the film surface, while the thick film shows that M is clearly in the plane of the film. Also interesting is the broad line Mössbauer spectrum of the 2.4 ML Fe film superlattice which clearly shows evidence of at least two overlapping six line sextets. The authors of this work interpret this as evidence of two magnetic sites having different hyperfine fields (358 and 344 kg) and different isomer shifts. They do not offer a detailed explanation of these two sextets, but later work by another group [5.9] shows a quite similar spectrum. In this case the Ag(1 0 0) substrate was grown epitaxially on mica by MBE and the Mössbauer spectrum was taken in transmission geometry instead of by CEMS. The nearly identical forms of the spectra indicates that the two "sites" seen in the spectra are real, and not trivial artifacts of the method of superlattice growth.

Koon et al. also mention the presence of line broadening above 50 K associated with magnetic relaxation. It is not clear what causes magnetic relaxation in thin film systems. One cannot be sure that it follows from the same causes as three dimensional super-paramagnetic relaxation of small ferromagnetic particles. However, this is, an interesting phenomenon which is studied easily and well by Mössbauer spectroscopy.

(III) C.J. Gutierrez, Z.Q. Qiu, M.D. Wieczorek, H. Tang, J.C. Walker: "The Observation of A 3-D to 2-D Crossover In the Magnetism of Epitaxial Fe(1 1 0)/Ag(1 1 1) Multilayers," J. Magn. Magn. Mat., **93**, 326 (1991)

These experiments were carried out by conventional transmission Mössbauer spectroscopy (TMS) on a series of Fe(1 1 0)/Ag(1 1 1) superlattices grown

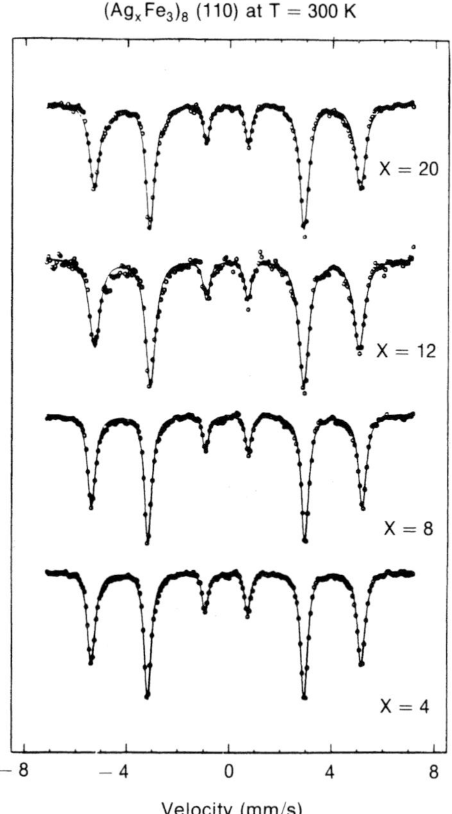

Fig. 5.7. The Mössbauer spectra of (Fe$_3$ Ag$_x$)$_8$ multilayers measured at room temperature

by Molecular Beam Epitaxy (MBE). The samples had 3 ML Fe(1 1 0) bilayer components and Ag(1 1 1) bilayer component thickness equal to 4, 8, 12 and 20 ML. The TMS spectra of each of these samples consisted of a single magnetically-split sextet with no features, even at temperatures of 300 K, which could be associated with magnetic relaxation (Fig. 5.7). The sample with 20 ML Ag(1 1 1) separation layers showed a linear temperature dependence of the magnetic hyperfine field. For a 3-d ferromagnetic the temperature dependence is well described by $H_{hf}(T)/H_{hf}(0) = 1 - BT^{3/2}$ as predicted from spin wave theories. Several theoretical works predict that for sufficiently thin films a quasi-linear magnetization (hyperfine field) temperature dependence should occur because of the difficulty in exciting spin waves normal to the film plane.

Previously a two-dimensional (2D) quasi-linear temperature dependence was obscured by the linear temperature dependence due to magnetic relaxation effects. The results of this work show no indications of magnetic relaxation. Also the slope of the linear temperature dependence of the hyperfine field did not change with the application of an external magnetic field as has been demonstrated for samples showing magnetic relaxation. The conclusion is inescapable that the 3 ML Fe(1 1 0) films in this work show 2D spin wave characteristics.

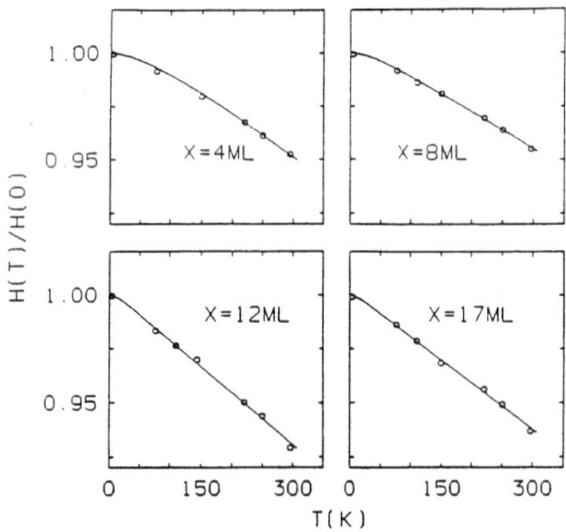

Fig. 5.8. The hyperfine field temperature dependence of the $(Ag_xFe_3)_8$ multilayers, where $x = 4, 8, 12,$ and 17 ML

A further interesting result is found when the Ag separation layers are made thinner. For Ag(1 1 1) layers of 4 ML the temperature dependence of the hyperfine field again follows a $T^{3/2}$ form characteristic of 3D behavior. For Ag(1 1 1) thicknesses between these two extremes the temperature dependence are neither linear nor $T^{3/2}$ but show an intermediate form well-explained by calculations of *Maccio* et al. [5.10], who used a spin wave model with surface anisotropy and *two* exchange couplings: a strong J_0 in the Fe film and a much weaker J_1 representing the interlayer magnetic couplings through the Ag spacer layers. The data fit by this more complex spin wave model is shown in Fig. 5.8.

Here the importance of Mössbauer spectroscopy for thin films and super-lattice studies shows most strikingly: the narrow spectral lines with no relaxa-tional broadening or central unsplit line features show the quality, flatness and continuity of the 3 ML Fe(1 1 0) film components, in agreement with the RHEED analysis during the growth. The linear temperature dependence of the hyperfine field seems to have no remaining explanation other than the effectively 2D nature of the Fe films. This conclusion is strengthened by the changes in the hyperfine field temperature dependence which occur when samples are made with thinner Ag separation layers. By providing interlayer magnetic inter-actions, the 3D nature of the spin wave excitations is restored. Recently a great deal of interest in these interlayer magnetic couplings has developed because of their possible role in the giant magneto-resistance often seen in such magnetic superlattice systems.

5.8 Conclusions

Mössbauer spectroscopy, particularly using ^{57}Fe, is an important tool in the structural characterization and in the study of the magnetic properties of thin magnetic films, magnetic superlattices, and surfaces. Information comes from the magnetic hyperfine splitting, electric quadrupole splittings and isomer shifts. The temperature dependence of the magnetic hyperfine field which tracks the magnetization in Fe gives interesting information about the thermal spin wave behavior of thin film and superlattice samples. Mössbauer spectroscopy has contributed significantly to our understanding of ultrathin films, superlattices, and magnetic surfaces. It will, no doubt, continue to do so.

References

5.1   For example, J.M. Siman: *Electrons and Phonons* (Oxford University Press, London, 1960)
5.2   APD Cryogenics, Inc., 1833 Vultee St., Allentown, PA
5.3   J. Tyson, A. Owens, J.C. Walker: J. Magn. Magn. Mat. **35**, 126–129 (1983)
5.4   G.T. Rado, H. Suhl: *Magnetism* (Academic Press, New York, 1966)
5.5   C.J. Gutierrez, Z.Q. Qui, M.D. Wieczorek, H. Tang, J.C. Walker: Phys. Rev. B **44**, 2190 (1991–I)
5.6   J.C. Walker, R. Droste, G. Stern, J. Tyson: J. Appl. Phys. **55**, 2500 (1984)
5.7   S. Onnishi, M. Weinert A.J. Freeman: Phys. Rev. B., **30**, 36 (1984)
5.8   J.G. Gay, R. Richter: Phys. Rev. Lett. **56**, 2728 (1986)
5.9   C.S. Gutierrez, M.D. Wieczorek, H. Tang, Z.Q. Qiu, J.C. Walker: J. Magn. Magn. Mat. **99**, 215 (1991)
5.10 M. Maccio, M.G. Pini, P. Politi, A. Rettori: Phys. Rev. B. **44**, 2190 (1991–I)

Subject Index

Printing: Krips bv, Meppel
Binding: Litges & Dopf, Heppenheim